ION-EXCHANGE MEMBRANE SEPARATION PROCESSES

Membrane Science and Technology Series

Volume 1: Pervaporation Membrane Separation Processes
Edited by R.Y.M. Huang (1991)
Volume 2: Membrane Separations Technology, Principles and Applications
Edited by R.D. Noble and S.A. Stern (1995)
Volume 3: Inorganic Membranes for Separation and Reaction
By H.P. Hsieh (1996)
Volume 4: Fundamentals of Inorganic Membrane Science and Technology
Edited by A.J. Burggraaf and L. Cot (1996)
Volume 5: Membrane Biophysics
Edited by H. Ti Tien and A. Ottova-Leitmannova (2000)
Volume 6: Recent Advances in Gas Separation by Microporous Ceramic Membranes
Edited by N.K. Kanellopoulos (2000)
Volume 7: Planar Lipid Bilayers (BLMs) and their Applications
Edited by H.T. Tien and A. Ottova-Leitmannova (2003)
Volume 8: New Insights into Membrane Science and Technology: Polymeric and Biofunctional Membranes
Edited by D. Bhattacharyya and D.A. Butterfield (2003)
Volume 9: Ion-Exchange Membrane Separation Processes
By H. Strathmann (2004)

Membrane Science and Technology Series, 9

ION-EXCHANGE MEMBRANE SEPARATION PROCESSES

Heiner Strathmann

Institut für Chemische Verfahrenstechnik
Universität Stuttgart, Stuttgart, Germany

2004

ELSEVIER

Amsterdam - Boston - Heidelberg - London - New York - Oxford
Paris - San Diego - San Francisco - Singapore - Sydney - Tokyo

ELSEVIER B.V.
Sara Burgerhartstraat 25
P.O.Box 211, 1000 AE
Amsterdam, The Netherlands

ELSEVIER Inc.
525 B Street, Suite 1900
San Diego, CA 92101-4495
USA

ELSEVIER Ltd
The Boulevard, Langford Lane
Kidlington, Oxford OX5 1GB
UK

ELSEVIER Ltd
84 Theobalds Road
London WC1X 8RR
UK

First edition 2004

Library of Congress Cataloging in Publication Data
A catalog record is available from the Library of Congress.

British Library Cataloguing in Publication Data
A catalogue record is available from the British Library.

ISBN: 978-0-444-50236-0
ISSN: 0927-5193

Preface

Ion-exchange membranes and their application are not a recent development. Processes such as electrodialysis have been used on a large industrial scale for more than half a century. The literature covering various processes involving ion-exchange membranes and their applications is very broad, extremely fragmented, and contained in a number of different scientific journals and in various patents. This makes it difficult to gain a reasonably accurate overview of ion-exchange membranes and their applications. While other membrane processes such as reverse osmosis or micro- and ultrafiltration are described in great detail in a number of excellent text and reference books there is hardly any comprehensive treatment of ion-exchange membrane separation processes. Therefore, the time seems to be appropriate for a book that emphasizes the most relevant aspects of this today already widely used technology.

The main difficulty confronted in such an endeavor is to find a balanced selection between the fundamental theoretical background and practical applications necessary to provide the reader with a sound understanding of the function of ion-exchange membranes and their importance for solving difficult separation problems without getting lost in too many details. Therefore, the objectives of this book are to give a comprehensive overview of ion-exchange membrane separation processes covering the fundamentals as well as recent developments of the different products and processes and their applications for a quite heterogeneous interest group including plant managers and process engineers as well as research scientists and graduate students.

Reading this book the plant engineer might complain that too much emphasis is put on the discussion of the theoretical background information, while on the other hand those in superior command of thermodynamics and electrochemistry might express their concern about the simplified approach of treating very complex phenomena. To facilitate the reading of the book I have tried to separate the different topics. In the first chapter a very general overview of the relevant electromembrane processes is given. In the second chapter some thermodynamic and physicochemical fundamentals are described. This chapter can cheerfully be skipped by those interested only in the various applications of the ion-exchange membrane separation processes and their technical and commercial relevance compared to competing technologies. This is discussed in the last chapter, which certainly can be missed by all those interested in the more scientific aspects. The third chapter gives some information about ion-exchange membrane preparation techniques, while in the fourth and fifth chapter the processes are discussed as unit operations giving examples for the design of specific plants. These two chapters are probably of interest mainly to the plant engineer and plant manager.

In a book of this kind it is impossible to mention all those who have contributed to the development and understanding of ion-exchange membrane separation processes. Quotations and literature references are restricted to publications, which are apt to illustrate the discussions in the text. The choice of the literature references is arbitrary and a matter of personal preference, and I apologize to all scientists and engineers whom I have not given due credit for their contribution to the field.

My sincere thanks are due to many of my students and colleagues who have contributed to this book with their work and many helpful suggestions. However, I am particularly indebted to Harm Schmoldt for his valuable and constructive suggestions concerning the content of the book, my wife Monika who struggled valiantly with my English spelling and style while proofreading the manuscript, and Andrej Grabowsky who spent many hours checking all the equations and drawings very thoroughly and pointing out mistakes. His sound scientific background and complete command of the computer was extremely helpful in writing this book. I thank both of them for the many hours they have spent to improve the manuscript in the technical content and in style.

Heiner Strathmann

Contents

Chapter 3 Preparation and Characterization of Ion-Exchange Membranes

Chapter 4 Operating Principle of Electrodialysis and Related Processes

Chapter 5 Ion-Exchange Membrane Process and Equipment Design

Chapter 1

Overview of Ion-Exchange Membrane Processes

1 INTRODUCTION

Ion-exchange membranes are used in a number of processes which are rather different in their basic concept, their practical application, and their technical relevance. However, in all these processes membranes carrying electrical charges are used to control the transport of ionic species and to separate them selectively from a mixture with neutral components.

Based on their application ion-exchange membrane separation processes can be classified in three mayor categories:

- Mass separation processes such as electrodialysis, Donnan and diffusion dialysis, and electrodialysis with bipolar membranes.
- Chemical synthesis processes such as the chlorine-alkaline electrolysis and the hydrogen and oxygen production by water electrolysis.
- Energy conversion and storage processes such as fuel cells and electrical batteries.

All ion-exchange membrane separation processes are based on the same fundamental principle, which is the coupling of the transport of electrical charges, i.e. an electrical current with a transport of mass, i.e. cat- or anions through a perm-selective membrane due to an externally applied or internally generated electrical potential gradient. The technical and commercial relevance and their present state of development of the various processes is rather different. Electrodialysis, for example, is a mature process with a multitude of interesting applications in water desalination and in the food and drug industry. The chlorine-alkaline electrolysis is also a state-of-the-art process. Its application, however, is limited to a rather small market segment. Fuel cells are still in an early state of development. Their future technical and commercial relevance as an efficient and convenient energy source with a multitude of potential applications is considered to be very large.

In this book, however, only ion-exchange membrane processes that are used in mass separation such as electrodialysis, diffusion dialysis, Donnan dialysis, and the application of bipolar membranes are discussed in some detail. Electromembrane processes used in chemical synthesis and energy conversion devices will only be briefly described in their fundamental principle.

2 HISTORICAL DEVELOPMENT

The first ion-exchange membrane separation process that was used on a large industrial scale is electrodialysis which was introduced in the late 1950's and early 1960's as an efficient and economic technique for the desalination of brackish water [J.R. Wilson, 1960]. The principle of the process has been known for more than 100 years [W. Oswald, 1890]. Between 1880 and 1900 Ostwald studied the properties of semipermeable membranes and discovered that a membrane is impermeable for an electrolyte if it is impermeable either for its cations or its anions [W. Ostwald, 1890]. To illustrate this, he postulated the existence of the so-called "membrane potential" at the boundary between a membrane and an aqueous solution as a consequence of a difference in ion-concentration. Twenty years later Donnan confirmed this postulate membrane potential between an ion-exchange membrane and the surrounding solution [F.G. Donnan, 1911]. Simultaneously, he developed a mathematical equation describing the concentration equilibrium which resulted in the so-called "Donnan exclusion potential" [F.G. Donnan et al., 1932].

The first use of electrodialysis in mass separation dates back to 1903, when Morse and Pierce introduced electrodes into two solutions separated by an ordinary dialysis membrane and found out that electrolytes could be removed more rapidly from a feed solution with the application of an electrical potential [H.W. Morse et al., 1903]. About 20 years later the first three compartment electrodialysis cell to be used in practical mass separation problems was designed. Here, the process was applied mainly to remove last traces of electrolytes from colloid solutions. Membranes used in the early electrodialysers were uncharged and thus not ion-selective [P.H. Prausnitz et al., 1931]. This limited the process basically to a three compartment cell design.

The first basic studies related to ion-selective membranes had already been carried out in 1925 [L. Michaelis et al., 1925] with homogeneous, weak acid collodium membranes. With the advent of ion-selective membranes which could effectively block the transport of coions, i.e. ions carrying the same electrical charge as the membrane, it became feasible to transport ions against a concentration gradient in a multi-cell arrangement. Such a multi-cell electrodialysis process was suggested in 1940 by Meyer and Strauß, who designed an electrodialysis apparatus, in which anion-selective and cation-selective membranes were arranged in alternating series to form many parallel solution compartments between two electrodes [K.H. Meyer et al., 1940]. With such a multi-compartment electrodialyser, demineralization or concentration of solutions could be achieved in many compartments with only one pair of electrodes. Thus the irreversible energy losses represented by the decomposition potentials at the electrodes could be distributed over many demineralizing compartments and therefore be minimized.

Around about 1940, the interest in industrial applications of electrodialysis led to the development of more effective synthetic ion-exchange membranes on the basis of resins prepared by polycondensation of phenol and formaldehyde [K. Haagen et al., 1959]. Later polystyrene cross-linked with divinylbenzene became more and more the basis of ion-exchange membranes [F. Helfferich, 1962]. After the importance of the multi-cell stack arrangement for the economy of electrodialysis was recognized, and with the development of stable, highly selective ion-exchange membranes of low electric resistance in the late 40's by Juda and McRae of Ionics Inc. [W. Juda et al., 1950] and by Winger et al. at Rohm and Haas [A.G. Winger et al., 1953], electrodialysis rapidly became a relevant industrial process for demineralizing and concentrating electrolyte solutions [K.S. Spiegler, 1956].

The main use envisaged for electrodialysis in the United States and Europe was the desalination of brackish water to produce potable water of sufficient quality. The membranes to be used in this application should have high selectivity and low osmotic and electro-osmotic water transfer in contact with dilute solutions. The electrical resistance of the membranes, however, was not very important because in water desalination the electrical resistance in an electrodialysis stack is mainly determined by the conductivity of the dilute stream. The membranes which fulfilled these requirements were manufactured as so-called heterogeneous structures by the dispersion of a fine ion-exchange resin powder within the solution of a matrix polymer followed by the evaporation of the solvent [F. Bergsma et al., 1961]. Generally, the membranes are reinforced by a screen which contributes to the form stability of the membranes. Compared to reverse osmosis or ultra- and microfiltration membranes ion-exchange membranes have much better mechanical as well as chemical and thermal stability. Especially cation-exchange membranes can be operated over a wide pH-range and at elevated temperatures without changing their properties.

In the early stage of developing electrodialysis as an industrial process a substantial amount of basic research on ion-exchange membranes and electrodialysis stack design was carried out in Europe, Israel, and South Africa. The first large scale application of electrodialysis for the economical production of potable water from brackish water sources, however, was carried out in the United States. With the financial support of the Office of Saline Water which was introduced by the United States government to promote the development of efficient water desalination techniques the first pilot plants were built, and very soon commercial plants were built and sold in the Middle East and the Persian Golf region.

A significant step towards the efficient application of electrodialysis was the introduction of a new mode of operation referred to as electrodialysis reversal by Ionics Inc. In this operation mode the polarity and the flow streams in an electrodialysis stack are reversed in certain time intervals [W.E. Katz, 1979], i.e. the anode becomes the cathode and the diluate stream becomes the concentrate stream. In this operation membrane fouling and scaling could be reduced to a minimum and costly and time-consuming membrane cleaning procedures became obsolete.

A completely different use of electrodialysis was envisaged in Japan. Here, electrodialysis was used for concentrating sodium chloride from seawater to produce table salts [T. Nishiwaki, 1972]. In this application the electrical resistance of the membrane was of prime importance for the economics of the process. These requirements have led to the development of homogeneous membranes with very low electrical resistance but less mechanical strength. With the introduction of electrodialysis into the food and drug industry and especially into the treatment of certain industrial effluents again further improvements of both the cell system design and the membrane properties, especially their chemical and thermal stability, became necessary [W.G. Grot, 1973].

With the introduction of bipolar membranes for the recovery of acids and bases from the corresponding salts on an industrial scale in the early 80´s [K.J. Liu et al., 1977] a completely new area of application of electrodialysis was opened up. In combination with mono-polar ion-exchange membranes bipolar membranes can be used to convert salts into the corresponding acids and bases. With the development of energy efficient and chemically stable bipolar membranes, a large number of applications in the chemical process industry, the food and drug industry, and in waste water treatment were identified for bipolar membrane electrodialysis.

The combination of electrodialysis with conventional ion-exchange technology and the use of conducting spacers were suggested in the 70's [Z. Matejka, 1971], [E. Korngold, 1975]. The process is now referred to as continuous electrodeionization and is commercialized on a large scale for the production of ultra-pure deionized water.

With the development of the highly stable perfluorinated ionomer membranes by DuPont [W.G. Grot, 1974, 1975] and Asahi Glass Co. [Y. Oda et al., 1977] the electrolytic production of chlorine and caustic soda was introduced on a large scale and has become the standard technique for the production of caustic soda and chlorine.

The development of an electrochemical cell for a highly efficient conversion of chemical into electrical energy by avoiding the Carnot-Cycle was envisioned already by the end of the19th century [W. Oswald, 1894]. The principle of the fuel cell, however, without the use of a membrane was already demonstrated 50 years earlier [W. R. Groves, 1839]. The modern fuel cell using ion-exchange membranes as a "solid electrolyte" made its first appearance in the middle of the last century [L.W. Niedrach et al., 1963]. With the need for a more efficient and less polluting energy source for the automotive industry and for replacing inefficient energy storage systems for portable devices such as computers of flash lights extensive research is concentrated today on the further development of a reliable and efficient solid polymer electrolyte fuel cell.

3 THE STRUCTURE OF ION-EXCHANGE MEMBRANES

The key components in electrodialysis and related processes are the ion-exchange membranes which resemble highly swollen ion-exchange resins in sheet form. There are two different types of ion-exchange membranes:

- cation-exchange membranes which contain negatively charged groups fixed to the polymer matrix, and
- anion-exchange membranes which contain positively charged groups fixed to the polymer matrix.

In a cation-exchange membrane, the fixed negative charges are in electrical equilibrium with mobile cations in the interstices of the polymer as indicated in Figure 1.1, which shows schematically the matrix of a cation-exchange membrane with fixed anions and mobile cations, the latter are referred to as counterions.

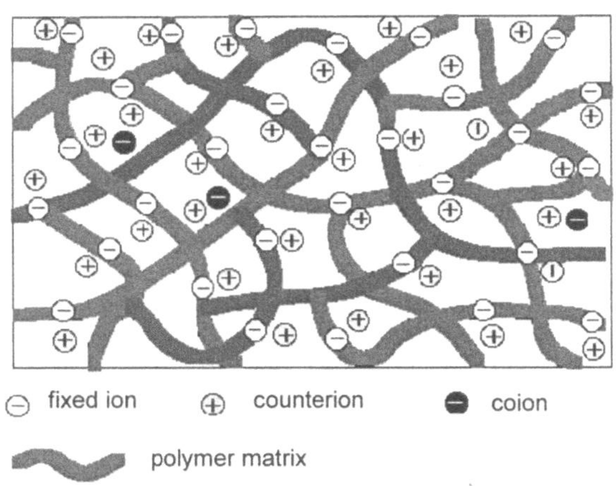

Fig. 1.1 Schematic drawing illustrating the structure of a cation-exchange membrane.

The mobile anions are referred to as coions. They are more or less completely excluded from the polymer matrix because of their electrical charge which is identical to that of the fixed ions. This type of exclusion is called Donnan exclusion in honor of his pioneering work [F.G. Donnan, 1911], [F.G. Donnan et a.,1932]. Due to the exclusion of the coions, a cation-exchange membrane permits transfer of cations only. Anion-exchange membranes carry positive charges fixed on the polymer matrix. Therefore, they exclude all cations and are permeable to anions only.

Thus, the selectivity of ion-exchange membranes results from the exclusion of coions from the membrane phase. To which extent the coions are excluded from an ion-exchange membrane depends on membrane as well as on solution properties. Furthermore, cation- and anion-exchange membranes are classified as strong acid and strong base, or weak acid and weak base membranes depending on the degree of dissociation of the fixed charged group in an aqueous solution. Strong acid membranes contain sulfonic acid as charged groups. In weak acid membranes carboxylic acid is the charge carrying group. Quaternary and tertiary amines are the fixed charges in strong and weak base cation-exchange membranes, respectively [F. Bergsma et al., 1961].The properties of ion-exchange membranes are determined by different parameters such as the density of the polymer network, the hydrophobic or hydrophilic character of the matrix polymer and the type and concentration of the charges in the polymer, and the morphology of the membrane itself [F. Helfferich, 1962]. The most desired properties of ion-exchange membranes are:

- high permselectivity
- low electrical resistance
- good mechanical and form stability
- high chemical and thermal stability

Many of today's available membranes meet most of these requirements. Especially the Nafion-type® cation-exchange membrane has quite satisfactory properties for applications in the chlorine-alkaline electrolyses as well as in electrodialysis [A. Eisenberg et al., 1982]. Anion-exchange membranes often show lower stability in strong alkaline solutions than cation-exchange membranes.

4 ION-EXCHANGE MEMBRANE PROCESSES

Ion-exchange membrane processes can conveniently be divided into three types. The first type of ion-exchange membrane processes involves the separation of components such as salts or acids and bases from electrolyte solutions. In these processes the driving force for the transport of ions from one phase to another is an electrical potential as in conventional electrodialysis or a concentration gradient as in diffusion dialysis and Donnan dialysis.

The second type of processes involves an electrochemical reaction producing certain chemicals such as chlorine, acids and bases, or inorganic and organic compounds. The chlorine-alkaline process is the most prominent electrochemical production process having an ion-exchange membrane as a key component.

The third type of ion-exchange membrane processes involves the conversion of chemical into electrical energy and vice versa. The solid polymer electrolyte fuel cell is today the most significant application of ion-exchange membranes in energy conversion devices.

4.1 Ion-Exchange Membranes in Separation Processes

In ion-exchange membrane separation processes such as diffusion dialysis, Donnan dialysis, and electrodialysis many aspects resemble ordinary dialysis with neutral membranes. In all four processes certain low molecular weight components are removed from a feed solution through a semipermeable membrane. In dialysis, however, the driving force for the transport of the components is a concentration difference between the two solutions separated by a membrane. In diffusion dialysis and Donnan dialysis the driving force for the process is also a concentration gradient, but the membrane separating the two solutions is carrying positive or negative electrical charges in form of fixed ions, and the permeability of the membrane for different ions is determined by the charges of the membrane and the permeating ions. The two processes will be discussed in more detail later. In electrodialysis the driving force for the transport of certain components from a feed solution to a receiving solution is an electrical potential difference across a membrane separating the two solutions. Modifications of the classical electrodialysis process, which was limited to the separation and concentration of salts, acids, and bases from aqueous solutions, have led to many new applications. Combined with electrically forced water dissociation in bipolar membranes electrodialysis is also used to produce acids and bases from the corresponding salts [K.N. Mani, 1991]. Diffusion dialysis utilizing anion-exchange membranes only is used today on a large scale to recover acids from pickling solutions [Y. Kobuchi et al., 1987]. Donnan dialysis is used for water softening or for recovering organic acids from their salts.

4.1.1 Electrodialysis

The principle of electrodialysis is illustrated in Figure 1.2 which shows a schematic diagram of a typical electrodialysis cell arrangement consisting of a series of anion- and cation-exchange membranes arranged in an alternating pattern between an anode and a cathode to form individual cells. A cell consists of a volume with two adjacent membranes. If an ionic solution such as an aqueous salt solution is pumped through these cells and an electrical potential is established between the anode and cathode, the positively charged cations migrate towards the cathode and the negatively charged anions towards the anode. The cations pass easily through the negatively charged cation-exchange membrane but are retained by the positively charged anion-exchange membrane. Likewise, the negatively charged anions pass through the anion-exchange membrane, and are retained by the cation-exchange membrane. The overall result is an increase in the ion concentration in alternate compartments, while the other compartments simultaneously become depleted. The depleted solution is generally referred to as the diluate and the concentrated solution as the brine or the concentrate. The driving force for the ion transport in the electrodialysis process is the applied electrical potential between the anode and cathode. Figure 1.2 shows only three cation- (C) and three anion-exchange membranes (A). An actual electrodialysis unit which is referred to as a stack may have a few hundreds of such membranes [M.S. Mintz, 1963]. The total space occupied by the diluate solution between two contiguous membranes, the concentrated solution between two contiguous membranes next to the diluate chamber, and the two contiguous anion- and cation-exchange membranes make up a cell pair.

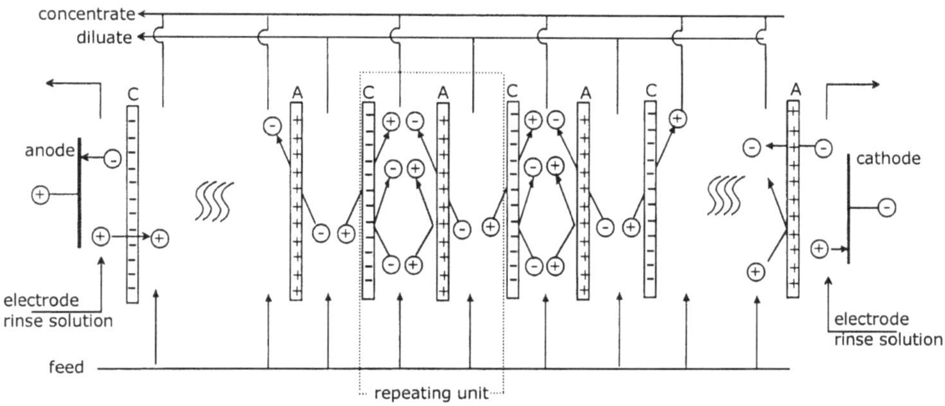

Fig. 1.2 Schematic diagram illustrating the principle of desalination by electrodialysis in a stack with cation- and anion-exchange membranes in alternating series between two electrodes.

A device composed of individual cells in alternating series with electrodes on both ends is referred to as an electrodialysis stack. The cell pair is a repeating unit in an electrodialysis stack. The electrode containing cells are often rinsed with a separate solution which does not contain Cl^--ions to avoid chlorine formation.

Although there are many different components necessary for the proper and efficient operation of an electrodialysis plant such as the electrical power supply, pumps, and control and monitoring devices, the stack is the key element. There are various stack design concepts described in the literature [L.H. Schaffer et al., 1966]. In praxis only two different stack designs are used on a large scale. One is the so-called sheet-flow and the other the tortuous path-flow concept. Both stack designs will be discussed in more detail later.

Electrodialysis is used mainly today for desalination of saline solutions either to produce concentrated brines and salt depleted waters for industrial use or potable water mainly from brackish water sources [R.E. Lacey, 1972]. Although of major importance, water desalination is by no means the only significant application. Stimulated by the development of new ion-exchange membranes with better selectivity, lower electrical resistance, and improved thermal, chemical, and mechanical properties, other uses of electrodialysis, especially in the food, drug, and chemical process industry as well as in biotechnology and waste water treatment, have recently gained a broader interest [T. Sata, 1986]. In many applications, electrodialysis is in direct competition with other separation processes such as distillation, ion-exchange, reverse osmosis, and various chromatographic procedures. In other applications, there are very few other economic alternatives to electrodialysis. Because of the inherent features of the process, electrodialysis plays an important role today in large scale industrial mass separation.

4.1.2 Electrodialysis with Bipolar Membranes

The conventional electrodialysis can be combined with bipolar membranes and utilized to produce acids and bases from the corresponding salts. In this process cation- and anion-exchange membranes are installed together with bipolar membranes in alternating series in an electrodialysis stack as illustrated in Figure 1.3. A cation-exchange, an anion-exchange, and a bipolar membrane form a repeating unit in the stack between two electrodes. Thus, a repeating unit is composed of three separate flow streams, i.e. the feed solution flow stream containing one salt and two product solution flow streams containing an acid and a base, respectively.

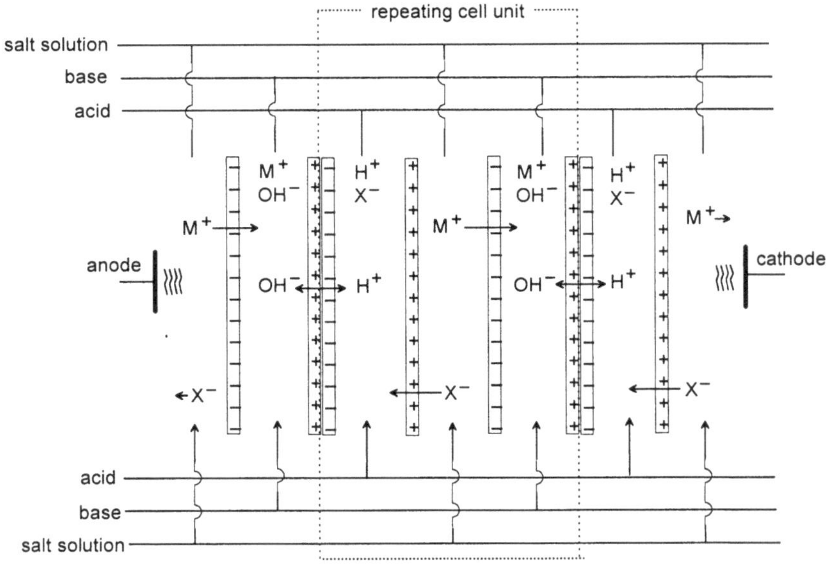

Fig. 1.3 Schematic drawing illustrating the principle of electrodialytic production of acids and bases from the corresponding salts with bipolar membranes.

The utilization of electrodialysis with bipolar membranes is economically very attractive and has a multitude of interesting potential applications. The key component in this process is the bipolar membrane. Its function and properties determine to a large extent the overall efficiency of the process. A bipolar membrane should have adequate water dissociation capability, low electrical resistance, high permselectivity, and a long useful life under operating conditions, which means it must be stable in highly concentrated acid and alkaline solutions. A bipolar membrane consists of a laminate of a cation- and an anion- exchange layer. If an electrical potential difference is established across the membrane charged species are removed from the interphase between the two ion-exchange layers. When this interphase contains only water the transport of electrical charges can be accomplished only by protons and hydroxide ions, which are available in a concentration of ca. 1×10^{-7} moles per liter. The protons and hydroxide ions removed from the interphase between the cation- and anion-exchange membranes are replenished due to the water dissociation equilibrium [K.N. Mani, 1991]. The process of electrodialytic water dissociation and the structure of a bipolar membrane will be discussed in more detail later.

4.1.3 Continuous Electrodeionization

Continuous electrodeionization is very similar to conventional electrodialysis. However, the cell which contains the desalted water, i.e. the diluate flow stream is filled with a mixed-bed ion-exchange resin. The principle of the process is illustrated in Figure 1.4. The mixed bed ion-exchange resin in the diluate cell of an electrodialysis stack binds the ions of a feed solution. Due to an applied electrical field the ions migrate through the ion-exchange bed towards the adjacent concentrate cells. The ion-exchange resin increases the conductivity in the diluate cell substantially and at very low salt concentrations in the feed solution water is dissociated at the contact point of the cation- and anion-exchange resin beads generating protons and hydroxide ions which further replace the salt ions in the resins [Z. Matejka, 1971]. The result is completely deionized water as a product.

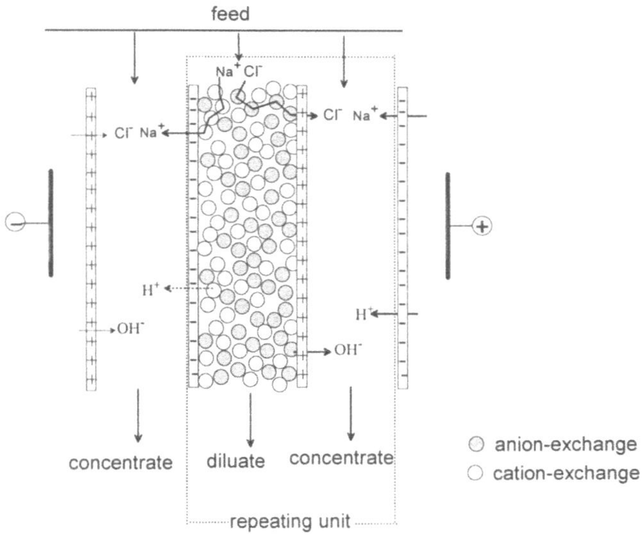

Fig. 1.4 Schematic drawing illustrating the principle of the continuous electrodeionization process.

A disadvantage of the process is the relatively poor current utilization which can in most practical applications, however, be tolerated. Compared to the deionization by a conventional mixed bed ion-exchange resin the continuous electrodeionization has several advantages since no chemicals are needed for the regeneration of the ion-exchange resins which is time consuming, labor intensive, and generates a salt containing waste water [A.J. Giuffrida et al., 1986].

4.1.4 Diffusion Dialysis

Diffusion dialysis is used mainly today to recover acids or bases from a mixture with salt ions. Its principle is illustrated in Figure 1.5 which shows a schematic diagram of a typical diffusion dialysis cell arrangement consisting of a series of anion-exchange membranes arranged in parallel to form individual cells. A cell consists of a volume with two adjacent membranes. If a feed solution containing a salt and an acid and pure water as so-called stripping solution are fed into adjacent cells the individual ions will diffuse from the salt and acid containing feed solution through the ion-exchange membrane into the cell with the stripping solution due to a concentration difference. The process is very similar to conventional dialysis with the difference that ion-exchange membranes instead of uncharged membranes are installed in the stack. The driving force for the transport of the different ions from a feed to a receiving solution is their concentration difference between the two solutions. Since the membranes carry positive or negative fixed charges they show different permeabilities for cations and anions. If the cell system contains anion-exchange membranes only as indicated in Figure 1.5 the anion can permeate the membrane freely but the salt cations will be retained by the membrane. The protons, however, can pass the anion-exchange membrane in spite of their positive charge because they are transported by a tunneling mechanism as will be discussed later in more detail. Thus, the acid will be removed from the salt solution. Correspondingly, a base can be removed from mixtures with salts if cation-exchange membranes are used.

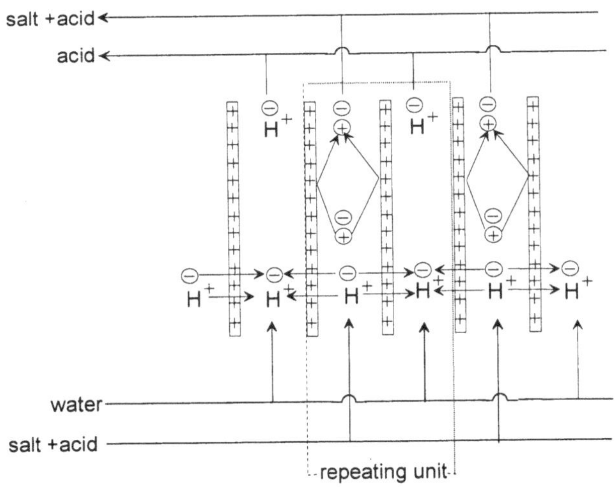

Fig. 1.5 Schematic drawing illustrating the principle of diffusion dialysis used to recover an acid from a mixture with salt in a stack of anion-exchange membranes only.

Diffusion dialysis is used today on a large scale to recover acids from pickling solutions in the metal surface treating industry [Y. Kobuchi et al., 1987]. However, its technical and also commercial relevance is still rather limited. The main reason for the reluctant use of diffusion dialysis are the costs. Since the diffusion through the relatively thick ion-exchange membranes is a rather slow process, large membrane areas are required to remove a significant amount of ions from a feed solution resulting in high investment costs for a given capacity plant.

4.1.5 *Donnan Dialysis*

The principle of Donnan dialysis is very similar to that of diffusion dialysis. Here again only cation- or anion-exchange membranes are installed in a stack as shown in Figure 1.6. The driving force for the transport of ions through the membranes is their concentration difference in the two phases separated by the membranes. A typical application of Donnan dialysis is the removal of divalent ions such as Ca^{++} from a feed stream by exchange for monovalent ions such as Na^+ in water softening. This is illustrated in Figure 1.6.

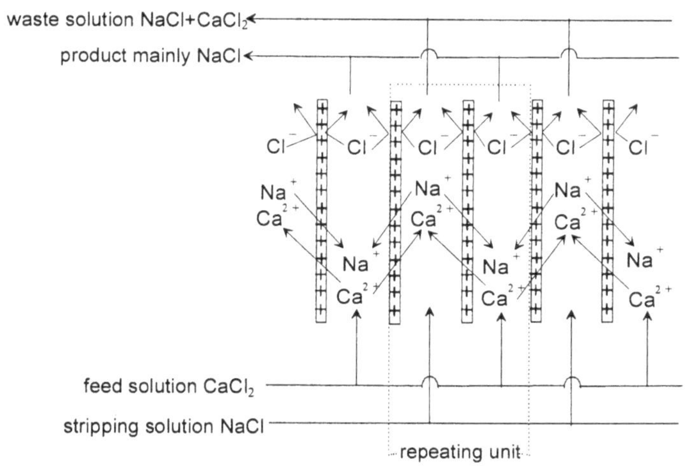

Fig. 1.6 Schematic drawing illustrating the principle of the Donnan dialysis water softening process by the exchange of Na^+- and Ca^{++}-ions in a stack with cation-exchange membranes only.

A feed solution containing $CaCl_2$ in relatively low concentration and a stripping solution containing NaCl in relatively high concentration is passed into alternating cells of a stack of cation-exchange membranes. Because of the concentration difference in the feed and the stripping solution Na^+ -ions diffuse from the stripping solution through the cation-exchange membrane into the feed solution. Since the Cl^--ions cannot permeate the negatively charged cation-exchange membrane an electrical potential is generated between the two solutions which acts as driving force for the transport of Ca^{++}-ions from the feed to the stripping solution. Because of the required electroneutrality the identical charges are exchanged between the two solutions, i.e., for two Na^+ -ions diffusing from the stripping into the feed solution one Ca^{++} -ion is removed from the feed solution if the membrane is completely impermeable for Cl^- -ions. The ion transport in Donnan dialysis is referred to as counter current transport. It will be discussed in more detail later.

As a major application of Donnan dialysis water softening was envisioned. However, the process is in direct competition with conventional ion-exchange technology and apparently not economic. There are several other interesting applications in waste water treatment and the recovery of heavy metals from industrial effluents which have been studied on a laboratory scale [T.A. Davis et al., 1971]. But up to today there is very little large scale commercial use of Donnan dialysis.

4.1.6 Electrolytic Processes Using Ion-Exchange Membranes

Electrolytic processes are used for the production of chlorine and caustic soda [D. Pletcher et al., 1990], or hydrogen and oxygen, or special organic compounds [T. Sata, 1991]. The principle of the chlorine/alkaline process is illustrated in the schematic drawing of Figure 1.7 which shows the electrolytic production of sodium hydroxide and chlorine.

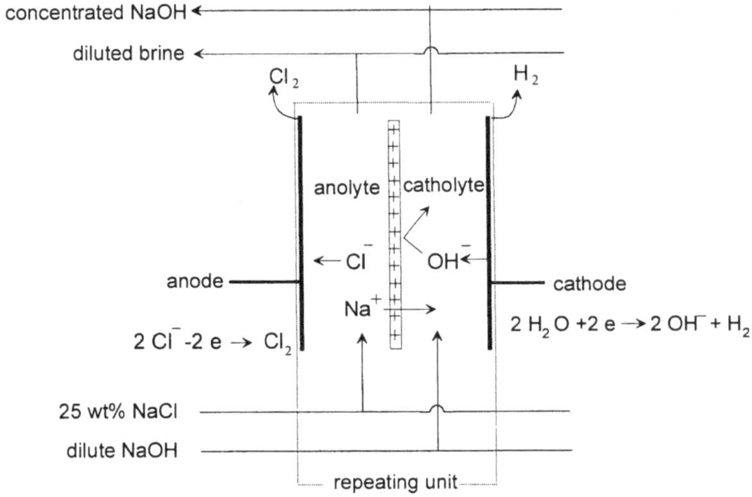

Fig. 1.7 Schematic drawing illustrating the principle of the electrolytic production process of chlorine and sodium hydroxide.

The cell unit used in the process is composed of two electrodes separated by a cation-exchange membrane forming two compartments. The compartment with the anode contains the anolyte, i.e. a 25 wt % NaCl solution, and the other compartment with the cathode contains the catholyte, i.e. dilute sodium hydroxide. When an electrical potential between the electrodes is applied the chloride ions in the anode compartment migrate towards the anode where they are oxidized and form chlorine which is released as gas. The sodium ion from the salt solution migrates through the cation-exchange membrane towards the cathode where it is reduced to sodium metal, which immediately reacts with the water to sodium hydroxide and hydrogen which is then released as gas. A multitude of the cell units shown in Figure 1.7 is usually integrated in a stack using bipolar electrodes [K. Motani et al., 1980].

The cation-exchange membrane used in the electrolytic chlorine alkaline production process should have a low resistance for the transport of Na^+-ions, it should have a low water transport, and it should retain the OH^--ions as complete as possible. Furthermore, it must have excellent chemical and thermal stability to withstand the environment of concentrated sodium hydroxide and chlorine at elevated temperature. With the development of a high performance perfluorinated membrane the process has become the standard technique for the production of caustic soda and chlorine.

Another application of the electrolytic process which might become interesting is the production of hydrogen and oxygen using electrical energy generated by photo-voltaic energy conversation. However, in this application there is no need for an efficient ion-exchange membrane. A diaphragm that has a high ion permeability and prevents the mixing of the two gases is sufficient for separating the solutions in the compartments of the cathode and anode.

The chlorine-alkaline and the hydrogen-oxygen production as well as organic synthesis processes with ion-exchange membranes will not further be described in this book which will discuss separation processes only.

4.1.7 The Solid Polymer Electrolyte Fuel Cell

A fuel cell is an electrochemical reactor in which chemical energy is converted into electrical energy. There are different types of fuel cells operated at different temperatures between 100 and 1000 °C. The low temperature fuel cell utilizes an ion-exchange membrane as an ion-conducting layer between two electrodes. In fuel cells operated at high temperature, i.e. in excess of 700 °C solid oxide ceramic is used for the proton transport. The principle of the so-called polymer electrolyte fuel cell is illustrated in Figure 1.8 which shows a single unit cell consisting of two porous electrodes, two catalyst layers, and a cation-exchange membrane in between.

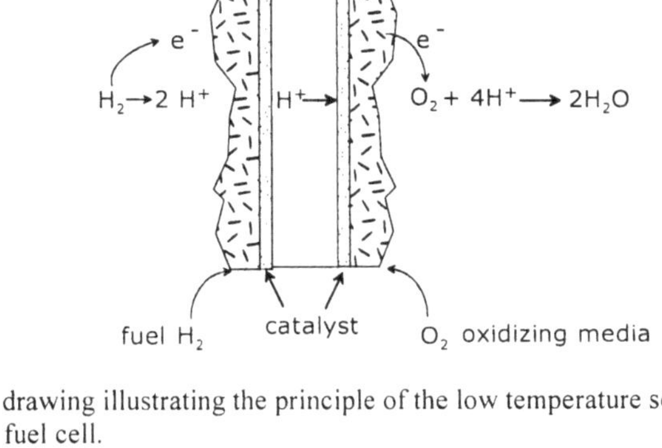

Fig. 1.8 Schematic drawing illustrating the principle of the low temperature solid polymer electrolyte fuel cell.

A fuel such as hydrogen gas is passed through the porous anode. It reacts in the catalyst layer forming protons and releasing electrons at the anode to an electric circuit. The protons diffuse through the ion-exchange membrane and react in the catalyst layer at the surface of the porous cathode with an oxidizing media such as oxygen gas to form water and take electrons from the electric circuit. The overall reaction in a fuel cell is the oxidation of hydrogen by oxygen to water. The reaction has a negative Gibb's free energy. That means work is done by the system and can be utilized as electrical energy. Fuel cells are free from the limitations of the Carnot process and thus a very efficient energy source.

The low temperature fuel cell with the ion-exchange membrane as proton conductive layer is of prime importance as energy source for mobile applications such as in automobiles, computers, cellular phones, etc. Future developments are focusing on the replacement of hydrogen by a more readily transportable liquid fuel such as methanol. A more detailed discussion of fuel cells is beyond the scope of this book and must be referred to the corresponding text books and literature [K.R. Williams, 1966.].

There are many more very interesting electrolytic processes with ion-exchange membranes as key elements used today in electro-organic synthesis and environmental oriented electrochemistry. But a discussion of these processes also goes far beyond the scope of this book, which is concentrated on the treatment of ion-exchange membrane separation, i.e. dialytic processes such as electrodialysis and closely related processes.

4.2 Technical Relevance of Ion-Exchange Membrane Processes

Today electrodialysis is the most important separation process using ion-exchange membranes and an electrical potential driving force as far as economic relevance is concerned. It is used on a large industrial scale in two major areas of application. One is the desalination of salts containing solutions such as brackish water or industrial effluents, and the second is the concentration of electrolytes such as table salt or certain acids and bases. Both applications have experienced a steady growth over the recent years and have developed into significant industrial processes. Water desalination is an application of electrodialysis that is utilized all over the world for the production of potable water from brackish water sources. In Japan and Korea electrodialysis is used today as pre-concentration step for the production of table salt from sea water. But more recently electrodialysis has found new applications both in its capacity of concentrating electrolyte solutions and in water purification. In the treatment of certain industrial effluents electrodialysis is used to recycle the water as well as valuable water constituents. In combination with other processes such as ion-exchange or reverse osmosis electrodialysis plays an increasingly important role today in the production of high quality industrial water.

The continuous electrodeionization is replacing more and more the conventionally used mixed bed ion-exchange resin in the production of ultra pure water. The process has significant technical and commercial advantages and will most likely replace to a large extent the conventional ion-exchange technology.

Electrodialysis with bipolar membranes is still in a state of development. In spite of the identification of a large number of potential applications there is hardly any large scale commercial use. This is mainly due to the inefficiency of today's bipolar membranes. One promising potential application of electrodialysis with bipolar membranes is the production of caustic soda replacing the conventional electrolytic alkaline-chlorine process. But insufficient permselectivity of the presently available bipolar membranes leads to a contamination of the produced acids and bases with salt. There are other applications mainly in waste water treatment and in the recovery of organic acids from fermentation processes were electrodialysis with bipolar membranes may play a more important role in the future.

Diffusion dialysis is used on a small scale today to recover acid from salt solutions obtained as effluents from metal surface treatment processes. But until now there are very few large scale installations of the process mainly for economic reasons.

The same is true for Donnan dialysis. The process has been studied extensively on a laboratory scale. But there are no real large scale industrial applications.

4.2.1 Desalination by Electrodialysis

The production of potable water from brackish water sources is certainly the most important application of electrodialysis. The capacity of the worldwide installed electrodialysis desalination plants is difficult to estimate. Some data concerning the production of potable water in the western world including the Middle East and Japan are available from sources such as the desalination inventory report published by the International Desalination Association in their magazine *Desalination &Water Reuse*. Substantial desalination capacity, however, is also installed in the Commonwealth of Independent States of the former Soviet Union [W.A. McRae, 1992] and in China [Gao Congjie, 1992]. A worldwide installed capacity of 1 to 2 million m^3 per day seems to be a realistic estimation [K. Wangnick, 2000]. The capacity of the individual desalination plants varies widely from a few 100 $m^3 day^{-1}$ to 20.000 $m^3 day^{-1}$. The brackish water treated by electrodialysis has generally a relatively low salt concentration in the range of 2000 to 5000 mg L^{-1} total dissolved solids.

Electrodialysis is competing in brackish water desalination with other processes such as low pressure reverse osmosis and nanofiltration. For certain feed water compositions it has, however, significant technical advantages over the competing processes. There are several features which make electrodialysis especially suited for the desalination of brackish waters. Electrodialysis is less sensitive to membrane fouling and scaling than reverse osmosis. Therefore, higher recovery rates can be achieved and brine disposal problems can be minimized. Ion-exchange membranes can be operated at elevated temperatures in excess of 50 °C without any effect on their chemical and mechanical stability or performance. The membranes are also quite stable at high and low pH-values and less sensitive to oxidizing agents than most reverse osmosis membranes. Brackish waters obtained from deep wells have often a high concentration of divalent ions and come at elevated temperature. These raw waters are difficult to desalt by reverse osmosis without significant pre-treatment including cooling. For electrodialysis, however, they exhibit no problems and can be processed without any or a minimum of pre-treatment. Even if electrodialysis has in some cases significantly higher investment and operating costs than low pressure reverse osmosis it is often the preferred process in brackish water desalination because of clear technical advantages.

By now electrodialysis has 25 years of proven history in desalination of brackish water for the production of potable water. This application will certainly continue into the future. But electrodialysis becomes more and more important for the production of boiler feed water to be used in power stations, for the treatment of waste water as well as cooling tower blow-downs, for the removal of nitrate and fluoride from potable water sources, for the recovering of heavy metals from industrial effluents, for the recycling of rinse waters from electroplating industry and for the treatment of dump leach water. In addition, electrodialysis is used today in the food industry for the demineralization of whey and in the chemical and pharmaceutical industry for the desalination of protein or dextran solutions or for recovering organic acids from fermentation processes. Although many of these applications require relatively small plant capacities they do provide a substantial future market for electrodialysis which might well exceed that of today's potable water production.

4.2.2 Concentration of Mineral Salts and Other Applications of Electrodialysis

The pre-concentration of sea water for the production of table salt is presently by far the largest application of electrodialysis in the field of brine concentration [T. Kawahara, 1992]. Several 100,000 tons of table salts are produced from sea water per year using electrodialysis as a pre-concentration step. The application is limited mainly to Japan, Korea and several other places in the world where there are no or limited native salt deposits and where large amounts of rainfall make solar evaporation of sea water a relatively expensive means of producing salts. In principle the electrodialytic concentration of brine does not differ from the desalination of brackish waters. There are, however, some specific problems which must be addressed. In the production of table salt a preferential transfer of sodium chloride is desirable to avoid precipitation of divalent ions in the concentrate. Therefore, membranes with a certain transport selectivity for univalent ions are needed. To minimize power consumption these membranes should also exhibit a low electrical resistance and low water transfer which are caused by osmotic or electroosmotic effects. These membranes have been developed in recent years and are very effectively used today. But the production of table salt is not the only relevant application of electrodialytic concentration. The process can also be used to recover valuable salts from sea water or other feed solutions such as industrial effluents or process streams. Here, the concentration of salts is generally combined with a recycling of the water and the water constituents. In certain areas reverse osmosis brine disposal is a problem and electrodialysis is used for further concentration to minimize environmental pollution.

4.2.3 The Continuous Electrodeionization and the Use of Bipolar Membranes

The combination of conventional electrodialysis with ion-exchange resins or with bipolar membranes have extended the application again substantially. Today, the use of electrodialysis goes far beyond the production of potable water from brackish water sources or the production of table salt from sea water. The continuous electrodeionization is today a proven and reliable technology. The main application of the process is in the production of ultra pure water for the semi-conductor industry and for analytical and chemical laboratories. The process and its application is described in great detail in various patents and company brochures [A.J. Giuffrida et al.,1986], [T.R.E. Kressman,1961].

While the continuous electrodeionization is already well established and a commercially available process, is the application of electrodialysis with bipolar membranes still in a state of development. There is a large variety of potential applications of the process in the chemical and the environmental protection industry as well as in biotechnology. Typical applications of the process that have been studied on a laboratory or pilot plant scale are the production of alkali hydroxides such as sodium or potassium hydroxides and mineral acids such as hydrochloric, hydrofluoric or sulphuric acid from the corresponding salt solutions [K.J. Liu et al., 1977]. The recovery of organic acids such as formic, acetic, citric, itaconic acid or certain amino acids from a fermentation broth is another interesting application of bipolar membranes. The adjustment of the pH-values in surface and sea water to control scaling, or in fermentation and chemical production processes, without increasing the ion potential can also be done by the electrodialytic water dissociation with bipolar membranes.

The large scale use, however, is effected by shortcomings of the today's available bipolar and monopolar membranes [H. Strathmann et al., 1993]. Electrodialysis with bipolar membranes is studied today mainly on a laboratory scale. There are still a large number of problems which must be solved before this very interesting process can be applied reliably on a large industrial scale.

4.2.4 Diffusion Dialysis and Donnan Dialysis

The main application of diffusion dialysis is in the recovery of mineral acids such as HCl, HF, HNO_3, H_2SO_4, etc. from a solution containing these acids in a mixture with metal salts. Mineral acids are used widely in the metal processing industry for surface finishing in a so-called pickling procedure. During this process the acids get contaminated with metal ions. When the metal ion concentration exceeds a certain value the acid becomes ineffective and the salt-acid mixture must be disposed as waste. For environmental and economic reasons it is desirable to recover the acids from the waste stream. This can be done very energy efficiently by diffusion dialysis [Y. Kobuchi et al., 1987]. However, there are several problems related to the process. One is the stability of the membranes in the very aggressive environment of strong acids such as fluoric or nitric acid. Another problem is the relatively low membrane flux that is obtained in diffusion dialysis. Since the transfer of the acids or bases from the feed solution into the receiving flow stream is due to diffusion through a relatively thick membrane with the concentration difference between the two solutions being the only driving force, the specific fluxes are small and large membrane areas are required for a certain capacity plant. Special property membranes which meet the requirement for diffusion dialysis have been developed in Japan and they are used there on a large scale [J. Sato et al., 1985].

Donnan dialysis is used for the treatment of certain industrial effluents to recover valuable or toxic heavy metal ions. In this application Donnan dialysis is competing directly with conventional ion-exchange technology. For economic reasons Donnan dialysis is not used on a large industrial scale today in spite of obvious advantages.

4.3 Relevance of Electrolytic Production Processes and Fuel Cells

Ion-exchange membranes are used on a large scale in organic synthesis and chemical production processes [J. Jörissen, 1996], [Z. Ogumi et al., 1981], [E. Raoult et al., 1984] and in energy storage and conversion systems. Today, these processes have reached a significant technical and commercial relevance which is rapidly increasing and which may in the near future exceed that of electrodialysis and related processes. The most technical relevant ion-exchange membrane processes involving electrochemical reaction or a conversion of energy are the electrolytic production of chlorine and alkaline and the fuel cell. Both processes are extensively described in the literature [T.C. Wellington, 1993], [K.V. Kordesch et al.,1989] and the treatment of these processes is beyond the scope of this book. Several literature references discussing these processes are recommended for further reading.

5 RECOMMENDED FURTHER READING:

5.1 Ion-Exchange Membrane Preparation

Bergsma, F., Kruissink, Ch.A., 1961. Ion-exchange membranes, Fortschr. Hochpolym.-Forsch. 21: 307-362.

Davis, T.A., Genders, J.D., Pletcher, D., 1997. Ion permeable membranes, Alresford Press Ltd. Alresford, Hants, England.

Eisenberg, A., Yeager, H. L., 1982. Perfluorinated ionomer membranes, ACS Symposium Series 180, Washington, DC, American Chemical Society.

Flett, D.S., 1983. Ion-exchange membranes, E. Horwood Ltd., Chichester, U.K.

Helfferich, F., 1962. Ion-exchange, McGraw-Hill Book Co., New York.

5.2 Electrodialysis

Schaffer, L.H., Mintz, M.S., 1966. Electrodialysis, in: Principles of desalination, edt.: Spiegler, K.S., pp. 3-20, Academic Press, New York.

Spiegler, K.S., 1956. Electrochemical operations, in: Ion-exchange technology, edts.: Nachod, F.C., Schubert, J., Academic Press, New York.

Wilson, J.R., edt., 1960. Demineralization by electrodialysis, Butterworth Scientific Publications, London.

Huffmann, E.L., Lacey, R.E., 1972. Engineering and economic considerations in electromembrane processing, in: Industrial processing with membranes, edts.: Lacey, R.E., Loeb, S., pp. 39-55, John Wiley & Sons, New York.

Strathmann, H., 1995. Electrodialysis and related processes, in: Membrane separation technology, edts.: Nobel, R.D., Stern, S.A., pp. 213-281, Elsevier, Amsterdam

5.3 Bipolar Membrane Electrodialysis

Liu, K.J., Chlanda, F.P., Nagasubramanian, K.J., 1977. Use of bipolar membranes for generation of acid and base: An engineering and economic analysis, J. Membrane Sci., 2: 109-124.

Kemperman, A.J.B., edt., 2000. Handbook on bipolar membrane technology, Twente University Press, Endschede , The Netherlands, ISBN 9036515203.

Simon, R., 1984. Electric field effects on proton transfer between ionisable groups and water in ion exchange membranes, Electrochim. Acta, 29: 151-158.

Strathmann, H., Krol, J.J., Rapp, H.J., Eigenberger, G. 1997. Limiting current density and water dissociation in bipolar membranes, J. Membrane Sci., 125: 123-142.

5.4 Continuous Electrodeionization

Giuffrida, A.J., Andover, N., Jha, A.D., Ganzi, G.C., 1986. Electrodeionization apparatus, US Patent, 4,632,745, December 10.

Giuffrida, A.J., Andover, 1990. Electrodeionization method and apparatus, US Patent, 4,931,160, June 5.

Matejka, Z., 1971. Continuous production of high purity water by electrodeionization, J. Appl. Chem. Biotechnol., 21: 117-120.

Ganzi,G.C., 1988. Electrodeionization for high purity water production, in: New membrane materials and processes for separation, edts.: Sirkar, K.K., Lloyd, D.R., AIChE Symposium Series, 84: 73-83.

5.5 Electrolytic Chlorine and Alkaline Production

Eisenberg, A., Yeager, H. L., edts., 1982. Perfluorinated ionomer membranes, ACS Symposium Series 180, American Chemical Society, Washington, DC.

Coulter, M.O., edt., 1980. Modern chlor-alkali technology, Society of Chemical Industry, Vol. 1, Ellis Horwood, Ltd. Publ., Chichester.

Jackson, C., edt., 1983. Modern chlor-alkali technology, Society of Chemical Industry, Vol. 2, Ellis Horwood, Ltd. Publ., Chichester.

Wall, K. edt., 1986. Modern chlor-alkali technology, Society of Chemical Industry, Vol. 3, Ellis Horwood, Ltd. Publ., Chichester.

Prout, N.M., Moorhouse, J.S., edts., 1990. Modern chlor-alkali technology, Society of Chemical Industry, Vol. 4, Ellis Horwood, Ltd. Publ., Chichester.

Wellington, T.C., edt., 1993. Modern chlor-alkali technology, Society of Chemical Industry, Vol. 5, Ellis Horwood, Ltd. Publ., Chichester.

5.6 Fuel Cells and Electrochemical Synthesis

Gould, R.F., 1965. Fuel cell systems, Advances in Chemistry Series No. 47, American Chemical Society, Washington, DC.

Kordesch, K.V., Tambasco de Oliveira, J.C., 1989. Fuel cells, in: Ullmann's Encyclopedia of Industrial Chemistry, 5. Edition, Vol. A12, 55–83, Verlag Chemie, Weihnheim.

Mitchell, W., edt., 1963. Fuel cells, Academic Press, New York.

Williams, K.R., 1966. An introduction to fuel cells, Elsevier Publ., London.

Jörissen, J., 1996. Ionenaustauscher-Membranen in der Elektrolyse und elektroorganischen Synthese, VDI-Fortschritt-Berichte Verfahrenstechnik, 442, VDI Verlag, Düsseldorf, Germany, ISBN 3-18-344203-5.

LITERATURE:

Bergsma, F., Kruissink, Ch.A., 1961. Ion-exchange membranes, Fortschr. Hochpolym.-Forsch., 21: 307-362.

Donnan, F.G., 1911. The theory of membrane equilibrium in presence of a non-dialyzable electrolyte, Z. Electrochem., 17: 572-581.

Donnan, F.G., Guggenheim, E.A., 1932. Exact thermodynamics of membrane equilibrium, Z. Phys. Chem. A, 162: 346-360.

Davis, T.A., Wu, J.S., Baker, B.L., 1971. Use of Donnan equilibrium principle to concentrate uranyl ions by an ion-exchange membrane process, AIChE Journal, 17: 1006-1008.

Eisenberg, A., Yeager, H. L., edts., 1982. Perfluorinated ionomer membranes, ACS Symposium Series 180, American Chemical Society, Washington, DC.

Gao Congjie, 1992. Electrodialysis in China, International Desalination &Water Reuse, Vol. 2/4, pp. 37-38.

Giuffrida, A.J., Andover, N., Jha, A.D., Ganzi, G.C., 1986. Electrodeionization apparatus, US Patent, 4,632,745, December 10.

Grot, W.G., 1973. Laminates of support material and fluorinated polymer containing pendant side chains containing sulfonyl groups, U.S. Patent 3, 770,567.

Grot, W.G., 1974. Films of fluorinated polymer containing sulfonyl groups with one surface in the sulfonamide or sulfonamide salt form and a process for preparing such, US Patent, 3,784,399, January 8.

Grot, W.G., 1975. Perfluorierte Kationenaustauscher Polymere, Chem.-Ing.-Tech., 47: 617.

Groves, W. R., 1839. On a gaseous voltaic battery, Phil. Mag. S. 3, 21: 417-420.

Haagen, K., Helfferich, F., 1959. Ionenaustauschermembranen, German Patent 971 729.

Helfferich, F., 1962. Ion-exchange, McGraw-Hill Book Co, New York.

Jörissen, J., 1996. Ionenaustauscher-Membranen in der Elektrolyse und elektroorganischen Synthese, VDI-Fortschritt-Berichte Verfahrenstechnik, 442, VDI Verlag, Düsseldorf, Germany, ISBN 3-18-344203-5.

Juda, W., McRae, W.A., 1950. Coherent ion-exchange gels and membranes, J. Am. Chem. Soc., 72: 1044.

Kawahara, T., 1992. Industrial applications of ion-exchange membranes, International Desalination &Water Reuse, Vol. 2/4, pp. 26-30.

Katz, W.E., 1979. The electrodialysis reversal (EDR) process, Desalination, 28: 31-40.

Kressman, T.R.E., 1961. Process for concentrating dissolved salts by electrodialysis, US patent 3,014,855, December 26.

Kobuchi, Y., Motomura, H., Noma, Y., Hanada, F., 1987. Application of ion-exchange membranes to recover acids by diffusion dialysis, J. Membrane Sci., 27: 173-179.

Korngold, E., 1975. Electrodialysis processes using ion-exchange resins between membranes, Desalination, 16: 225–233.

Lacey, R.E., 1972. Basis of electro membrane processes, in: Industrial processing with membranes, edts.: Lacey, R. E., Loeb, S., John Wiley & Sons, New York.

Liu, K.J., Chlanda, F.P., Nagasubramanian, K.J., 1977. Use of bipolar membranes for generation of acid and base: an engineering and economic analysis, J. Membrane Sci., 2: 109-124.

Mani, K.N., 1991. Electrodialysis water-splitting technology, J. Membrane Sci., 58: 117-138.

Matejka, Z., 1971. Continuous production of high-purity water by electrodeionization, J. Appl. Chem. Biotechnology, 21: 117-120.

McRae, W.A., 1992. Electrodialysis in the Commonwealth of Independent States of the former Soviet Union, International Desalination & Water Reuse, Vol. 2/4, pp. 37-38

Meyer, K.H., Strauss, W., 1940. La perméabilité des membranes VI, Sur le passage du courant électrique a travers des membranes sélective, Helv. Chim. Acta, 23: 795-800.

Michaelis, L., Fujita, A., 1925. The electric phenomena and ion permeability of membranes II. Permeability of apple peel, Biochem., Z., 158: 28-37.

Mintz, M.S., 1963. Electrodialysis: Principles of process design, Ind. Eng. Chem., 55: 18-28.

Morse H.W., Pierce, G. W., 1903. Diffusion und Übersättigung in Gelantine, Z. Phys. Chem., 45: 589-607.

Motani, K., Sata, T., 1980. The Tokuyama soda membrane–caustic-chlorine process, in: Modern chlor-alkali technology, Society of Chemical Industry, edt.: Coulter, M.O., Vol. 1, pp. 223-234, Ellis Horwood, Ltd. Publ., Chichester.

Niedrach, L.W., Grubb, W.T., 1963. In: Fuel cells, Edt.: Mitchell, W., Academic Press, New York.

Nishiwaki, T., 1972. Concentration of electrolytes prior to evaporation with an electromembrane process, in: Industrial processing with membranes, edts.: Lacey, R.E., Loeb, S., Wiley & Sons, New York.

Oda, Y., Suhara, M., Endo, E., 1977. Process for producing alkali metal hydroxide, US Patent 4 065 366, December 27.

Ogumi, Z., Nishio, K., Yoshizawa, S., 1981. Application of solid polymer electrolyte method to organic electrochemistry II. Electrochemical hydrogenation of olefinic double bonds. Electrochim. Acta, 26: 1779–1782.

Ostwald, W., 1890. Elektrische Eigenschaften halbdurchlässiger Scheidewände. Z. Physik. Chemie, 6: 71-82.

Pletcher, D., Walsh, F.C., 1990. Industrial electrochemistry, Chapman & Hall, London.

Prausnitz, P.H., Reitstotter, J., 1931. Elektrophorese, Elektroosmose, Elektrodialyse in Flüssigkeiten, Steinkopff Verlag Dresden.

Raoult, E., Sarrazin, J., Tallec, A., 1984. Use of ion-exchange membranes in preparative organic electrochemistry: Anodic methoxylation of some olefins, J. Appl. Electrochem., 14: 639–643.

Sata, T., 1986. Recent trends in ion-exchange research, Pure & Appl. Chemistry, 58: 1613-1626.

Sata, T., 1991. Ion-exchange membranes and separation processes with chemical reactions, J. Appl. Electrochem., 21: 283-294.

Sato, J., Onuma, M., Motomura, H., Noma, Y., 1985. Metal Finishing Society of Japan, 32: 220-226.

Schaffer, L.H., Mintz, M.S., 1966. Electrodialysis, in: Principles of desalination, edt.: Spiegler, K.S., pp. 3-20, Academic Press, New York.

Spiegler K.S., 1956, Electrochemical operations, in: Ion-exchange technology, edts.. Nachod; F.C., Schubert, J., Academic Press, New York.

Strathmann, H., Bauer, B., Rapp, H.J., 1993. Better bipolar membranes, CHEMTECH, June, pp: 17-24.

Wangnick, K., 2000. IDA Desalting plants inventory report No.16, International Desalting Association.

Williams, K.R., 1966. An introduction to Fuel Cells, Elsevier Publ., London.

Winger, A.G., Bodamer, G.W., Kunin, R., 1953. Some electrochemical properties of new synthetic ion-exchange membranes, J. Electrochem. Soc., 100: 178-184.

Chapter 2

Electrochemical and Thermodynamic Fundamentals

1 INTRODUCTION

The performance of ion-exchange membrane separation processes is determined by the transport rate of different components in the membranes and the adjacent electrolyte solutions. The transport rate of a component is governed by its mobility and concentration in a given environment and by the driving force or forces acting on the component. The mobility and concentration of a component are determined by its interaction with other components in its surrounding. In electrolyte solutions the interaction between components is defined to a large extent by long-range electrostatic forces with the consequence that on a macroscopic scale electro neutrality prevails, i.e. in a certain volume all positive electrical charges must be counter-balanced by negative charges. The driving forces acting in electrolyte solutions are gradients in the chemical, electrical, or the electrochemical potential. For applying an electrical potential in an electrolyte solution an electrochemical cell is needed. Such a cell consists of two metallic, i.e. electron conductors in contact with an ionic conductor, i.e. an electrolyte. At the electrode/electrolyte interface the electron conductance is converted to an ionic conductance by an electrochemical reaction. Thus, ion-exchange membrane separation processes are determined by electrochemical, thermodynamic, and kinetic parameters, and before entering into a more detailed discussion of these processes and their technical applications, some fundamental electrochemical and thermodynamic relations as well as mass transport phenomena in electrolyte solutions and ion-exchange membranes shall be reviewed.

2 BASIC ELECTROCHEMICAL RELATIONS

In ion-exchange membrane separation processes the ions of an electrolyte solution are transported due to an externally applied or by activity gradients induced electrical potential through a membrane from one electrolyte solution into another or to the surface of an electrode. The membrane may carry positively or negatively charged fixed ions. The negatively charged ions in the solution, i.e. the anions move towards the anode where they are oxidized and the negative charges, i.e. the electrons are transferred in an electrochemical reaction to the electrode. Thus, at the electrode/electrolyte solution interface the transport of electrical charges changes from a transport by ions to a transport by electrons.

Likewise, the positively charged cations are transported due to an electrical potential gradient towards the cathode where they are reduced by receiving electrons in an electrochemical reaction at the electrode/electrolyte solution interface. Thus, the transport of ions in an electrolyte solution between electrodes results in a transport of electrical charges, i.e. an electrical current and can be described by the same relation as the transport of electrons in a metallic conductor, i.e. by Ohm's law.

2.1 Electron and Ion Conductivity and Ohm's Law

The transport of electric charges, i.e. an electric current can be achieved by two characteristic modes [G. Kortüm, 1957]:

- the transport of electrons through solid material such as metals and
- the transport of ions through liquids such as electrolyte solutions.

The conductivity of electrons, i.e. metal conductors is generally 3 to 5 orders of magnitude higher than the conductivity of electrolyte solutions. Furthermore, the conductivity of metals is decreasing with increasing temperature while the conductivity of electrolyte solutions is increasing with temperature. The most important difference between electron and ion conductivity, however, is the fact that ion conductivity is always coupled with a transport of mass while due to the very small mass of an electron virtually no mass is transported in an electron conductor. In spite of the basic difference in the transport mode between electron and ion conductivity the electrical current can be described in both cases by the same mathematical relation which states that the electrical current is proportional to the electrical potential driving force. This relation is referred to as Ohm's law which is given by:

$$U = R\,I \tag{2.1}$$

Here U is the electrical potential between two electron sources, e.g. between two electrodes in an electrolyte solution expressed in Volt, I is the electrical current between the electron sources expressed in Ampere, and R the electrical resistance expressed in Ohm.

The resistance R is a function of the specific resistance of the material, the distance between the electron sources, and the cross section area of the material through which the electric current passes. It is given by:

$$R = \rho \frac{l}{q} \tag{2.2}$$

Here R is the overall resistance, ρ is the specific resistance, l is the length, and q the cross-section area of the conducting material.

The reversal of the resistance and of the specific resistance, respectively, is the conductivity and the specific conductivity, thus is:

$$S = \frac{1}{R} \text{ and } \kappa = \frac{1}{\rho} \tag{2.3}$$

Here S is the conductivity and κ the specific conductivity.

The conductivity of electrolyte solutions depends on the concentration and the valence of the ions in the solution. It is expressed as the equivalent conductivity or molar conductivity [H.S. Harned et al., 1958]. For a solution of a single electrolyte the equivalent conductivity is given by:

$$\Lambda_{eq} = \frac{\kappa}{C(z_a \nu_a + z_c \nu_c)} \tag{2.4}$$

Here Λ_{eq} is the equivalent conductivity, C is the molar concentration of the electrolyte in the solution, z_a and z_c are the valences of the anion and cation, respectively, and ν_a and ν_c are the stoichoimetric coefficients of the anion and cation, respectively.

The stoichoimetric coefficient gives the number of anions and cations in a mole electrolyte and the valence gives the number of charges related to an ion. For example for NaCl ν_c and ν_a are identical and 1 and also z_a and z_c are 1. However, for $MgCl_2$ ν_c is 1 and ν_a is 2, and z_c is 2 and z_a is 1.

The equivalent conductivity can be expressed as the sum of the contributions from its individual ions. Thus is:

$$\Lambda_{eq} = \lambda_a + \lambda_c \tag{2.5}$$

Here Λ_{eq} is equivalent conductivity, λ_a and λ_c are the equivalent conductivities of anion and cation, respectively.

In addition to the equivalent conductivity the molar conductivity can be used. The molar conductivity is related to the equivalent conductivity by:

$$\Lambda_{mol} = \nu_a z_a \lambda_a + \nu_c z_c \lambda_c \tag{2.6}$$

Here is Λ_{mol} the molar conductivity and z_a and z_c are the valences, and ν_a and ν_c the stoichoimetric coefficients of anion and cation, respectively.

The electrical current passing through an electrolyte solution under the driving force of an electrical potential gradient is proportional to the concentration of the ions in the solution, the stoichiometric coefficients for cat- and anions, the number of electrical charges carried by one ion, i.e. its valence and the ion mobility in the electrolyte solution. The number of electrical charges carried by all the ions of an electrolyte under the driving force of an electrical potential gradient through a certain area A is given by:

$$J_e = \sum_i z_i u_i \nu_i C\, e\, N_A \operatorname{grad}\varphi \tag{2.7}$$

Here J_e is the flux of electrical charges, z, u, and ν are the valence, the ion mobility, and the stoichiometric coefficient, respectively; C is the concentration of the electrolyte; e the charge of an electron, N_A the Avogadro number, φ is an electrical potential, and the subscript i refers to the ions in the solution.

The ion mobility u has the dimension $m^2s^{-1}V^{-1}$. The charge of an electron is $e = 1.6019 \times 10^{-19}$ C, and the Avogadro number is $N_A = 6.0232 \times 10^{23}$ molecules per mol or in this case electrical charges per mole. The product of e and N_A is called the Faraday constant which is $F = 96485$ C eq^{-1} [P.W. Atkins, 1990].

The flux of electrical charges represents an electrical current, which is for a solution of a single electrolyte according to Ohms's law given by:

$$I = \frac{U}{R} = \frac{\kappa\, A\, \Delta\varphi}{l} \equiv \frac{\sum_i z_i u_i \nu_i C\; F\; A\, \Delta\varphi}{l} \tag{2.8}$$

Here I is the current, U is the applied voltage, F is the Faraday constant, R is the resistance, κ is the specific conductivity, A is the area through which the current passes, $\Delta\varphi$ is the voltage difference between two points, and l is the distance between the two points; furthermore z, u, and ν are the valence, the mobility, and the stoichoimetric coefficient of an ion, respectively; the subscript i refers to cation and anion and C is the concentration of the electrolyte.

From Equation 2.8 follows:

$$\kappa = \sum_i z_i u_i \nu_i C F \tag{2.9}$$

The equivalent conductivity of an ion i is:

$$\lambda_i = u_i F \tag{2.10}$$

Equation 2.10 applies to all cations and anions. The molar conductivity of an electrolyte can also be expressed by:

$$\Lambda_{mol} = (z_+ \nu_+ u_+ + z_- \nu_- u_-) F \tag{2.11}$$

2.2 The Electric Field and the Permittivity

Electrical charges carried by molecules interact with each other. Opposite charges attract each other and identical charges repel each other. The interaction force between two point electrical charges q_1 and q_2 in a vacuum separated by a distance r is given by:

$$K = \frac{q_1 q_2}{4\pi\varepsilon_0 r^2} \tag{2.12}$$

Here K is the interaction force between two point electrical charges, q_1 and q_2 are electrical charges, r is the distance between the two point electrical charges, and ε_o is the permittivity of vacuum.

Equation 2.12 is referred to as Coulomb's law and the electrical charge is expressed in Coulomb.

The permittivity depends on the medium between the electrical charges. In a vacuum the permittivity is a fundamental constant denoted as ε_0. It has the value 8.854×10^{-12} [$C^2 J^{-1} m^{-1}$]. If the medium is not a vacuum the permittivity can be expressed in terms of a relative permittivity ε_r which is also referred to as dielectric constant of the medium [P.W. Atkins, 1990]. It is given by:

$$\varepsilon_r = \frac{\varepsilon}{\varepsilon_o} \tag{2.13}$$

Here ε, ε_r and ε_o are the permittivity of a medium, the relative permittivity, and the permittivity in the vacuum, respectively.

2.2.1 *The Electric Field*

Coulomb's law as defined by Equation 2.11 involves the interaction between two point charges. In some cases it is desirable to know the force exerted by one charge q in a certain distance r to its surrounding. This force is referred to as electric field. It is given by:

$$E = \frac{K}{q} \tag{2.14}$$

Here K is the force, E is the electric field and q is an electrical charge.

In determining the electric field the direction of the field must be taken into account. For a point electrical charge q_i the electric field is radial symmetric [M.E. Starzak, 1984], [G. Kortüm, 1957] and given by:

$$E = \frac{q}{4\pi\varepsilon_o\ r^2} \tag{2.15}$$

Here E is the electric field around a point electric charge q, r is the distance from the electrical charge, and ε_o is the vacuum permittivity.

If instead of a vacuum the point charge is surrounded by another media the permittivity of the media must be taken into account. The electric field has the units of Newton per Coulomb or Volt per distance.

The electrical field around a point charge is illustrated in Figure 2.1 a) which shows the field strength or field density symbolized by density of the lines which is decreasing with the distance from the point charge. The electric field between two point charges of opposite charge are illustrated in Figure 2.1 b). In this case the two charges will attract each other. If they carry the same charge they will repulse each other. The decrease of the electric field density with the distance from the point charge is expressed by Equation 2.15.

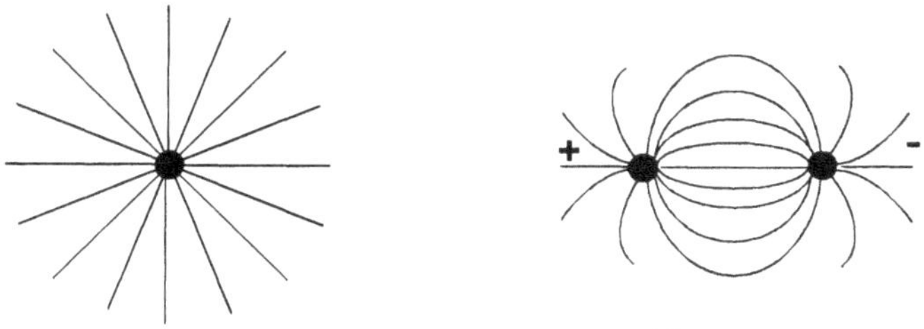

a) b)

Fig. 2.1 Schematic drawing illustrating the electric field lines a) around a point charge and b) between a positive and a negative point charge.

Equation 2.15 can be rewritten as:

$$E\,(4\,\pi\,r^2) = \frac{q}{\varepsilon_o} \tag{2.16}$$

Furthermore, this electric field through the surface $4\pi r^2$ of a sphere around a point charge is proportional to the total charge within the volume surrounded by the surface. Using a vector notation, this result can be generalized to the surface of any shape by:

$$\oint E\,dS = 4\pi\frac{q}{\varepsilon_o} = 4\pi\frac{\sum_i q_i}{\varepsilon_o} \tag{2.17}$$

Here $\oint$ indicates an integral over the entire surface S and q is the total charge within the surface enclosed region. Equation 2.17 is referred to as Gauss's law [G. Kortüm, 1957].

Using the divergence theorem the surface integral can be converted to a volume integral. Thus is:

$$\oint EdS = \int^{V} divEdV \tag{2.18}$$

which converts the surface integral into a volume integral. The total charge within a volume is $\int^{V}\rho dV$, where ρ is the space charge density. Gauss's law is then:

$$\oint EdS = \int^{V} divEdV = \frac{4\pi}{\varepsilon_o}\int^{V}\rho\,dV \tag{2.19}$$

Since the two volume integrals are equal, their arguments must be equal. Thus is:

$$divE = \frac{4\pi}{\varepsilon_o}\rho \tag{2.20}$$

This is the three dimensional version of the Poisson equation.

The electric field can be expressed by an electrical potential gradient. If the electrical potential varies over a certain distance the electrical field associated with this potential is given by the derivative in the direction of the electric field. For the three dimensional case is:

$$E = -\,grad\,\varphi \tag{2.21}$$

Combining Equations 2.20 and 2.21 leads to:

$$div\,grad\,\varphi = -\frac{4\pi}{\varepsilon_o}\rho \tag{2.22}$$

The divergence of the grad φ can also be written as:

$$div\,grad\varphi = \frac{\partial^2\varphi}{\partial x^2} + \frac{\partial^2\varphi}{\partial y^2} + \frac{\partial^2\varphi}{\partial z^2} = \nabla^2\varphi = -\frac{4\pi}{\varepsilon_o}\rho \tag{2.23}$$

Here x, y, and z are the directional coordinates.

Equations 2.22 and 2.23 are different forms of the Poisson differential equation which provides a relation between the electrical potential and the space charge density in an electric field.

2.2.2 *The Debye-Hückel Theory of Ionic Solutions*

The consequent application of the Poisson equation for an electrolyte solution is also the basis for the Debye-Hückel theory of electrolyte solutions [G. Kortüm, 1957], [P.W. Atkins, 1990], which will here only qualitatively be described. The Debye-Hückel theory provides a relation between the electrical potential around a single ion and oppositely charged ions. It describes the distribution of oppositely charged ions around a central ion as a function of the distance and their effect on the decay of the electrical potential with distance from the central ion. Since oppositely charged ions attract each other and equally charged ions repulse each other cations and anions are not uniformly distributed in a solution which is overall neutral. Near any given ion there is an excess of counter ions. The result is that any ion is surrounded by an atmosphere of oppositely charged ions. This means that within the ion atmosphere around a center ion there is an excess of oppositely charged ions. The total space charge density of the ion atmosphere is equal to that of the center ion. The potential around a center ion is related to the space charge density of the center ion and the surrounding ion atmosphere by the Poison equation. The distribution of ions in the ion atmosphere follows the Boltzmann statistic, i.e. the concentration of the excess charges is decreasing exponentially with distance from the center ion. The distribution of the ion atmosphere around the center ion and the decay of the electrical potential are illustrated in Figure 2.2 a) and b).

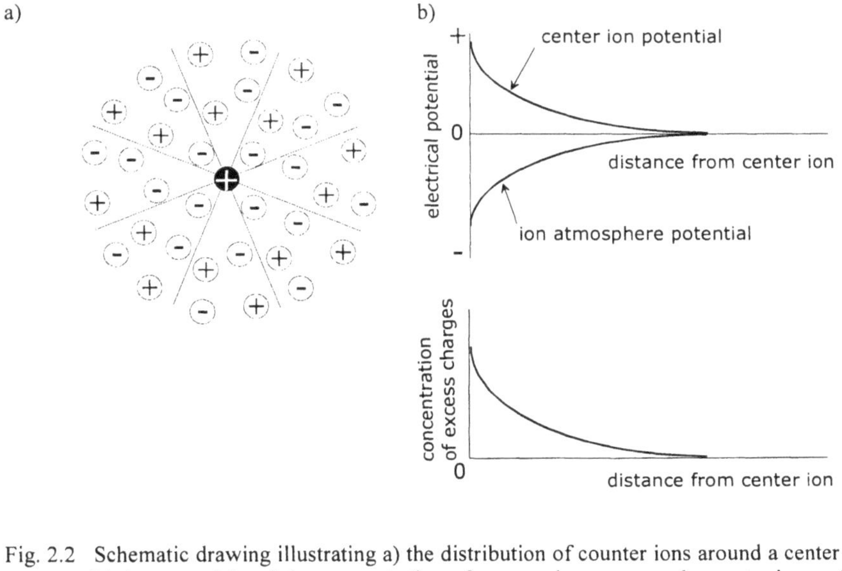

Fig. 2.2 Schematic drawing illustrating a) the distribution of counter ions around a center ion and b) the potential and the concentration of excess charges around a center ion as function of distance.

The Debye-Hückel theory of ionic solutions leads to two significant conclusions:

- Within the ion atmosphere the potential of the center ion decays much faster than in an atmosphere of neutral molecules.
- The electrochemical potential of the center ion is lowered as a result of its interaction with the oppositely charged ion of the ion atmosphere.

Since the electrical potential surrounding a center ion is decreasing with the distance from the center ion it will reach zero at a certain distance. This distance is called the Debye length and marks the outer boundary of the ion atmosphere. Within an area around the center ion determined by a radius which is smaller than the Debye length there are uncompensated electrical charges. Outside this area there are no uncompensated charges. The Debye length is a function of the permittivity of the solution and its ion concentration.

The Debye length is given by:

$$r_D = \left(\frac{\varepsilon RT}{2\rho F^2 J} \right)^{\frac{1}{2}} \tag{2.24}$$

Here r_D is the Debye length, ρ is the space charge of the center ion, F is the Faraday constant, ε is the permittivity, and J is the ionic strength of the solution, which is given by:

$$J = \frac{1}{2} \sum_i z_i^2 C_i \tag{2.25}$$

Here z_i is the valence of the ion i and C_i its concentration.

Equations 2.24 and 2.25 show that the radius of the ion atmosphere around a center ion, i.e. the Debye length is decreasing with the square root of the ionic strength of the solution, i.e. with the square root of the overall ion concentration. But since the ionic strength is proportional to the square of the valence of the ions in the solution the Debye length is decreasing linearly with the valence of the ions in the solution.

The experimentally determined radii of the ion atmosphere for different electrolytes and ion concentrations at 25°C are given in Table 2.1.

Table 2.1 Radii of the ion atmospheres around a center ion in electrolyte solutions of different concentrations and salts with different valences [G. Kortüm, 1957].

concentration [mol L^{-1}]	radius of ion atmosphere in aqueous solutions of ions with different valence [nm]			
	$z^+=1, z^-=-1$	$z^+=1, z^-=-2$	$z^+=2, z^-=-2$	$z^+=1, z^-=-3$
10^{-4}	30,4	17,6	15,2	12,4
10^{-3}	9,6	5,5	4,8	3,9
10^{-2}	3,0	1,8	1,5	1,2
10^{-1}	1,0	0.6	0.5	0,4

These data show that the size of the ion atmosphere changes from less than 0.5 nm in concentrated electrolyte solutions of multivalent ions to more than 30 nm in highly dilute electrolyte solutions of monovalent ions.

Since the energy, and thus the chemical potential of the center ion is changed as a result of its interaction with the oppositely charged ion of the ion atmosphere its activity coefficient in the solution is also changed and according to the Debye-Hückel theory the activity coefficient in dilute solutions can be calculated by:

$$\lg \gamma_{\pm} = -\left|z^{+}z^{-}\right| A \sqrt{J} \qquad (2.26)$$

Here $\gamma_{\pm}$ is the activity coefficient of the electrolyte, A is a constant which depends on the temperature and the relative permittivity in a solution, and J is the ionic strength of the solution which is given by Equation 2.25.

In an aqueous solution at 25°C the constant A is 0.509 [mol $kg^{-1}]^{0.5}$ [P.W Atkins, 1990].

Equation 2.26 shows that the activity coefficient is a function of the concentration and the valence of all ions in the solution.

2.2.3 Electrical Dipoles and Intermolecular Forces

An electrical dipole consists of two opposite charges attached to a molecule and separated by a certain distance. An attraction of the charges is prevented by the rigidity of the molecular structure. The potential produced by the dipole will be different from that produced by a single charge and will depend on the directional orientation of the dipole towards another charge. This is illustrated in Figure 2.3 which shows the potential an isolated charge experiences in a certain distance from a dipole.

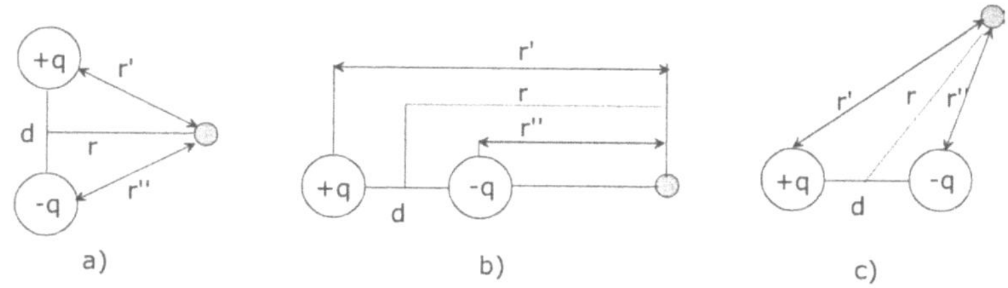

Fig. 2.3 Schematic drawing illustrating the net dipole potential on an isolated charge with a distance r from the center of the dipole: a) the isolated charge is perpendicular to the dipole axis, b) the isolated charge is in line with the dipole axis and c) the isolated charge aligned at a certain angle to the dipole axis.

Figure 2.3 a) illustrates the net potential of a dipole with a distance d between the two charges that an isolated charge experiences in a distance r from the center of the dipole perpendicular to the dipole axis d. r' and r'' are the distances between the isolated charge and the center of dipole charges. In this case the net potential an isolated charge experiences is zero [M.E. Starzak, 1984]. Figure 2.3 b) shows that the isolated charge is in line with the dipole axis. In this case the potential the isolated charge experiences is at a maximum. If the dipole is aligned at a certain angle as shown in Figure 2.3 c) the potential the isolated charge experiences is between zero and the maximum.

The general expression for the dipole potential an isolated charge experiences is given by [M.E. Starzak, 1984]:

$$\varphi = -\frac{\cos\theta \; q\, d}{4\pi\varepsilon r^2} = \frac{-\cos\theta \; \mu}{4\pi\varepsilon r^2} \tag{2.27}$$

Here φ is the potential, θ is the angle between the dipole axis d and the line r between the dipole and the isolated charge, and μ is the so-called dipole momentum given by the product of d and q.

If the angle θ is 90 degrees, i.e. the isolated charge is perpendicular to the dipole axis, the potential is zero, and if θ is zero degrees, i.e. the isolated charge is in line with the dipole axis the potential reaches a maximum as illustrated in the Figure 2.3 a) to c).

The potential of a dipole with a single charge is decreasing with the distance by the square of the distance, i.e. $\varphi \propto \frac{1}{r^2}$.

Dipoles interact not only with single charges such as ions, but also with other dipoles. However, in dipole-dipole interaction the potential drops by the third power with the distance between the dipoles, and the interaction energy is significantly lower as indicated by the data given in the following table.

Table 2.2 Potentials and energies between charged components [P.W. Atkins 1990]

type of interaction	distance dependence potential	typical energy [kJ mol^{-1}]
ion-ion	proportional to: r^{-1}	ca. 250
ion-dipole	proportional to: r^{-2}	ca. 15
dipole-dipole	proportional to: r^{-3}	ca. 2

The dipole momentum of molecules is measured in Debye [D] with D = 3.33×10^{-30} A s m. The dipole momentum varies substantially depending on the polarity of the molecules. It is 0 for a completely none-polar molecule such as CCl_4 or H_2 and ca. 1.80 D for a polar molecule such as water [P.W. Atkins, 1990]. The high dipole momentum of the water is the result of the water molecule structure in which the hydrogen atoms are attached to one side of the oxygen atom as illustrated in Figure 2.4.

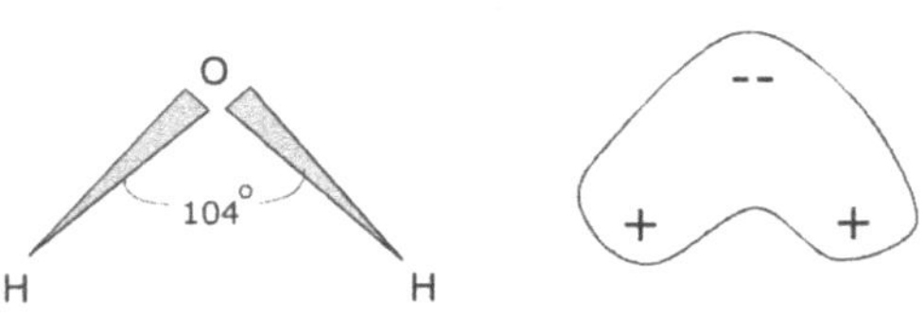

Fig. 2.4 Schematic drawing illustrating the structure of a water molecule indicating the dipole character due to the uneven distribution of the electrical charges within the molecule.

The strong dipole momentum of water is also the reason for its good solubility properties for ionic crystals such as salts. The water dipoles tend to break the generally strong Coulomb forces between the negatively and positively charged ions of a crystal by surrounding the ions. Thus, in an aqueous solution each ion is surrounded by a shell of water molecules as indicated in Figure 2.5 a) showing water molecules surrounding a cation and b) water molecules surrounding an anion.

Another important result of the molecular interaction of water dipoles is the formation of the hydronium ions from protons and water. In water the proton recombines readily with the water molecule to the hydronium ion in which the three H-atoms are no longer distinguishable and carry the positive charge together. The hydronium ion resembles a tripod with the three H-atoms on one side and all equally apart from the oxygen atom. The rapid formation of the hydronium ion is the reason for the much faster transport of protons compared to other ions in an aqueous solution.

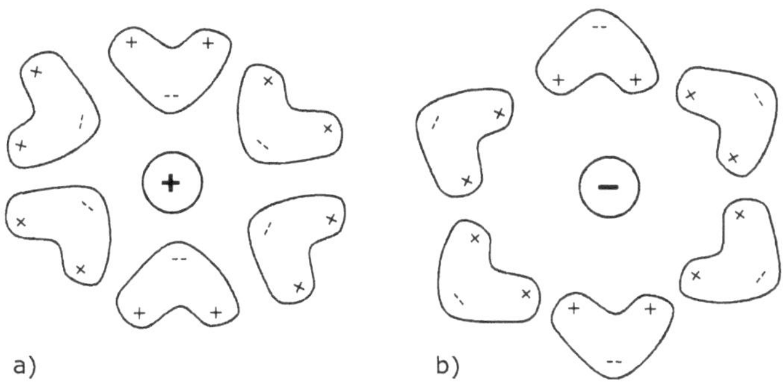

a) b)

Fig. 2.5 Schematic drawing illustrating the distribution of water molecules a) around a cation and b) around an anion.

2.3 The Double Layer at the Surface of Ion-Exchange Membranes

When an ion-exchange membrane is immersed into an electrolyte solution the fixed charges at the surface of membrane create an electric field which will attract oppositely charged ions from the solution. This leads to a layer on the membrane surface in which the concentration of the oppositely charged ions is higher than in the bulk solution. In many aspects an ion-exchange membrane in an electrolyte solution behaves very similar to a parallel plate capacitor.

2.3.1 The Principle of a Capacitor

The capacitance measures the ability of a system to store electrical charges. It is expressed by charges that can be stored on a surface of a capacitor per voltage, i.e.:

$$C = \frac{q}{U} \tag{2.28}$$

Here is C the capacitance with units [Farad], q is the electrical charge and U the voltage.

The most simple configuration of a capacitor is the parallel plate capacitor which is shown in Figure 2.6.

If a potential difference is established between the two plates positive charges will flow to one plate and the same amount of negative charges to the other. Since all charges are accumulated at the surface of the plates the electric field is directed only in direction to the opposite plate.

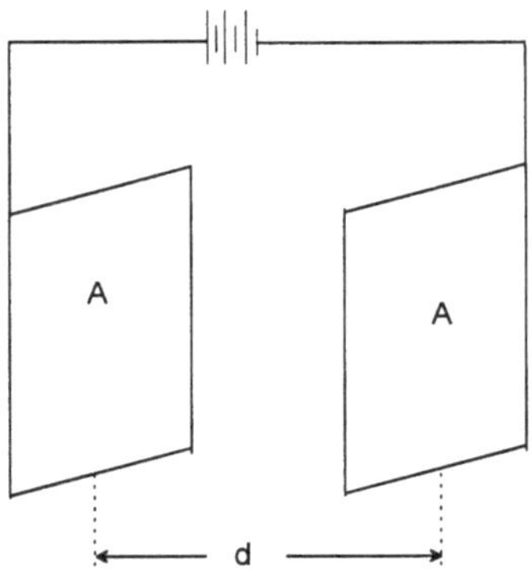

Fig. 2.6 Parallel plate capacitor with plate area A separated by a distance d.

The solution of the Poisson differential equation which gives a relation between the charge density, the electric field, and the electrical potential for a parallel plate capacitor leads to:

$$C = \varepsilon \frac{A}{4 \pi d} \tag{2.29}$$

Here is ε the permittivity of the medium between the plates, A the area of plates, and d the distance between the parallel plates.

For other configurations instead of flat plates the capacitance can be expressed by similar relations by taking the surfaces of the charged device, the distance between the charged surfaces and the permittivity into account.

2.3.2 The Electrical Double Layer

The fixed charges on the surface of an ion-exchange membrane will generate an electric field directed into an adjacent solution and attract due to the Coulomb forces ions with the opposite charge. These ions will form a layer at a certain distance from the membrane surface which neutralize the fixed charges at the membrane surface. This double layer at the interface between the ion-exchange membrane and the adjacent solution is kept in place by the attractive force of the electrical potential generated by the fixed ions of the membrane, a chemical potential gradient, and the dispersing force of the thermal motion. This double layer at the interface between the membrane and the electrolyte solution, which is important for certain electro-kinetic phenomena such as electro-osmosis or streaming potential, can be described by several models [P.W. Atkins, 1990], [M.E. Starzak, 1984]. The most simple one is the Helmholtz model.

The Helmholtz Double Layer Model

Helmholtz postulates that all counterions necessary to neutralize the surface charge of an ion-exchange membrane are aligned in a single layer in parallel at a certain distance from the membrane surface. This distance is determined by the radius of the hydrated or unhydrated counterion as indicated in Figure 2.7 which shows a) the closest approach of a hydrated positive ion to a membrane surface with negative fixed ions and b) the potential dissipation in the double layer.

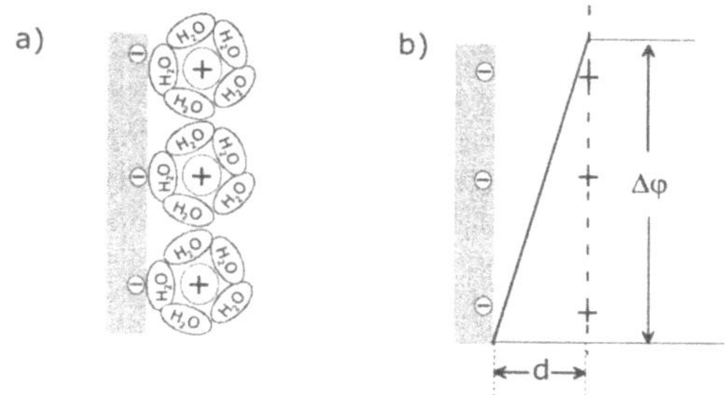

Fig. 2.7 Schematic drawing illustrating the a) closest approach of hydrated cations to the negatively fixed ions of the surface of a cation-exchange membrane and b) the potential dissipation in the double layer according to the Helmholtz model.

The Helmholtz model resembles a parallel plate capacitor where the charge is determined by the number of charges at the surface and in a plane parallel to the membrane surface. The relation between the surface charges, the potential between the negatively and positively charged layers and their distance is given by the same equation which describes the capacitance of a parallel plate capacitor [M.E. Starzak, 1984], i.e.

$$C = \varepsilon \frac{A}{4\pi d} = \frac{q}{U} \tag{2.30}$$

Here C is the capacitance, A is the area of the charge surface, d is the thickness of the Helmholtz layer, q is the charge.

Introducing the surface charge density σ which is given by:

$$\sigma = \frac{q}{A} \tag{2.31}$$

and replacing the voltage U by the potential difference $\Delta\varphi$ between the two layers

$$U = \Delta\varphi \tag{2.32}$$

gives a relation between the thickness of the double layer d, the surface charge, and the potential difference in the double layer:

$$\Delta\varphi = \frac{4\pi d}{\varepsilon_r\,\varepsilon_o}\sigma \qquad (2.33)$$

Here ε_r is the dielectric constant or the relative permittivity given by $\varepsilon/\varepsilon_o$, and σ is the surface charge density.

The Gouy-Chapman Double Layer Model

The Helmholtz model contains a number of rather drastic simplifications. For example it assumes that the fixed ions form a homogeneous sheet and that the counter ions form a complementary parallel sheet not effected by the thermal motion of the solution. In reality this is not the case.

In the Gouy-Chapman model the counter ions in the solution are no longer assumed to be aligned in a sheet but will occupy a finite volume in a mixture of cat- and anions. The total number of counter ions in this volume is identical to the fixed charges of the membrane surface. Their distribution, however, is such that the area close to the membrane surface contains the largest excess of counter ions as indicated in Figure 2.8 and is very similar to the ion atmosphere around a single charge.

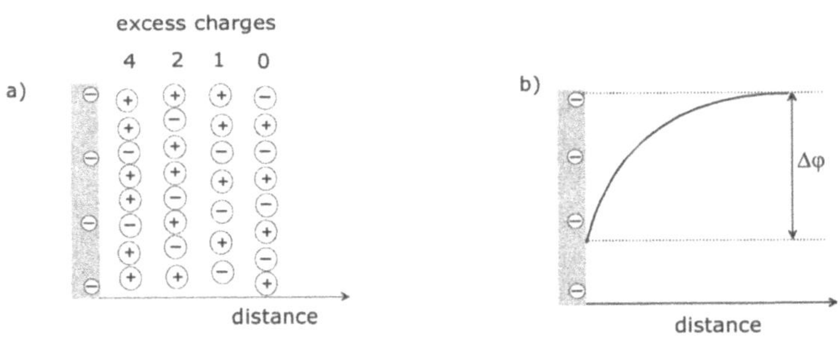

Fig. 2.8 Schematic drawing illustrating a) the excess counter ions in the double layer as function of the distance from the membrane surface and b) the potential dissipation as function of the distance from the surface of the membrane.

Applying the Poisson equation which provides a relationship between the potential and the excess charges at a given point in the solution and under the assumption that the system obeys the Boltzmann statistic [M.E. Starzak, 1984], the potential at a distance x in the solution is given by:

$$\varphi(x) = \frac{\sigma\, x_D}{\varepsilon}\exp\left[-\frac{1}{x_D}(x-a)\right] \qquad (2.34)$$

Here $\varphi(x)$ is the potential at a certain distance from the membrane surface in the solution, σ is the surface charge density of the membrane, a is the closest approach of a counterion to the membrane surface and x_D is the Debye length which is identical with the Debye radius of the ion atmosphere around a center ion. It determines the distance from the membrane surface into the solution where no longer excess charges are available that contribute to the compensation of the surface charges. The potential difference between the solution and the membrane surface is reached when x is equal to a. Then Equation 2.34 becomes:

$$\Delta\varphi = \frac{\sigma\, x_D}{\varepsilon} \tag{2.35}$$

The Debye length x_D plays the same role as the distance d in Equation 2.33 of the Helmholtz model, i.e. it determines the thickness of the double layer. Since the Debye length is inverse proportional to the square root of the ionic strength of the solution which will be shorter when the ion concentration in the solution is higher, because then there are sufficient ions available to compensate the surface charges of the membrane at a shorter distance.

The Stern Model

In the Gouy-Chapman model a distance a is defined for the closest approach of counter ions to the membrane surface. At the distance a the mobility of the counter ions is more restricted because of the strong interaction with the surface charges of the membrane, and the distribution of these ions in a diffuse double layer due to the thermal motion as assumed in the Gouy-Chapman model is not possible. Therefore, the Gouy-Chapman model is modified by Stern by suggesting a combination of the Helmholtz and the Gouy-Chapman model. In this model the double layer is characterized by two distinct types of potential gradients. In an area between the membrane surface and the so-called Helmholtz plane the potential changes linearly with the distance from the membrane surface as postulated in the Helmholtz model, while in an area from the Helmholtz plane into the solution the potential changes exponentially as postulated in the Gouy-Chapman model.

3 BASIC THERMODYNAMIC RELATIONS

Transport of heat, mass, electrical charges, and individual components between two systems separated by a semipermeable membrane will occur only when the systems are not in equilibrium, i.e. when the two systems are in different thermodynamic states. The driving force for any transport of matter, heat, or electrical charges through a membrane is based on differences or gradients in temperature, pressure, concentration or electrical potential across the membrane. Properties such as temperature, pressure, or composition are thermodynamic parameters which describe the state of a system. They are referred to as state variables. A key for describing the different membrane transport phenomena is the understanding of the thermodynamic relations that govern the process. A brief discussion of some fundamental thermodynamic relations is therefore very helpful for a further detailed description of the various ion-exchange membrane processes and their application.

The basic concept of thermodynamics is the description of a system and its interaction with its surrounding. A "system" is defined as a part of the universe which is under special consideration. A system is referred to as "open" when matter can be transferred through its boundary from or to the surrounding. Otherwise it is referred to as "closed". A system is referred to as "isolated" when neither matter nor heat is exchanged with the surrounding. The state of a system is determined by a number of state variables describing the extensive and intensive properties of the system. Extensive state variables are the mass, the volume, and the total energy of a system. Extensive state variables depend on the size of the system under consideration. They are additive and the total quantity consists of the sum of its parts. Intensive state variables such as temperature, pressure, or molar properties are independent of the size of the system. The relation between the different state variables is described by the state functions. The four fundamental state functions are: the internal energy U, the enthalpy H, the Helmholtz free energy A, and the Gibbs free energy G [P.W. Atkins, 1990]. For a system without exchange of matter with its surrounding the state functions are given by:

$$U = q + w \tag{2.36}$$

$$H = U + pV \tag{2.37}$$

$$A = U - TS \tag{2.38}$$

$$G = H - TS \tag{2.39}$$

Here is U the internal energy, H the enthalpy, A the Helmholtz free energy, and G the Gibbs free energy; q is the heat or energy of a system, w is the work done by the system or done on the system, T is the absolute temperature and S is the entropy of the system.

The internal energy describes the total energy of a system which is composed of the heat q of the system and the work w done by the system or done on the system. The enthalpy H takes into account that some of the heat of the system is converted into expansion work if the system changes its pressure or volume. The Helmholtz free energy A and the Gibbs free energy G are a relation determining the part of the internal energy U, or the enthalpy H can freely be used to provide work and which part is bound to the system in form of entropy.

The entropy of a system can be defined by taking the statistical or thermodynamic approach. The statistical definition describes the entropy as a degree of disorder in a system. It is based on the Boltzmann statistic and takes into account the number of different ways in which the energy in a system can be achieved by rearranging the atoms or molecules in their available states. For the discussion of membrane phenomena the thermodynamic definition of the entropy is more useful.

The thermodynamic approach to defining the entropy is based on the change of the entropy during a process. The entropy change of a system in a completely reversible process is defined as:

$$dS = \frac{dq_{rev}}{T} = 0 \tag{2.40}$$

and for a spontaneous, i.e. irreversible process:

$$dS \geq \frac{dq_{irrev}}{T} \geq 0 \tag{2.41}$$

Here is dq_{rev} the energy exchanged by the system with the surrounding in form of heat in a completely reversible process, dq_{irrev} is the energy exchange in an irreversible process, and T is the temperature at which the process occurs.

In a reversible process the entropy change is 0 and in a spontaneous, i.e. irreversible process the entropy is increasing.

The interaction of a system with its surrounding occurs by an exchange of energy, i.e. heat, mass, or work with the surrounding and leads to a change of the system properties. The change is counted positive if something is added to the system or done on the system.

The change is expressed by a negative term if something is taken from the system or if work is done by the system. This is referred to as the "egoistic principle".

The change of a system is given by the derivative of the state functions. If only mechanical work is considered, i.e. only work due to expansion or compression of a system, and if no exchange of matter with the surrounding takes place and the thermodynamic concept of entropy is introduced the changes of the state functions are given by their differential:

$$dU = dw + dq = - pdV + TdS \tag{2.42}$$

$$dH = dU + d(pV) = Vdp + TdS \tag{2.43}$$

$$dA = dU - d(TS) = - pdV - SdT \tag{2.44}$$

$$dG = dU - d(TS) = Vdp - SdT \tag{2.45}$$

Here pdV is the work of expansion at constant pressure and Vdp the expansion at constant volume, TdS is the entropy change in a reversible process at constant temperature, and SdT the temperature change at constant entropy.

The interaction of the system with its surrounding is governed by the first and second law of thermodynamics which state:

- The internal energy of a system is constant unless it is changed by interaction with the surrounding.
- The entropy of an isolated system increases in the course of a spontaneous, i.e. irreversible process.

For membrane processes the change of the Gibbs free energy is the most important state function for describing the state of a system since the Gibbs free energy describes the part of the total energy of a system which is available to do work, i.e. the energy which is available to act as driving force for a process.

Equation 2.45 describes the change of the Gibbs free energy for a closed system, i.e. a system that has no exchange of matter with its surrounding. However, in membrane processes generally a transport of matter is obtained and the change of the state functions due to the exchange of matter must be considered. The change of the Gibbs free energy of a system due to an exchange of matter with the surrounding is given by:

$$dG = Vdp - SdT + \sum_i \mu_i dn_i \tag{2.46}$$

Here μ_i is referred to as the chemical potential of a component i. It is a partial molar quantity and given by:

$$\mu_i = \left(\frac{\partial G}{dn_i} \right)_{p,T,n_j} \tag{2.47}$$

$\left(\frac{\partial G}{\partial n_i}\right)_{p,T,n_j}$ is the partial molar Gibbs free energy. It should be noted that the partial molar Gibbs free energy is an intensive property and thus independent of the size of the system, while the total Gibbs free energy is an extensive property given by the sum of the chemical potential of all components in the system.

The Gibbs free energy is the summation of its partial molar properties. This holds true for all other state properties, too. Thus is:

$$\left(\frac{\partial V}{\partial n_i}\right)_{p,T,n_j} = \overline{V}_i \tag{2.48}$$

and

$$\left(\frac{\partial S}{\partial n_i}\right)_{p,T,n_j} = \overline{S}_i \tag{2.49}$$

Here i refers to the component under consideration and j refers to all other components in the mixture.

Furthermore is:

$$V = \sum_i n_i \overline{V}_i, \tag{2.50}$$

$$S = \sum_i n_i \overline{S}_i, \tag{2.51}$$

$$G = \sum_i n_i \mu_i \tag{2.52}$$

Here $\overline{V}_i$ is the partial molar volume, and $\overline{S}_i$ is the partial molar entropy.

The chemical potential μ_i is also a function of temperature, pressure, and composition, and its change can be expressed by:

$$d\mu_i = \overline{S}_i dT + \overline{V}_i dp + \sum_j \frac{\partial \mu_i}{\partial n_j} dn_j \tag{2.53}$$

The last term in Equation 2.53 describes the chemical potential of a component i as a function of other components in a mixture. In analogy to an ideal gas mixture the composition dependent part of the chemical potential can be expressed by a standard potential referring to the pure component and a composition dependent term. At constant pressure and temperature the chemical potential of a component i in an ideal mixture is given by:

$$(\mu_i)_{p,T} = \mu_i^{\circ} + RT \ln X_i \tag{2.54}$$

Here μ_i^o is the chemical potential of the pure component i under standard conditions of temperature and pressure, and X_i is the mole fraction of the component i in a mixture which is defined as the number of moles n_i in a mixture, divided by the number of all moles in the mixture:

$$X_i = \frac{n_i}{n_i + \sum_j n_j} \tag{2.55}$$

Equation 2.54 is correct only for an ideal mixture. For real mixtures the mole fraction X_i must be replaced by the activity a_i.

The mole fraction and the activity are related by the activity coefficient γ:

$$a_i = \gamma_i X_i \tag{2.56}$$

The activity coefficient describes the deviation of a real mixture from an ideal solution. For ideal mixtures the activity coefficient is 1.

For a mixture composed of a solvent and various solutes the activity coefficient of the solvent approaches 1 if its mole fraction approaches 1. For a solvent s in a mixture is: $\gamma_s \rightarrow 1$ as $X_s \rightarrow 1$. Thus, the standard state for a solvent is defined as that of the pure solvent.

For a solute in a mixture the activity coefficient approaches 1 if its mole fraction approaches 0. For a solute i in a mixture therefore is: $\gamma_i \rightarrow 1$ as $X_i \rightarrow 0$.

Thus, for a solute in a mixture the standard state, i.e. the ideal behaviour is defined for an infinitely low concentration.

The change of the chemical potential of a component i in a real mixture at isothermal conditions is obtained by introducing the activity concept in Equations 2.53 and 2.54:

$$d\mu_i = \overline{V}_i dp + RT\, d \ln a_i \tag{2.57}$$

Here μ_i, $\overline{V}_i$, and a_i are the chemical potential, the partial molar volume, and the activity of the component i.

Since most ion-exchange membrane separation processes are operated at constant temperature Equation 2.57 is a key relation for describing the mass transport between two systems separated by a membrane.

The gradient in the chemical potential across a membrane between two mixtures is the driving force for the transport of a component through the membrane from one mixture into the other. In membrane processes it is assumed that the membrane is in equilibrium with the adjacent mixture. Thus, the chemical potential in a mixture at the membrane surface is identical to the chemical potential in the membrane at the surface which is in contact with the mixture.

The gradient in the chemical potential and the activity of a component can, therefore, be expressed by the difference in the chemical potentials and the activities of the two mixtures separated by the membrane. Integration of Equation 2.57 leads to:

$$\Delta\mu_i = \mu_i' - \mu_i'' = \overline{V}_i\,(p' - p'') + RT \ln \frac{a_i'}{a_i''} \tag{2.58}$$

Here μ_i' and μ_i'' are the chemical potentials and a_i' and a_i'' the activities of a component i in the two mixtures (') and ('') separated by the membrane, and Δp is the pressure difference between the two mixtures. Thus, $\Delta\mu_i = \mu_i^{'} - \mu_i^{''}$ is the driving force for the transport of the component i from one phase to the other.

Equation 2.58 is an important relation which not only describes the driving force for the transport of a component i from one mixture or phase to another through a membrane as a function of a pressure and activity difference, it also indicates that equilibrium between the two phases for the component i can be achieved, i.e. $\mu_i^{'} = \mu_i^{''}$ even if it has different activities in the two different phases when this activity difference is compensated by a pressure difference as will be shown later in the discussion of the osmotic pressure.

If the membrane is permeable for the component i and μ_i' and μ_i'' are not identical the number of moles n_i transported from one mixture to the other is identical to the change of the number of moles in the two mixtures separated by the membrane. Thus is::

$$dn_i^{'} = -dn_i^{''} \qquad (2.59)$$

Here $dn_i^{'}$ and $dn_i^{''}$ are the changes of the number of moles in the phases (') and ('').

The flux of the component i through the membrane is given by the change of the number of moles of the component i per unit time:

$$\frac{dn_i^{''}}{dt} = A\,J_i \qquad (2.60)$$

Here A is the membrane area, t is the time, and J_i is the flux per unit area.

The chemical potential μ_i of the component i describes the partial molar Gibbs free energy for a system that contains neutral components only.

If a system contains charged components, i.e. ions, the partial molar Gibbs free energy depends on the chemical potential and on an electrical potential. The partial molar Gibbs free energy of a mixture containing ions is expressed by the so-called electrochemical potential [H.S. Harned et al., 1958]. The change in the electrochemical potential of an ion is given by:

$$d\eta_i = d\mu_i + z_i F\,d\varphi = \overline{V}_i dp + RT\,d\ln a_i + z_i F d\varphi \qquad (2.61)$$

Here η_i is the electrochemical potential, z_i is the electrochemical valence of the ion i, F is the Faraday constant, and φ is the electrical potential. R, T, p, μ_i, a_i, and $\overline{V}_i$ have their normal meaning.

The valence of an ion indicates the number of electrical charges carried by this ion. For example, a monovalent ion such as the Na^+-ion carries one positive charge. Thus, its valence is z = +1. For a negatively charged ion such as the Cl^--ion is z = -1, and for a divalent ion such as Ca^{++}-ion is z = +2.

The ion concentrations in an electrolyte solution are related by the electroneutrality requirement which postulates that the number of positive and negative charges in a macroscopic system must be equal. The electroneutrality condition for an electrolyte solution composed of a single electrolyte is given by:

$$\sum_i z_i C_i = \sum_i z_i \nu_i C_s = 0 \qquad (2.62)$$

Here C_i is the concentration, z_i is the valence of an ion i, C_s is the concentration of the electrolyte, e.g. a salt, an acid, or a base, and ν_i is a stoichiometric coefficient which gives the number of ions formed when one mole of an electrolyte dissociates in a solution.

For a monovalent electrolyte such as NaCl is $\nu_{Na^+} = \nu_{Cl^-} = 1$ for both ions. For an electrolyte that dissociates into more than two ions such as $CaCl_2$, is $\nu_{Ca^{++}} = 1$ and $\nu_{Cl^-} = 2$.

3.1 Chemical and Electrochemical Equilibrium

Equilibrium between two systems is obtained when the systems are in the same thermodynamic state, i.e. when the Gibbs free energy in both systems is identical. For systems containing electrically charged components such as salts, acids, or bases two systems are in equilibrium when the electrochemical potential of all components which can be exchanged between the systems is identical in both systems. The Gibbs free energy of a system is, therefore, a function of various state variables such as temperature, pressure, electrical potential and the activity of individual components and differences in these state variables between two systems can lead to fluxes of heat, momentum, electrical charges or individual components, and to chemical reactions as pointed out earlier. In ion-exchange membrane separation processes the equilibrium between an ion-exchange membrane and the adjacent electrolyte solution as well as the equilibrium between two phases separated by an ion-exchange membrane are of main interest. But in many applications of ion-exchange membrane separation processes the water and electrolyte dissociation equilibrium is also of importance.

3.1.1 The Water Dissociation and the pH- and pK-Values of Acids and Bases

Water is dissociated to a certain extent into hydronium- and hydroxide-ions, i.e. H_3O^+ and OH^- ions according to following equilibrium relation:

$$2H_2O \Leftrightarrow H_3O^+ + OH^- \qquad (2.63)$$

The water dissociation equilibrium constant K is:

$$\frac{[H_3O^+]\,[OH^-]}{[H_2O]^2} = K \qquad (2.64)$$

The brackets [ ] denote concentrations of the various components in equilibrium.

In pure water the concentration of water is 55.6 mol L^{-1}. The product of K× $[H_2O]^2$ is constant and referred to as the water dissociation product K_w which is 1.008×10^{-14} [$mol^2\ L^{-2}$] at 25°C [P.W. Atkins, 1990].

It is:

$$[H_3O^+]\,[OH^-] = K\,[H_2O]^2 = K_w = 10^{-14} \quad (\text{at } 25\ ^\circ\text{C}) \qquad (2.65)$$

The K_w value is a function of temperature and changes from 0.1139×10^{-14} at 0°C to 9.610×10^{-14} at 60°C.

However, at constant temperature and in electrolyte solutions with low to moderate ion concentrations the water concentration does not change very much and K_w can be considered constant, and in pure water the concentrations of H_3O^+- and OH^--ions are identical and are 10^{-7} [mol L^{-1}] at 25°C.

In electrolyte solutions containing an acid or a base the concentrations of H_3O^+- and OH^--ions are no longer identical and their concentrations are expressed by the so-called pH-value (potentia hydrogenii). The pH-value is defined as the negative logarithm of the H_3O^+-ion concentration to the basis of 10. Thus is:

$$[H_3O^+] = 10^{-pH} \qquad (2.66)$$

and

$$[OH^-] = 10^{-(14-pH)} \qquad (2.67)$$

Pure water or neutral electrolyte solutions at 25°C have a pH-value of ca 7. Acid exhibit pH-values < 7 and bases >7.

An important parameter for acids and bases is the dissociation constant expressed in the so-called pK-value. The dissociation of acids and basis follows the general scheme:

acids:

$$HA + H_2O \overset{K_1}{\Leftrightarrow} H_3O^+ + A^- \qquad (2.68)$$

bases:

$$BOH \overset{K_2}{\Leftrightarrow} B^+ + OH^- \qquad (2.69)$$

Here K_1 and K_2 are the dissociation equilibrium constants, and HA and BOH represent the undissociated acid and base.

The acid and base equilibrium can be expressed by the following relations:

$$\frac{[H_3O^+]\,[A^-]}{[HA]} = K_1[H_2O] = K_a \qquad (2.70)$$

and

$$\frac{[B^+]\,[OH^-]}{[BOH]} = K_2 = K_b \qquad (2.71)$$

Here K_a and K_b are the dissociation constants for the acid and the base.

The dissociation constant between strong and weak acids or bases can vary by several orders of magnitude. Therefore, the dissociation constant is generally expressed by the so-called pK_a- and pK_b-values which are defined similarly to the pH-value as the negative logarithm of the dissociation constant., i.e.:

$$pk_a = -\log K_a = 10^{-pK_a} \qquad (2.72)$$

and

$$pk_b = -\log K_b = 10^{-pK_b} \qquad (2.73)$$

Multivalent acids or bases, such as H_2SO_4 or $Ca(OH)_2$ which have more than one dissociable proton or hydroxide ion dissociate in consecutive steps and, therefore, have more than one pK_a- or pK_b-value, respectively.

In practical applications in addition to the pK_a- and pK_b-values the degree of dissociation of a component such as a weak acid is of interest. The degree of dissociation of an electrolyte is defined as the ratio of dissociated molecules to the total number of molecules present in a solution. For an electrolyte composed of two monovalent ions A^- and B^+ the dissociation equilibrium is given by:

$$AB \overset{K_{diss}}{\Leftrightarrow} A^- + B^+ \qquad (2.74)$$

Here AB is the part of the electrolyte that is not dissociated, K_{diss} is the dissociation equilibrium constant, and A^- and B^+ represent the an- and cations of the electrolyte under consideration.

The concentration of the electrolyte and the concentrations of the un-dissociated part of the electrolyte and the dissociated ions are related as follows:

$$[AB] = (1-\alpha)C \qquad (2.75)$$

and

$$[A^-] = [B^+] = \alpha C \qquad (2.76)$$

Thus is:

$$\alpha = \frac{[A^-]}{[AB] + [A^-]} = \frac{[B^+]}{[AB] + [B^+]} \qquad (2.77)$$

Here α is the dissociation degree and C is the total concentration of the electrolyte in the solution.

The degree of dissociation α has values between 1 for a totally dissociated electrolyte and 0 for a completely un-dissociated electrolyte. The dissociation constant K_{diss} is related to the degree of dissociation α by:

$$K_{diss} = \frac{[A^-][B^+]}{[AB]} = \frac{\alpha^2}{1-\alpha} C \qquad (2.78)$$

and

$$\alpha = \frac{K_{diss}}{2C}\left(\sqrt{1 + \frac{4C}{K_{diss}}} - 1\right) \qquad (2.79)$$

Equation 2.79 shows that the degree of dissociation α depends on the dissociation constant and on the total concentration of the electrolyte in the solution. At very low concentrations α approaches 1 and at very high concentrations it approaches 0.

It should be noted, however, that while the degree of dissociation is decreasing with increasing concentration, the total number of ions in the solution is increasing with increasing concentration since the dissociation equilibrium constant is independent from the total concentration [R.A. Robinson et al, 1970].

From practical experience it is well known that the degree of dissociation of weak acids or bases depends on the pH-value of the solution. In solutions with high pH-values weak acids have a higher degree of dissociation than in solutions with low pH-values. With weak bases the situation is reversed. This can easily be shown by introducing the pH-value for the proton concentration into the dissociation equilibrium relation given by Equation 2.70 and by replacing the concentrations of the anion and the un-dissociated molecule as well as the degree of dissociation according to Equation 2.77.

Rearranging Equation 2.70 leads to:

$$\log K_a = \log [H_3O^+] + \log [A^-] - \log[HA] \qquad (2.80)$$

Introducing Equations 2.66 and 2.72 into Equation 2.80 leads to:

$$-pK_a = -pH + \log [A^-] - \log[HA] \qquad (2.81)$$

or

$$[A^-] = [HA]\ 10^{(pH-pKa)} \qquad (2.82)$$

Introducing the degree of dissociation from Equation 2.77 gives α as a function of the pH- and pK-value:

$$\alpha = \frac{[A^-]}{[HA]+[A^-]} = \frac{1}{1+10^{(pK_a-pH)}} \qquad (2.83)$$

The degree of dissociation α calculated according to Equation 2.83 for a weak acid with a pK_a-value of 5 is shown in Figure 2.9 as a function of the pH-value.

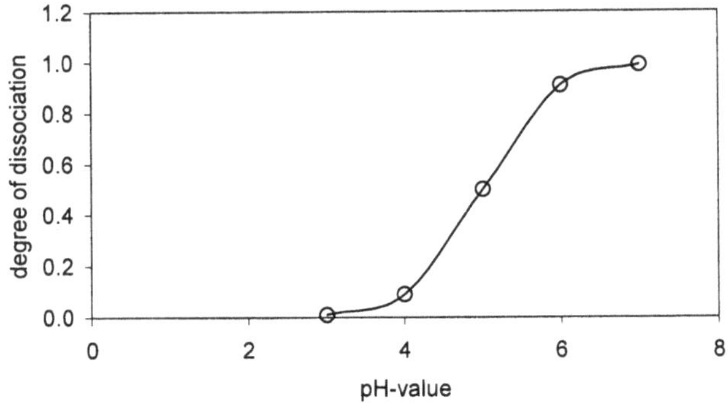

Fig. 2.9 Degree of dissociation of a weak acid with a pK_a-value of 5 as the function of the pH of the solution calculated according to Equation 2.83.

The diagram shows that for a weak acid the degree of dissociation is increasing with increasing pH and decreasing with decreasing pH and is 0.5 when the pH- and the pK_a-values are the same.

3.1.2 The Ion-Exchange Equilibrium and the Donnan Potential

In any electrolyte solution which may be a liquid such as an aqueous mixture or a solid phase like an ion-exchange membrane electroneutrality is required on a macroscopic scale. This means that every positive or negative charge is counter-balanced by an oppositely charged ion. The fixed negative charges of a cation-exchange membrane are counter-balanced by positively charged cations, the so-called counterions as indicated in Figure 1.1. This means, if an ion-exchange membrane is in equilibrium with an adjacent electrolyte solution counterions will be attracted into the membrane and coions will be expelled from the membrane by the strong Coulomb forces of the fixed ions [F. Helfferich, 1962].

Equilibrium between Different Counterions

The Coulomb forces of the fixed membrane charges act on counterions with certain preference due to distances and charge number of the ions. However, there are also additional forces of the van der Waals-type that can be different for different counterions and can result in a preferential take-up of certain counterions by the membrane. The distribution of counterions between a membrane and an adjacent electrolyte solution is determined by the counterion exchange equilibrium. Considering an exchange of protons and sodium ions between a cation-exchange membrane and an adjacent solution leads to the following relation:

$$^{m}H_3O^+ + {}^{s}Na^+ \Leftrightarrow {}^{s}H_3O^+ + {}^{m}Na^+ \qquad (2.84)$$

The superscripts m and s refer to the membrane and the adjacent aqueous solution.

The equilibrium constant K_{Na^+,H^+} for the counterion distribution between the membrane and the solution is given by:

$$K_{Na^+,H^+} = \frac{[{}^{m}Na^+][{}^{s}H_3O^+]}{[{}^{s}Na^+][{}^{m}H_3O^+]} \qquad (2.85)$$

The brackets in Equation 2.85 depict concentrations.

Under the assumption that the concentration in the aqueous solution is substantially lower than the fixed ion concentration in the membrane coions will be more or less completely excluded and the counterion concentration is about equal to the fixed ion concentration of the membrane. Thus, to a first approximation is:

$$C^m_{fix} = C^m_{H_3O^+} + C^m_{Na^+} \qquad (2.86)$$

Here C is the concentration, the subscripts fix, H_3O^+, and Na^+ refer to fixed charges of the membrane, to protons, and to sodium ions, respectively.

The concentration of the sodium ions in the membrane can be determined as function of the ion-exchange equilibrium constant K, the fixed charge density of the membrane, and the concentration of protons and sodium ions in the solution by combining Equations 2.85 and 2.86:

$$\frac{C^m_{Na^+}}{C^m_{H_3O}} = \frac{C^s_{H_3O^+}}{C^s_{Na^+} K_{Na^+,H^+}} \tag{2.87}$$

Equation 2.87 shows that a cation-exchange membrane in the proton form will readily exchange the proton for the sodium ion from a binary solution containing sodium ions as long as the pH-value of the solution is high, which means that the proton concentration in the solution is low compared to that of the sodium ions. For equal concentrations of sodium and protons in the solution the concentration in the membrane is determined by the equilibrium constant.

3.1.3 The Electrochemical Equilibrium

Two systems may well be in equilibrium, even if the pressures or the electrical potentials or the activities of individual components in both systems are different, as long as these differences compensate each other and the Gibbs free energy in both systems is identical.

The electrochemical equilibrium between two phases is referred to as the Donnan equilibrium which is obtained when an electrical potential difference between two phases is compensated by activity and pressure differences of the ions in the two phases.

The Electrochemical Equilibrium and the Donnan Potential

An ion-exchange membrane and an adjacent electrolyte solution will be in equilibrium if the electrochemical potentials of all ions in the two phases are equal. Thus, for each ion in equilibrium is:

$$\eta_i^m = \eta_i^s = \mu_i^m + z_i F \varphi^m = \mu_i^s + z_i F \varphi^s \tag{2.88}$$

Here μ_i and η_i are the chemical and the electrochemical potential of the component i, φ is the electrical potential, and F is the Faraday constant. The superscripts m and s refer to the membrane and to the solution, respectively.

Thus, the electrochemical potential of an ion is composed of two additive terms, the first is the chemical potential and the second is the electrical potential multiplied by the Faraday constant and the valence of the ion. Introducing the chemical potential μ_i as defined in Equation 2.58 as a function of pressure and composition into Equation 2.88 and rearranging gives the electrical potential difference between the membrane and the adjacent solution to:

$$\varphi^m - \varphi^s = \frac{1}{z_i F}\left[RT \ln \frac{a_i^s}{a_i^m} + \overline{V}_i \left(p^s - p^m\right)\right] = \varphi_{Don} \tag{2.89}$$

Here φ is the electrical potential, a is the ion activity, $\overline{V}_i$ is the partial molar volume, z is the valence, F is the Faraday constant, p is the pressure, T is the absolute temperature, and R is the gas constant; the subscript i refers to individual ions and the superscripts m and s refer to the membrane and the electrolyte solution, respectively. The potential difference between the membrane and the solution is referred to as Donnan potential φ_{Don} [F.G., Donnan, 1932].

The Donnan potential between an electrolyte solution and an ion-exchange membrane cannot be measured directly. It can, however, be calculated from the ion activities in the solution and the membrane, and the pressure difference between the membrane phase and the adjacent solution, p^m-p^s, which is referred to as swelling pressure [F.G. Donnan et al., 1932]. The swelling pressure can be considered as the osmotic pressure difference between the solution and the membrane phase. The osmotic pressure difference $\Delta\pi$ can be calculated from the water activity in the membrane and in the solution as will be shown in detail later. The pressure difference is given by:

$$\Delta\pi = p^m - p^s = \frac{RT}{\overline{V}_w} \ln \frac{a_w^m}{a_w^s} \tag{2.90}$$

Introducing Equation 2.90 into Equation 2.89 and rearranging gives the Donnan potential as a function of the ion and the water activities in the membrane and the solution:

$$\varphi_{Don} = \varphi^m - \varphi^s = \frac{1}{z_i F}\left[RT \ln \frac{a_i^s}{a_i^m} - \overline{V}_i \frac{RT}{\overline{V}_w} \ln \frac{a_w^m}{a_w^s}\right] = \frac{1}{z_i F}\left[RT \ln \frac{a_i^s}{a_i^m} - \overline{V}_i \, \Delta\pi\right] \tag{2.91}$$

The Donnan potentials between a cat- and an anion-exchange membrane and an adjacent bulk solution and the concentrations of ions in the solution and co- and counterions in the membranes are illustrated in Figure 2.10 a) to c). The schematic drawing shown in a) illustrates the cation- and anion-exchange membranes with the corresponding counterions and a electrolyte solution in between. It is indicated that the ion concentration in the solution is significantly lower than in the membranes. The diagram shown in b) indicates the concentration profiles of the counter- and coions in the corresponding membrane and in the solution. The counterion concentration in the membrane is generally slightly higher than the fixed ion concentration because of a small number of coions that are not completely excluded from the membrane phase. At the membrane/solution interface there is a sharp change in the ion concentration. The coion concentration increases and the counterion concentration decreases. These steep gradients in the ion concentration at the interface act as driving force for the transport of coions into the membrane and counterions from the membrane into the solution. Since the solution and the membrane are in equilibrium no net transport of ions between the membrane and the solution takes place. This means that the concentration gradient driving force is counter-balanced by a driving force acting in opposite direction. This driving force is the Donnan potential as illustrated in Figure 2.10 c) which shows the electrical potentials in the membranes and the solution indicating a sharp change in the electrical potential at the membrane/solution interface. This potential gradient at the interface is the Donnan potential φ_{don}.

It should be noted that the value of Donnan potential is negative for the cation-exchange membrane and positive for the anion-exchange membrane in equilibrium with a dilute electrolyte solution.

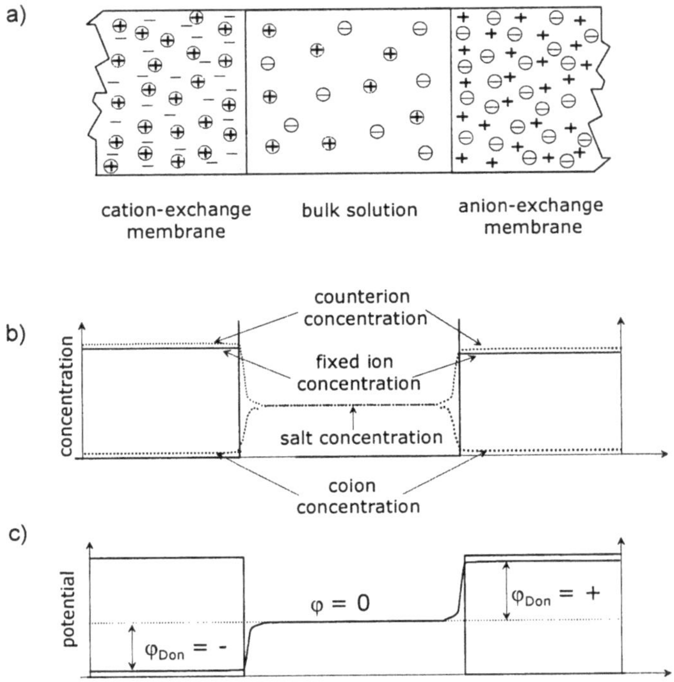

Fig. 2.10 Schematic drawing illustrating the distribution of ions between a bulk solution and a cat- and anion-exchange membrane and the Donnan potentials between the membranes and the bulk solution.

The numerical value of the Donnan potential φ_{don} can be calculated either from the cation or anion activities. For a single salt thus is:

$$\varphi_{Don} = \varphi^m - \varphi^s = \frac{1}{z_c F}\left[RT \ln \frac{a_c^s}{a_c^m} - \overline{V}_c \, \Delta \pi\right] \tag{2.92}$$

or

$$\varphi_{Don} = \varphi^m - \varphi^s = \frac{1}{z_a F}\left[RT \ln \frac{a_a^s}{a_a^m} - \overline{V}_a \, \Delta \pi\right] \tag{2.93}$$

Combination of Equations 2.92 and 2.93 leads to a general relation for the cation and anion distribution at the interface between a solution and an ion-exchange membrane.

$$\frac{1}{z_a}\left[RT \ln \frac{a_a^s}{a_a^m} - \overline{V}_a \, \Delta \pi\right] = \frac{1}{z_c}\left[RT \ln \frac{a_c^s}{a_c^m} - \overline{V}_c \, \Delta \pi\right] \tag{2.94}$$

Here z is the valence, F is the Faraday constant, R is the gas constant, T is the absolute temperature, $\overline{V}$ is the partial molar volume, and a is the activity; the subscripts a, c, and w refer to anion, cation, and water, respectively, and the superscripts s and m refer to the solution and the membrane, respectively.

To determine the sorption equilibrium between a strong electrolyte and an ion-exchange membrane by Equation 2.94 it is assumed that the electrolyte is completely dissociated. Thus, one mole of electrolyte gives ν_a moles of anions and ν_c moles of cations and the partial molar volume of the electrolyte is the sum of the partial molar volumes of the ions. Hence is:

$$z_a \nu_a = -z_c \nu_c \text{ and: } \overline{V}_s = \nu_a \overline{V}_a + \nu_c \overline{V}_c \qquad (2.95)$$

Here ν is the stoichiometric coefficient of the electrolyte, $\overline{V}$ the partial molar volume, and the subscript s, a, and c refer to the salt, anion, and cation, respectively.

Combination of Equations 2.94 and 2.95 leads to:

$$\left(\frac{a_a^s}{a_a^m}\right)^{\frac{1}{z_a}} \left(\frac{a_c^m}{a_c^s}\right)^{\frac{1}{z_c}} = e^{-\frac{\Delta\pi}{RT}\frac{\overline{V}_s}{z_c \nu_c}} \qquad (2.96)$$

Equation 2.96 is the general thermodynamic form of the Donnan equilibrium for the sorption of an electrolyte by an ion-exchange membrane from an adjacent solution containing a single salt.

The Donnan Exclusion of Coions

The Donnan equilibrium describes the relation between the concentrations of the different ions in the membrane and in the adjacent electrolyte solution. The ion concentration in the membrane and especially that of the coion, which is the ion carrying the same charge as the fixed ion of the membrane matrix, is of prime importance for the membrane permselectivity. To obtain a completely permselective membrane the coions should completely be excluded from the membrane phase. This exclusion is referred to as Donnan exclusion. The concentration of the coions in the membrane can be derived from Equation 2.96 by substituting the ion activities by the ion concentrations and the corresponding activity coefficients.

The ion activity is given by:

$$a_{co}^m = C_{co}^m \gamma_{co}^m;\; a_{co}^s = C_{co}^s \gamma_{co}^s;\; a_{cou}^m = C_{cou}^m \gamma_{cou}^m; \text{ and } a_{cou}^s = C_{cou}^s \gamma_{cou}^s \qquad (2.97)$$

Here a, C, and γ are the activity, the concentration, and the activity coefficient, respectively, the subscripts co and cou refer to coions and counterions, respectively, and the superscripts m and s refer to the membrane and the adjacent solution.

Introducing the electroneutrality condition which prevails in the membrane and the solution leads to:

$$|z_{cou}| C_{cou}^m = C_{fix}^m + |z_{co}| C_{co}^m \quad \text{and} \quad z_{cou} C_{cou}^s = -z_{co} C_{co}^s \qquad (2.98)$$

Here C_{fix} is the concentration of fixed ion groups of the ion-exchange membrane, and z is the valence of the ions.

Introducing Equations 2.97 and 2.98 into Equation 2.96 leads to:

$$\left(\frac{C_{co}^{s}}{C_{co}^{m}}\right)^{\frac{1}{z_{co}}}\left(\frac{C_{fix}+z_{co}C_{co}^{m}}{z_{cou}C_{cou}^{s}}\right)^{\frac{1}{z_{cou}}}=\left(\frac{\gamma_{co}^{m}}{\gamma_{co}^{s}}\right)^{\frac{1}{z_{co}}}\left(\frac{\gamma_{cou}^{s}}{\gamma_{cou}^{m}}\right)^{\frac{1}{z_{cou}}}e^{-\frac{\Delta\pi\,\overline{V}_s}{RT\,z_c\nu_c}} \tag{2.99}$$

Here C is the concentration of ion species, z is the electrochemical valence, ν is the number of ions in one mole, $\overline{V}$ is the partial molar volume, and γ the activity coefficient. The subscripts co, cou, fix, and s refer to coion, counterion, fixed ion of the membrane, and salt. The superscripts m and s refer to the membrane and the adjacent solution.

Equation 2.99 describes the distribution of the coions between an ion-exchange membrane and the adjacent solution as a function of the concentration of the fixed ions of the membrane and co- and counterions in the solution. Equation 2.99 is rather complex and contains several parameters such as the osmotic pressure or the activity coefficients of the various ions in the membrane which are difficult to determine in independent measurements. Therefore, the practical relevance is limited. A more simple relation for the calculation of the coion exclusion is obtained with certain assumptions. It can be showen that the osmotic pressure term is small compared to the energy term, i.e. $\Delta\pi \times \overline{V}_s << RT$. For an osmotic pressure of 300 bar and a molar volume of the electrolyte of 40 cm^3 mol^{-1} even for a monovalent electrolyte the term $e^{-\frac{\Delta\pi\,\overline{V}_s}{RT\,z_c\nu_c}} \approx 1$. For a monovalent electrolyte is $z_{cou} = -z_{co}$ and $\nu_i = z_i = 1$. Furthermore, the coion concentration in the solution is equal to the counterion and the salt concentration, and the activity coefficients of the individual ions in the membrane and the solution can be expressed by an average activity coefficient, i.e. $\left(\frac{\gamma_{co}^{s}\gamma_{cou}^{s}}{\gamma_{co}^{m}\gamma_{cou}^{m}}\right)=\left(\frac{\gamma_{\pm}^{s}}{\gamma_{\pm}^{m}}\right)^2$. With these assumptions Equation 2.99 is reduced to:

$$\left(\frac{C_{s}^{s}}{C_{co}^{m}}\right)\left(\frac{C_{s}^{s}}{C_{fix}-C_{co}^{m}}\right)=\left(\frac{\gamma_{\pm}^{s}}{\gamma_{\pm}^{m}}\right)^2 \tag{2.100}$$

Analytical solution of Equation 2.100 leads to:

$$C_{co}^{m}=\frac{C_{fix}}{2}+\sqrt[2]{\frac{C_{fix}}{4}+C_{s}^{s}\left(\frac{\gamma_{\pm}^{s}}{\gamma_{\pm}^{m}}\right)^2} \tag{2.101}$$

If it is further assumed that for very dilute solutions the coion concentration in the membrane is small compared to the fixed ion-concentration, i.e. $C_{co}^{m} << C_{fix}^{m}$ and the electrolyte is monovalent , Equation 2.100 can be solved to give a very simple relation for the coion concentration in the membrane:

$$C_{co}^{m} = \frac{C_s^s}{C_{fix}}\left(\frac{\gamma_\pm^s}{\gamma_\pm^m}\right)^2 \tag{2.102}$$

Here C is the concentration, γ is the activity coefficient, the subscripts co, fix, s, and $\pm$ refer to coion, fixed ion, electrolyte, and average value of the activity coefficient of cation and anion, respectively, and the superscripts m and s refer to the membrane and the solution, respectively.

Equations 2.101 and 2.102 describe the coion concentration in the membrane only to a first approximation under the above mentioned assumptions. Considerable deviation of measured concentrations in the membrane from those calculated by Equations 2.101 and 2.102 is often obtained. The differences between the observed and the calculated membrane behavior are mainly due to the non-uniformities in the distribution of the fixed ions in the membrane, which results from structural irregularities of the membrane on a molecular level and from the influence of the electric field on the ion distribution. The calculation of the Donnan equilibrium for ion-exchange membrane in a solution of multivalent ions or mixed salts is significantly more complex and requires a numerical solution. For a rough estimation of the permselectivity of an ion-exchange membrane as a function of the electrolyte solution Equations 2.101 and 2.102 are quite useful and often used in praxis [L.H. Shaffer et al.,1966].

In general, however, the Donnan equilibrium and thus the selectivity of an ion-exchange membrane depends on:

- the concentration of the fixed ions,
- the valence of the coions,
- the concentration of the electrolyte solution, and
- the affinity of the ion-exchange material with respect to the counterions.

The permselectivity of an ion-exchange membrane increases with increasing concentration of the fixed ions with decreasing electrolyte solution concentration and with decreasing chemical valence of the coions.

3.1.4 The Osmotic Equilibrium and the Osmotic Pressure

The water transport between solutions separated by an ion-exchange membrane due to osmotic pressure differences between the solutions can effect the efficiency of electrodialysis and related processes and must be considered in an overall plant design. The osmotic pressure is the result of different transport rates of water and salts through a membrane. It can be calculated from the osmotic equilibrium. Osmotic water transport occurs when two aqueous salt solutions of different concentrations are separated by a membrane which is permeable for the solvent, e.g. water, but significantly less permeable for the solute, e.g. salt. Then a transport of water from the more dilute solution into the more concentrated solution is observed. This phenomenon which is illustrated in Figure 2.11 is referred to as osmosis. The figure shows a schematic diagram of a membrane separating two solutions consisting of a solvent, e.g. water, and a solute, e.g. a salt. The two solutions are indicated by (') and ("). The membrane is assumed to be permeable to the solvent, but impermeable to the solute.

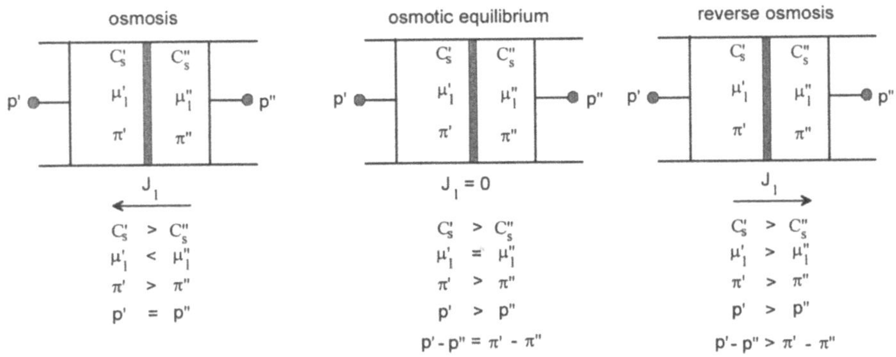

Fig. 2.11 Schematic drawing illustrating the osmotic phenomenon. It shows two phases (') and (") separated by a semipermeable membrane. The phases consist of a solvent and a solute indicated by the subscripts l and s. C and μ are the concentration and the chemical potential, p is the hydrostatic and π the osmotic pressure, respectively, J is the solvent flux across the membrane.

Depending on the concentrations and hydrostatic pressures in the two phases separated by the membrane three different situations can be distinguished:

- The hydrostatic pressures in the two phases separated by the membrane are equal but the solute concentration in solution (') is higher than the one in solution ("). In this case the osmotic pressure in solution (') is higher than that in solution (") and there will be a flow of solvent from the more diluted solution (") into the more concentrated solution (') because the chemical potential of the solvent is higher in the solution with the lower salt concentration. This situation is referred to as osmosis.

- The solute concentration in solution (') is higher than the one in solution ("). In this case the osmotic pressure in solution (') is higher than that in solution (") and the two phases have different hydrostatic pressures, but the difference in hydrostatic pressures is equal to the difference in the osmotic pressures between the two solutions. This situation is referred to as osmotic equilibrium and there is no flow of solvent through the membrane because the chemical potential of the solvent is identical in both phases although the concentrations in the two solutions are different. This situation is referred to as osmotic equilibrium.

- The two phases separated by the membrane have different hydrostatic pressures, and different salt concentrations and therefore also different osmotic pressures. But the difference in hydrostatic pressures between the two phases separated by the membrane is larger than that of the osmotic pressures which act in opposite direction. The solvent will flow through the membrane from the solution (') with the higher solute concentration into the solution (") with the lower solute concentration because the chemical potential of the solvent is now higher in the phase with the higher salt concentration. This situation is referred to as reverse osmosis.

The Osmotic Pressure

The osmotic pressure of a solution π is proportional to its solute concentration C_S. It can be calculated from the chemical potential of the solvent.

Assuming a solution of a solute in a solvent is separated from the pure solvent by a strictly semipermeable membrane which is only permeable to the solvent. For simplicity reasons the osmotic pressure shall be derived for a single component solute in a solvent. If the solvent is water and the solute is a salt the degree of dissociation and the stoichiometric coefficients of the salt must be considered for determining the osmotic pressure. Since in an aqueous solution salts are dissociated in two or more ions and since the osmotic pressure of the solution is determined by the total number of individual species in the solution the osmotic pressure must be expressed by the total ion concentrations.

To determine the osmotic pressure of a solution consisting of a solvent and a single solute osmotic equilibrium is assumed. This means there is no flux through the membrane and the hydrostatic pressure of the solution is higher than that of the solvent and in fact equal to the osmotic pressure of the solution. Under osmotic equilibrium conditions the chemical potential of the solvent in the solution, μ_l^s is equal to that of the pure solvent, i.e. μ_l^l. Thus is:

$$\mu_l^s = \mu_l^l \tag{2.103}$$

The chemical potential of the solvent in the solution is:

$$\mu_l^s = \mu_l^o + \overline{V}_l\, p^s + RT \ln a_l^s \tag{2.104}$$

Here $\overline{V}_l$ is the partial molar volume of the solvent, μ_l^o is the chemical potential of the solvent under standard conditions of pressure and temperature, a_l is the activity of the solvent, p is the hydrostatic pressure, and the superscript s refers to the solution.

The chemical potential of the solvent in the phase that contains only solvent is given by:

$$\mu_l^l = \mu_l^o + \overline{V}_l\, p^l \tag{2.105}$$

In osmotic equilibrium the chemical potential of the solvent in the solution is equal to that of the solvent. Thus, introducing Equations 2.104 and 2.105 into Equation 2.103 leads to:

$$\overline{V}_l\, p^s + RT \ln a_l^s = \overline{V}_l\, p^l \tag{2.106}$$

Rearranging Equation 2.106 leads to:

$$\overline{V}_l \left(p^l - p^s\right) = RT \ln a_l^s \tag{2.107}$$

The activity of the solvent in the solution a_l^s is proportional to its mole fraction in the solution X_l^s. Thus is:

$$\ln a_l^s = g \ln X_l^s \tag{2.108}$$

Here g is a proportionality factor which is referred to as osmotic coefficient [R.A. Robinson et al, 1970].

In dilute solutions the mole fraction of the solvent is much larger than that of the solute, i.e. $X_l^s >> X_s^s$. Under these conditions it can be assumed that:

$$g \ln X_l^s = g \ln\left(1 - X_s^s\right) \approx - g X_s^s \tag{2.109}$$

Here X_s^s is the mole fraction of the solute in the solution.

Introducing Equation 2.109 and 2.108 into Equation 2.107 gives the osmotic pressure of the solution as the function of the mole fraction of the solute:

$$\pi = \left(p^l - p^s\right) = -g \frac{RT}{\overline{V}_l} X_s^s \tag{2.110}$$

For dilute solutions it can be assumed that to a first approximation:

$$X_s^s \approx \frac{n_s^s}{n_l^s} \tag{2.111}$$

and

$$n_l^s \overline{V}_l \approx V \tag{2.112}$$

Further is:

$$\frac{n_s^s}{V} = C_s^s \tag{2.113}$$

Here n_s^s and n_l^s are the numbers of moles of the solute and the solvent, respectively, in the solution, V is the volume of the solution, $\overline{V}_l$ is the partial molar volume of the solvent and C_s^s the solute concentration in the solution.

Combination of Equations 2.110 to 2.113 describes the osmotic pressure as a function of the concentration of the solute in the solution:

$$\pi = - g\, RT\, C_s^s \tag{2.114}$$

If the solute is a salt, as is the case in many practically relevant applications, the concentration of the individual ions must be considered when determining the osmotic pressure, i.e. the dissociation of the salt must be taken into account.

In an aqueous solution the salt is dissociated in ν_c cations and ν_a anions as indicated in Equation 2.62 and the total concentration of dissolved components is:

$$C_c^s + C_a^s = \nu_c C_s^s + \nu_a C_s^s = \left(\nu_c + \nu_a\right) C_s^s \tag{2.115}$$

Here, C_c^s, C_s^s and C_a^s are the concentrations of the salt, the cations, and the anions in the solution, and ν_c and ν_a are stoichiometric coefficients which determine into how many cations and anions a salt will dissociate in the solution.

Introducing Equation 2.115 into Equation 2.114 gives the osmotic pressure of an aqueous salt solution as a function of the salt concentration:

$$\pi = - g\, RT\, (\nu_c + \nu_a)\, C_s^s \tag{2.116}$$

The osmotic pressure of concentrated solutions can be rather high. For a one molar sodium chloride solution the osmotic pressure is at 25°C close to 5 MPa or 50 bar. Therefore, the water transport from a dilute to the concentrate due to osmosis can effect the economics of ion-exchange membrane separation processes severely.

3.1.5 Electroosmosis and Streaming Potential

If an electrical potential gradient is applied to an electrolyte solution in a pore or thin capillary with fixed charges at the wall as indicated in the schematic drawing of Figure 2.12 a volume flow of electrolyte through the pore or capillary is observed. This phenomenon is referred to as electroosmosis [R.A. Wagenen et al, 1980], [K.S. Spiegler, 1958].

The causes and consequences of electroosmosis can be elucidated by the schematic drawing of Figure 2.12 which shows the cross-section of a capillary.

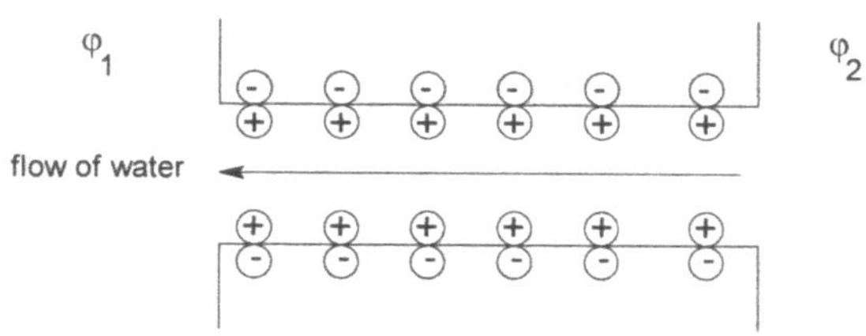

Fig. 2.12 Schematic drawing illustrating the flow of water through a capillary with negative surface charges filled with an electrolyte solution due to electroosmosis caused by an electrical potential difference ($\varphi_1 < \varphi_2$).

The wall of the capillary carries negative fixed charges and is in contact with an electrolyte solution. The fixed negative charges at the wall are compensated by positively charged mobile ions of the electrolyte solution. Thus, in a thin layer at the wall of the capillary there is an excess of positively charged ions. If an electrical potential is applied across the solution the positive excess ions in the double layer will move relative to the surface of the capillary wall towards the cathode dragging a certain amount of water in a thin layer to the exit of the capillary. The velocity with which the ions move is determined by the electrical potential driving force and the friction they encounter in the solution. This friction develops because the layers of water surrounding the charged particle move with different velocities. The layer closest to the ion will move at the same speed as the ion. The velocity of the water will decrease with increasing distance from the moving ion in the layer at the capillary wall and in a certain distance from the wall the water will be unaffected by the flow of the excess ions. The water transport by the excess ions in the layer at the capillary wall, i.e. the electroosmotic water flow can be calculated from the electrical potential driving force and the friction acting on the moving ions.

In the steady state the frictional force and the electrical potential force acting on the excess ions are equal and the flow of ions and thus of water through the capillary is constant. The frictional force of a moving plane of an unit area is given by:

$$F_{fr} = -\eta \frac{du}{dx} \qquad (2.117)$$

Here F_{fr} is the frictional force, η is the dynamic viscosity, u is the linear velocity, and x is the directional coordinate perpendicular to the capillary wall.

For simplicity reason it is assumed that the velocity change of the different water layers around the moving excess ions is decreasing linearly with the distance from the capillary wall and the frictional force can be expressed by:

$$F_{fr} = -\eta \frac{u}{d} \qquad (2.118)$$

Here d is the distance from the capillary wall into the solution where the water flow velocity changes from the speed of the ions to 0.

The electrical potential force of an unit area is given by:

$$F_{el} = -\sigma \frac{\Delta\varphi}{l} \qquad (2.119)$$

Here F_{el} is the electrical potential force, σ is the surface charge that must move relative to the solution, $\Delta\varphi$ is the voltage drop along the capillary, and l is the length of the capillary.

In the steady state the two forces are equal, thus is:

$$\eta \frac{u}{d} = \sigma \frac{\Delta\varphi}{l} \qquad (2.120)$$

The linear flow velocity u can be related to the surface charges at the capillary wall and the double layer potential generated by this charge by the Helmholtz double layer expression given in Equation 2.33.

Introducing Equation 2.33 into Equation 2.120 and rearranging gives an expression of the linear flow velocity in electroosmosis:

$$u = \frac{\varepsilon \, \Delta\varphi\zeta}{4\pi\eta \, l} \qquad (2.121)$$

Here u is the linear flow velocity, ε is the dielectric constant of the solution, $\Delta\varphi$ is the applied potential difference over the length of the capillary, η is the dynamic viscosity of the solution, l is the length of the capillary, and ζ is the so-called zeta potential

[J. Lyklema, 1995] which represents the Helmholtz double layer potential. The zeta potential is often used to determine the double layer potential from electro-kinetic measurements [R.J. Hunter, 1981].

In praxis, it is often more convenient to measure electroosmotic volume flow instead of the linear velocity u. The electroosmotic volume flow rate in a cylindrical capillary is given by:

$$\frac{dV}{dt} = u\,\pi\,r^2 = \frac{\varepsilon\,\zeta\,r^2\Delta\varphi}{4\eta\,l} \tag{2.122}$$

In analogy to the osmotic equilibrium an electroosmotic equilibrium with an electroosmotic pressure can be defined. In the electroosmotic equilibrium the electrical potential driving force is just counter-balanced by the hydrostatic pressure driving force.

The volume flow through a capillary due to a hydrostatic pressure difference is given by the Hagen-Poiseuille law [P.W. Atkins, 1990] to:

$$\frac{dV}{dt} = \frac{\Delta p\,\pi\,r^4}{8\,\eta\,l} \tag{2.123}$$

Introducing Equation 2.123 into Equation 2.122 and rearranging gives the electroosmotic pressure:

$$\Delta p = \frac{2\,\varepsilon\,\zeta\,\Delta\varphi}{\pi\,r^2} \tag{2.124}$$

Here Δp is the electroosmotic pressure, ζ is the zeta potential of the capillary wall, $\Delta\varphi$ is the applied voltage, and r is the radius of the capillary.

In electroosmosis a hydrostatic pressure difference is created by applying an electrical potential difference in the solution between the entrance and exit of a capillary. Alternatively, a hydrostatic pressure difference over the length of the capillary can create a flow of electrical charges, i.e. the so-called streaming current which then leads to a build-up of an electrical potential difference between the entrance and the exit of the capillary. This electrical potential is referred to as streaming potential.

The streaming current and the streaming potential, respectively, can also be derived from the Helmholtz relation in the same way the electroosmotic volume flow rate and the electroosmotic pressure were derived [G. Kortüm, 1957].

The streaming potential is given by:

$$\Delta\varphi_{str} = \frac{\varepsilon\,\zeta\Delta p}{\eta\,\kappa} \tag{2.125}$$

Here is $\Delta\varphi_{str}$ the streaming potential, κ is the conductivity of the solution in the capillary, Δp is the applied hydrostatic pressure, ζ is the zeta potential, η is the dynamic viscosity, and ε the dielectric constant of the solution.

The electroosmosis is of some relevance in ion-exchange membrane separation processes and specifically in electrodialysis since it effects the efficiency of the process by the transport of water from the diluate to the concentrate solution. This effect will be discussed in more detail later.

3.2 The Electrochemical Cell and the Electromotive Force

The transport of ions in electrodialysis is based on an external power source and two electrodes, i.e. a negatively charged cathode and a positively charged anode immersed in an electrolyte solution. This arrangement is referred to as an electrolytic cell. It is shown schematically in Figure 2.13.

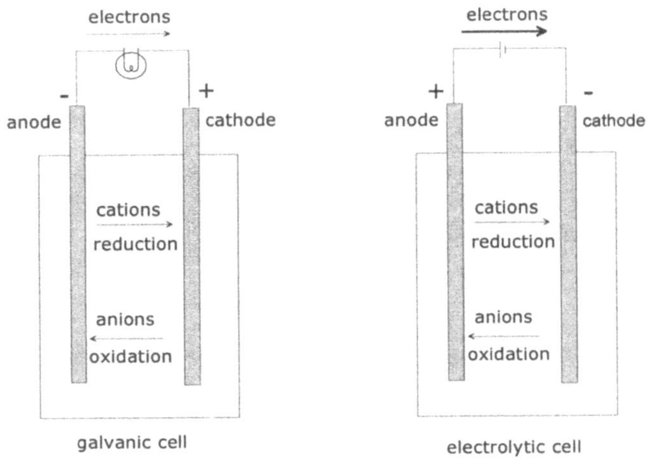

Fig. 2.13 Schematic diagram illustrating the functions of an electrochemical cell.

An electrolytic cell consists of two electrodes (electron conductors) dipping into an electrolyte (ion conductor). There are two types of electrochemical cells:

- a galvanic cell that produces electricity as a result of a spontaneous reduction and oxidation reaction in the electrolyte solution at the electrode surfaces generating electrical energy, and
- an electrolytic cell with a reduction and an oxidation reaction in the electrolyte solution at the electrode surfaces driven by an external electrical power source.

Thus, in an electrochemical cell a redox reaction takes place with a transfer of electrons from one substance to another. The reducing agent is the electron donor and the oxidizing agent is the electron acceptor. In case of a spontaneous redox reaction the result is a flux of electrons, i.e. an electric current from the anode to the cathode in the external electron conducting system, and the electrochemical cell represents a galvanic cell which can produce electrical energy. If the redox reaction is not spontaneous energy must be provided for the transfer of electrons from the reducing agent to the oxidizing agent. In this case, the electron flux from a power source in the outside electron conducting system is from the cathode to the anode.

A typical redox reaction is the combustion of hydrogen in a galvanic cell and the production of oxygen and hydrogen in the electrolytic cell. Thus, is:

- galvanic cell reaction: $2\,H_2 + O_2 \rightarrow 2\,H_2O$ + electrical energy
- electrolytic cell reaction: electrical energy + $2H_2O \rightarrow 2H_2 + O_2$

In an electrochemical cell the overall redox reaction is composed of two half-reactions. One half-reaction, i.e. the oxidation takes place at the anode and the other half-reaction, i.e. the reduction is taking place at the cathode. In this way the reduction and oxidation processes are separated in space. Thus, is:

anode reaction (oxidation): $n_{red} \rightarrow n_{oxi} + \nu_e$ (2.126)

cathode reaction (reduction) $n_{oxi} + \nu_e \rightarrow n_{red}$ (2.127)

Here n_{red} and n_{oxi} are the components n in the reduced and the oxidized form, respectively, and ν_e is the number of electrons released to or withdrawn from the electrodes during the oxidation or reduction, respectively.

In a galvanic cell with a spontaneous reaction a component is oxidized and releases electrons at the anode which is gaining negative charges. In the reaction taking place at the cathode electrons are withdrawn from the electrode to reduce a component. Thus, in a galvanic cell the anode is negatively charged and has an excess of electrons which can travel through an external circuit to the cathode to provide electrons needed in the second half-reaction for the reduction. In a galvanic cell the cathode has a higher electrical potential than the anode. The current generated in the galvanic cell can be used to do work. Fuel cells and batteries represent Galvanic cells.

In an electrolytic cell the situation is reversed. Due to an external power source electrons are withdrawn from the component in reduced form at the anode, and at the cathode electrons are provided by the external power source to a component in the oxidized form. Therefore, the anode in the electrolytic cell is negatively charged and the cathode positively charged. The half-reactions, however, are identical with those in the galvanic cell, i.e. oxidation at the anode and reduction at the cathode. The production of caustic soda and chlorine is a typical application of an electrolytic cell.

The electrodes in an electrochemical cell have a key function by establishing an efficient transfer of the electrons from a metal conductor to an ion conducting electrolyte. There are a variety of electrodes used in praxis. One example is the hydrogen electrode which is shown schematically in Figure 2.14.

In a hydrogen electrode hydrogen gas is in equilibrium with hydrogen ions of the solution in the presence of an inert metal. The metal which often is platinum acts as a source or sink for electrons and has no part in the reaction itself which is $2H^+ \Leftrightarrow H_2$. The hydrogen electrode is denoted:

$$\mathrm{Pt}, H_2 \,|\, H^+(\mathrm{sol})$$

The vertical line denotes interfaces between the hydrogen gas in the metal and the solution containing hydrogen ions.

The hydrogen electrode may be either a cathode or an anode, i.e. it may have a higher or lower potential than the opposite electrode depending on the nature of the other electrode in the cell and the direction of the overall reaction. At 25°C, with a gas pressure of one bar, and an ion activity of one, the potential of a hydrogen electrode is zero by definition. The hydrogen electrode is used as reference for the characterization of other electrodes [G. Kortüm, 1957].

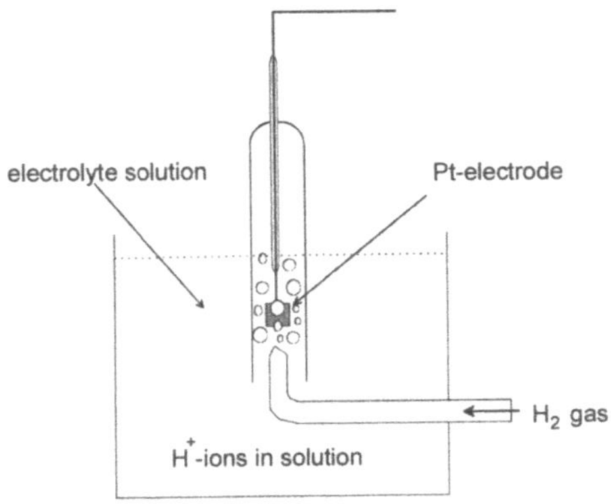

Fig. 2.14 Schematic drawing illustrating the construction of a hydrogen electrode.

The hydrogen electrode is referred to as an electrode of the first kind which means that the electrode material and the ions produced in the solution represent the same chemical species such as H_2 and H^+. Other electrodes of the first kind are the silver metal electrodes:

$$\mathrm{Ag(metal)}\,|\,\mathrm{Ag^+\ (sol)}$$

or the chlorine gas electrode:

$$\mathrm{Pt, Cl_2(gas)}\,|\,\mathrm{Cl^-\ (sol)}$$

In addition to the electrodes of the first kind there are the electrodes of the second or third kind. Important second kind electrodes are the Ag(metal),AgCl and the Hg(metal),Hg_2Cl_2 electrodes. These electrodes are composed of a porous layer of a virtually insoluble salt on metal surface, e.g. AgCl on a metal silver surface or Hg_2Cl_2 on mercury. The only chemical species that moves in and out of the solution is the Cl^--ion. The function of the Ag(metal),AgCl electrode as a cathode is such that electrons provided by the Ag (metal) reduce the Ag^+-ions in the salt layer on the surface of the metal part of the electrode leaving residual Cl^--ions. These ions drift into the solution. If the electrode is an anode electrons are withdrawn from the Ag(metal) and Ag^+-ions are generated. These ions are immediately precipitated by Cl^--ions from the solution as AgCl because of the low solubility of the AgCl salt. The net effect is a reversible exchange of Cl^--ions between the Ag(metal),AgCl electrodes without any other reaction.

The most common of the secondary type of electrodes is the calomel electrode consisting of Hg(metal),Hg_2Cl_2 salt in contact with a Cl^--ions containing solution. The calomel electrode has a significant property which makes it very useful as reference electrode. For any current applied to the electrode, the calomel electrode will always release or absorb the proper number of Cl^--ions. To an outside observer it would appear that transfer of charges across the interface occurs easily with very little resistance. Such an electrode is referred to as a highly non-polarizable electrode in contrast to an electrode where the transfer of charges is very difficult and the resistance of the interface is very high. Such an electrode is referred to as highly polarizable.

The calomel electrode is widely used in electrodialysis and related processes for characterizing electrolyte solutions or ion-exchange membranes. Its construction, however, is rather complex as the following schematic drawing of Figure 2.15 shows.

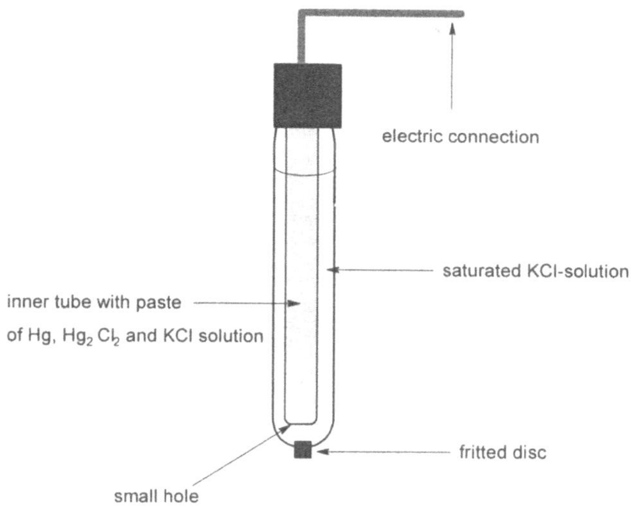

Fig 2.15 Schematic drawing illustrating the construction of a calomel electrode.

It consists of an inner tube filled with a paste of Hg, Hg_2Cl_2, and saturated KCl solution in contact with an electric plug on one side and through a small hole with a saturated KCl solution in an outside tube. The KCl solution is in contact with an outside test solution through a glass frit which avoids a convective flow of the salt solution. Since the mobility of K^+- and Cl^--ions are nearly identical no diffusion potential occurs at the calomel electrode test solution interface which could obscure a potential measurement.

3.2.1 The Electrochemical Cell Reaction and the Gibbs Free Energy

The current produced by a galvanic cell is the result of a spontaneous redox reaction taking place in the cell. This current can be utilized to do work which can be expressed in electrical energy. The transport of ions in an electrolytic cell requires an external power source and consumes energy. The work which can be obtained from a galvanic cell at constant temperature and pressure is the product of the current, the cell potential, and the time. The maximum work that the galvanic cell can do is given by the Gibbs free energy of the spontaneous reaction in the cell:

$$W_{e,max} = (\Delta G)_{p,T} \tag{2.128}$$

Here $W_{e,max}$ is the maximum electrical work and ΔG is the Gibbs free energy of the reaction.

The maximum work is obtained in a completely reversible reaction without any entropy production. In a galvanic cell this situation is achieved at zero current. A zero current situation is obtained when the cell potential is balanced by an opposing outside potential source or the system as an infinitely high resistance. The zero-current cell potential is referred to as the electromotive force of a cell. It is related to the work that can be provided by the cell and therefore to the Gibbs free energy of a reversible reaction. It can be derived from Equation 2.46:

$$(dG)_{p,T} = \sum_i \nu_i \mu_i d\xi = \Delta G_r d\xi, \tag{2.129}$$

Here $(dG)_{p,T}$ is the total Gibbs free energy change at constant pressure and temperature as a function of the reversible Gibbs free energy ΔG_r of a reaction and the extent of the reaction $d\xi$, which is given by $\sum_i dn_i = d\xi$, while ν_i and μ_i are the stoichiometric coefficient and the chemical potential of a component i.

Introducing Equation 2.129 into Equation 2.128 gives the electrical energy generated in a galvanic cell to:

$$dW_e = \Delta G_r \, d\xi \tag{2.130}$$

During the reaction a certain number of electrons are transported between the electrodes resulting in an external electrical current. The number of electrons that are transported between the electrodes and thus the electrical current is the product of the Faraday constant F and the extent of the reaction:

$$I = \nu F \xi. \tag{2.131}$$

Here I is the external current and the coefficient ν indicates the number of electrons generated in the reaction according to Equations 2.126 and 2.127.

The electrical work generated in the galvanic cell is:

$$W_e = -\nu F \xi \Delta\varphi \tag{2.132}$$

The maximum electrical work is given by:

$$W_{e,max} = \Delta G_r \xi = -\nu F \xi \varphi_{emf} \tag{2.133}$$

and

$$\Delta G_r = -\nu F \varphi_{emf} \tag{2.134}$$

Here φ_{emf} is the zero-current cell potential or electromotive force.

3.2.2 The Standard Cell Potential and the Nernst Equation

The electromotive force or zero-current potential can be related to the activities of participating components of the cell reaction. The Gibbs free reaction energy is:

$$\Delta G_r = \Delta G^o + RT \ln \prod_i a_i^{\nu_i} \qquad (2.135)$$

Here ΔG^o is a standard value of the Gibbs free reaction energy determined at 25° C and with the reaction component activities of 1, a is the activity of the reaction component, ν is the stoichiometric coefficient and the subscript i refers to a component i.

Introducing Equation 2.134 into 2.135 and rearranging leads to:

$$\varphi_{emf} = -\frac{\Delta G^o}{\nu F} - \frac{RT}{\nu F} \ln \prod_i a_i^{\nu_i} \qquad (2.136)$$

The first term in Equation 2.136 refers to standard conditions and is denoted φ^o and is called the standard cell potential. Introducing the standard cell potential into Equation 2.136 results in the so-called Nernst equation:

$$\varphi_{emf} = \varphi^o - \frac{RT}{\nu F} \ln \prod_i a_i^{\nu_i} \qquad (2.137)$$

The standard cell potential in the Nernst equation is defined in reference to the normal hydrogen electrode which is 0 by definition as indicated earlier. The Nernst equation is an important relation which is used to determine reaction equilibrium constants and concentration potentials as well as to calculate the minimum energy required in electrodialysis as will be shown later.

4 TRANSPORT OF IONS IN SOLUTIONS AND MEMBRANES

To achieve a separation of ions from a solution in ion-exchange membrane separation processes these ions must be transported from one solution through a cation- or anion-exchange membrane into another solution. The transport rate of the ions is determined by kinetic as well as by thermodynamic parameters. The kinetic parameters are expressed by the mobility or diffusivity of the ions in the membrane matrix and the electrolyte solution. The thermodynamic parameters are expressed by the driving forces which must be applied to overcome the friction a component experiences while being moved through the solution or the membrane. Thus, to describe the mass transport in a given system it must be characterized by thermodynamic state functions such as the Gibbs free energy and by its kinetic parameters such as diffusion coefficient, ion mobility or electric conductivity.

A very important boundary condition in treating the mass transport in electrolyte solutions is the electroneutrality requirement which postulates that on a macroscopic scale there is no excess in positive or negative charges. Although cations and anions can be considered as independent components in an electrolyte solution, their fluxes are related in such a way that there is no accumulation of positive or negative electrical charges on a macroscopic scale.

4.1 Mathematical Description of Transport Phenomena in Solutions

The mass transfer in electrolyte solutions is determined by the driving forces acting on the individual components of the solution and by the friction which the components have to overcome when being moved. The driving forces acting on the components of a system can be expressed by gradients in their electrochemical potential. The friction or the resistance that has to be overcome by the driving force to achieve the transport of the component is generally expressed by a hydrodynamic permeability, a diffusion coefficient or in electrolyte solutions also by the electrical resistance of the electrolyte. To describe the mass transport in a system the thermodynamic and kinetic parameters must be mathematically related. Several relations are described in the literature. Most are semi-empirical such as Fick's law, Ohm's law, Hagen-Poisseuille's law, etc.. To describe mass transport phenomena in electrolyte solutions often the extended Nernst-Planck equation is used [R. Schlögl, 1964]. This equation considers three modes of mass transport in ion-exchange membranes:

- diffusion,
- migration, and
- convection.

Diffusion is a movement of molecular components due to a local gradient in the chemical potential. Gradients in the chemical potential of a component under isothermal conditions can arise from gradients in concentration or pressure. The dominant term for molecules or ions transported through an ion-exchange membrane or in a solution by diffusion is the concentration gradient. For the diffusion of ions it is mandatory that the electroneutrality requirement is fulfilled, i.e. cations and anions have to diffuse in the same direction.

Migration is a movement of ions due to an electrical potential gradient. An electrical potential gradient is usually established by applying a voltage difference between two electrodes dipping into an electrolyte solution. The positively charged cations migrate towards the negatively charged cathode and the negatively charged anions migrate towards the positively charged anode, i. e. cations and anions are transported in opposite directions.

Convection is a movement of mass due to a mechanical force, i.e. in general a hydrostatic pressure difference. Convective transport is generally less important in homogeneous solid materials such as ion-exchange membranes. It is, however, dominant in porous structures and stirred solutions.

The advantage of the relatively simple relations such as Fick's law or the Nernst-Planck equation is that only a limited number of parameters such as the diffusion coefficient, the ion mobility or the hydrodynamic resistance are required which must be determined in independent measurements. The disadvantage of Fick's law and the Nernst-Planck equation is that any kinetic coupling between individual fluxes is neglected. A more comprehensive description of transport phenomena in multi-component electrolyte solutions is based either on the so-called Stefan-Maxwell approach or on the thermodynamic of irreversible processes which relates the fluxes of heat, electricity, momentum, and individual components to the corresponding driving forces in a set of phenomenological equations. These two approaches will be reviewed briefly.

4.1.1 Phenomenological Description of Mass Transport

A flux of individual components, heat, and electrical charges can be expressed in the following general relation [A. Katchalsky et al., 1965]:

$$J_i = \sum_k L_{ik} X_k \qquad (i,k = 1, 2, 3, \ldots, n) \tag{2.133}$$

Here J_i refers to fluxes of heat, or electric charges as well as to that of individual components, X_k refers to the driving forces related to a given flux, and L_{ik} is a phenomenological coefficient relating flux and driving force.

For multi-component systems with fluxes of heat, electric charges, and neutral components Equation 2.133 can be written as a matrix. The diagonal coefficients relate the fluxes to the corresponding driving forces, and the cross-coefficients express the coupling of fluxes with indirectly conjugated driving forces [R. Haase, 1963] .

Membrane processes are generally spontaneous, i.e. irreversible. Therefore, the entropy during the process is increasing. The entropy change is given by:

$$T\frac{d_i S}{dt} = \Psi = \sum_k J_i X_k = \sum_i \sum_k L_{ik} X_i X_k \geq 0 \tag{2.134}$$

Here $T\frac{d_i S}{dt}$ is the entropy production, Ψ is the dissipation function, X is the driving force, L is a phenomenological coefficient, J is the flux, and the subscripts i and k refer to individual components, momentum, heat, and electrical charges.

It can be shown by introducing Equation 2.133 into Equation 2.134 that the diagonal coefficients are always positive, while the cross-coefficients might be positive or negative. Their number, however, is reduced by the Onsager relation [L. Onsager, 1931], which gives:

$$L_{ik} = L_{ki} \tag{2.135}$$

To describe the transport in a multi-component system such as an electrolyte solution with gradients in hydrostatic pressure, temperature, chemical potential, and electrical potential as driving forces a very large number of phenomenological coefficients must be known which may be a function of temperature, pressure or composition but which are independent of the driving forces [D.G. Miller, 1966]. This limits the practical use of Equation 2.133 substantially. However, under certain conditions many of the phenomenological coupling coefficients can be neglected and the rather complex set of equations can be reduced to simple relations such as Fick's or Ohm's law.

Another approach to describe the mass transport in membrane processes is based on a relation developed by Maxwell and Stefan around 1870 and which was extended by Spiegler [K.S. Spiegler, 1958]. In this relation the forces are expressed as a linear function of the fluxes:

$$X_i = \sum_k R_{ik} J_k \tag{2.136}$$

Here R is a phenomenological coefficient relating driving forces and fluxes.

The coefficient L of Equation 2.133 has the dimension of a generalized conductance and is expressed by a flux per unit force while the coefficient R in Equation 2.136 has the dimension of a generalized resistance and is expressed by a unit force divided by a flux. Thus is:

$$L_{ik} = \left(\frac{J_i}{X_k}\right)_{X_i} \quad \text{and} \quad R_{ik} = \left(\frac{X_i}{J_k}\right)_{J_i} \tag{2.137}$$

4.1.2 Description of Mass Transport by the Stefan- Maxwell Approach

The Stefan-Maxwell approach is based on the assumption that at a constant flux of a component i the driving forces acting on this component are equal to the sum of the friction forces between the component i and other components in the system [J.A.Wesselingh et al., 1990]. If viscous flow is excluded the Stefan-Maxwell equation is given by [G. Kraaijeveld, 1994]:

$$X_i = \sum_k \frac{f_{ik}}{C_i}(v_i - v_k) = \sum_k \frac{RT}{Ð_{ik} C_i}(v_i - v_k) \tag{2.138}$$

Here X is the driving force, C is the concentration, v is the linear velocity, f is the friction coefficient, and Ð is the Stefan- Maxwell diffusion coefficient. The subscripts i and k refer to individual components.

The flux of an individual component is always proportional to its velocity relative to the velocity of another component which is used as a frame of reference. Thus, the number of independent fluxes in a given system is equal to the total number of components in the system minus the one used as reference. A system containing a membrane with n components, including the membrane, has (n-1) independent fluxes.

The flux of the component i relative to that of the component k which serves as a frame of reference is given by:

$$J_i = C_i(v_i - v_k) \tag{2.139}$$

In membrane processes such as electrodialysis the fluxes through the membranes are of interest. Therefore, the membrane is used as frame of reference and its velocity is assumed to be zero.

Combination of Equations 2.133, 2.136, 2.138, and 2.39 provides the relation between the various terms of the phenomenological equation and the Stefan-Maxwell equation.

$$\sum_i L_{ik} = \frac{1}{\sum_i R_{ik}} = \sum_i \frac{C_i}{f_{ik}} = \sum_i C_i \frac{Ð_{ik}}{RT} \tag{2.140}$$

Equation 2.138, i.e. the Stefan-Maxwell relation provides the same complete description of transport processes through a membrane separating two homogeneous mixtures as does the phenomenological Equation 2.133 based on the thermodynamic of irreversible processes. The only further boundary condition relevant for a system which contains electrical charged components is the electroneutrality requirement and the conservation of electrical charges. All phenomena observed in membrane systems, such as osmosis, electro-osmosis, the diffusion of individual components, viscous flow of a bulk solution, electric current, or the build-up of an osmotic pressure, a streaming and a diffusion potential can be described by applying the Equations 2.133 or 2.138. However, as mentioned before, the practical value of these equations is rather limited since they are only applicable close to equilibrium because of the assumed linear relationships between fluxes and driving forces and the large number of different coefficients that must be known when multi-component systems with viscous flow, diffusion, and migration shall be treated. But even if the phenomenological treatment of mass transport in membrane processes is of little practical use it indicates how complex the processes in multi-component systems and heterogeneous membranes really are. It shows that through the coupling of fluxes a transport of components may be obtained without a directly related driving force. The magnitude of the transport of certain components obtained through a coupling with other fluxes depends on the coupling coefficient. In many cases the coupling between fluxes can be neglected. In some cases, however, it can be significant. For instance, in electrodialysis the coupling of water with an ion flux can lead to a significant water transport through a membrane.

4.1.3 Description of Mass Transport by the Nernst-Planck Equation

A more simplified approach to describe the mass transport in electrolyte solution is provided by the so-called Nernst-Planck flux equation which can be derived from the Equations 2.133 or 2.138 by introducing the electrochemical potential as driving force for the mass transport and by neglecting all other kinetic coupling between individual components.

The electrochemical potential η is given by the chemical potential μ and the electrical potential φ. Thus is:

$$\eta_i = \mu_i + z_i F\varphi = \mu_i^o + \overline{V}_i p + RT\ln a_i + z_i F\varphi \tag{2.141}$$

Here a is the activity, p is the hydrostatic pressure, $\overline{V}_i$ is the partial molar volume, z is the valence, F is the Faraday constant, R is the gas constant, and T is the absolute temperature, the subscript i refers to a component, and the superscript o refers to a standard state.

Introducing the activity coefficient and the ion concentration for the ion activity into Equation 2.141 and differentiation in transport direction x, i.e. in direction perpendicular to the membrane surface leads to:

$$\frac{d\eta_i}{dx} = \vec{V}_i \frac{dp}{dx} + RT\frac{d\ln(C_i\gamma_i)}{dx} + z_i F\frac{d\varphi}{dx} \tag{2.142}$$

Here C is the concentration and γ is the activity coefficient.

In electrodialysis and other ion-exchange membrane separation processes the pressure difference across the membrane is kept as low as possible to minimize viscous flow.

Therefore, the term $\vec{V}_i \frac{dp}{dx}$ in Equation 2.142 can generally be neglected and the term $RT\frac{d\ln(C_i\gamma_i)}{dx}$ can be expressed by $\frac{RT}{C_i}\left(1+C_i\frac{d\ln\gamma_i}{dC_i}\right)\frac{dC_i}{dx}$.

Thus is:

$$\frac{d\eta_i}{dx}=\frac{RT}{C_i}\left(1+C_i\frac{d\ln\gamma_i}{dC_i}\right)\frac{dC_i}{dx}+Fz_i\frac{d\varphi}{dx} \tag{2.143}$$

Introduction of Equation 2.143 into Equation 2.138 gives the anion and cation fluxes of a completely dissociated salt:

$$J_i=-Đ_i\left[\left(1+C_i\frac{d\ln\gamma_i}{dC_i}\right)\frac{dC_i}{dx}+\frac{z_iC_iF}{RT}\frac{d\varphi}{dx}\right]+v_kC_i \tag{2.144}$$

If the solution is assumed to be ideal, i.e. the activity of a component is identical to its concentration, the Stefan-Maxwell diffusion coefficient can be replaced by the Fick's diffusion coefficient and Equation 2.144 is simplified to:

$$J_i=-D_i\frac{dC_i}{dx}-D_i\frac{z_iC_iF}{RT}\frac{d\varphi}{dx}+v_kC_i \tag{2.145}$$

Equation 2.145 represents the extended Nernst-Planck flux equation where D_i is the Fick's diffusion coefficient of the component i in reference to the water in the membrane.

The first term $D_i\frac{dC_i}{dx}$ in Equation 2.145 represents the diffusion, the second term $D_i\frac{z_iC_iF}{RT}\frac{d\varphi}{dx}$ the migration, and the third term v_kC_i the convection.

The Fick's diffusion coefficient is related to the Stefan-Maxwell diffusion coefficient and the phenomenological coefficient. Thus, it is given by:

$$D_i=L_{ii}\frac{RT}{C_i}\left(1+\frac{d\ln\gamma_i}{d\ln C_i}\right) \tag{2.146}$$

and

$$D_i=Đ_i\left(1+\frac{d\ln\gamma_i}{d\ln C_i}\right) \tag{2.147}$$

The Nernst-Planck equation becomes identical with the phenomenological equation of the thermodynamic of irreversible processes, and the Stefan-Maxwell equation if all kinetic coupling between individual ions is neglected and if the activity coefficient is assumed to be 1. The Nernst-Planck flux equation is a very convenient approach to describe the transport of ions in ion-exchange membranes. It must, however, be realized that a series of assumptions has been made which in practical ion-exchange membrane separation processes are not always applicable.

4.1.4 Electrical Current and Ion Fluxes

In electrodialysis it is assumed that the total current through the membrane is carried by ions only, thus is:

$$i = \frac{I}{A} = F\sum_i z_i J_i \tag{2.148}$$

Here i is the current density, I the current, A the membrane surface, F the Faraday constant, J the flux, and z the valence, the subscript i refers to cations and anions.

The current density i can be related to the specific conductivity κ by:

$$i = \kappa \frac{d\varphi}{dx} \tag{2.149}$$

Here κ is the specific conductivity and φ the electrical potential.

The specific conductivity κ can be expressed by the specific electrical resistance, the equivalent conductivity, the ion migration velocity, or the ion diffusivity:

$$\kappa = \frac{1}{\rho} = \sum_i |z_i| C_i \lambda_i = F\sum_i |z_i| C_i u_i = F^2 \sum_i z_i^2 C_i \frac{D_i}{RT} \tag{2.150}$$

Here ρ is the specific resistance, C is the concentration, F is the Faraday constant, λ is the equivalent conductivity, u is the ion migration velocity, D is the ion diffusivity, R is the gas constant, T the absolute temperature, and the subscript i refers to anions and cations.

Introducing Equations 2.145 into Equation 2.149 and neglecting the convection term leads to:

$$i = F\sum_i z_i J_i = -F^2 \sum_i z_i^2 \frac{C_i D_i}{RT}\left(\frac{RT}{z_i C_i F}\frac{dC_i}{dx} + \frac{d\varphi}{dx}\right) \tag{2.151}$$

Here i is the current density, C is the concentration, F is the Faraday constant, φ is the electrical potential, z is the valence, D is the ion diffusivity, R is the gas constant, T is the absolute temperature, and the subscript i refers to anions and cations.

The term $\frac{RT}{z_i C_i F}\frac{dC_i}{dx}$ has the dimension of an electrical potential gradient and represents the concentration potential which is established between two electrolyte solutions of different concentrations. In electrodialysis this potential is established between the concentrate and the diluate solution. It represents an electromotive force which has to be overcome by the applied electrical potential.

The convection term $v_k C_i$ in Equation 2.145 has been neglected since in ion-exchange membrane separation processes pressure differences between two solutions separated by a membrane are kept as low as possible.

4.1.5 The Transport and Transference Numbers

In an electrolyte solution the current is carried by both ions. However, cations and anions usually carry different portions of the overall current. In ion-exchange membranes the current is carried preferentially by the counterions.

The fraction of the current that is carried by a certain ion is expressed by the ion transport or transference number which are given by:

$$T_i = \frac{z_i J_i}{\sum_i z_i J_i} \qquad (2.152)$$

and

$$t_i = \frac{T_i}{|z_i|} = \frac{|J_i|}{\sum_i z_i J_i} \qquad (2.153)$$

Here T_i and t_i are the transport and transference number of the component i, J_i is its flux, and z_i its valence.

The transport number T_i indicates the fraction of the total current that is carried by the ion i, and the transference number t_i indicates the number of moles of ion i transported per mole of electrons, i.e. per Faraday.

The sum of the transport number or the transference number of all ions in a solution is 1.

Unfortunately, in the literature the definition of the transport and the transference number is not used consistently and very often transference and transport number are mixed up.

The transport numbers of cations and anions in an aqueous salt solution do not differ very much. In an ion-exchange membrane, however, the situation is quite different. In addition to the ions of the electrolyte there are the "fixed" ions of the membrane. Since these ions are bound to the membrane matrix they are unable to contribute to the transport of electrical charges, i.e. the electrical current. Thus, their transport number is per definition 0. Furthermore, the concentration of the counterions, i.e. the ions which carry the opposite charge of the membrane is rather different from that of the coions, i.e. the ions which carry the same charge as the membrane. Their concentration depends on the fixed ion concentration of the membrane and the salt concentration of the solution in contact with the membrane as will be shown later. If the coions are completely excluded from the membrane the transport number of the counterion in the membrane is 1 and the membrane is strictly semipermeable.

The application of Equations 2.145 and 2.151 for determining the ion fluxes or the current in electrolyte solutions, however, is restricted by the electroneutrality requirement and the conservation of charges which must be fulfilled on a macroscopic scale at all times. For a system consisting of a membrane and solutions the electroneutrality condition requires for the solution:

$$\sum_i z_i C_i = 0 \qquad (2.154)$$

and for a membrane:

$$\sum_i z_i C_i^m + z_{fix} C_{fix}^m = 0 \qquad (2.155)$$

Here the superscript m refers to ions in an ion-exchange membrane, the subscript fix refers to charges fixed to the membrane matrix, the subscript i refers to cations and anions.

In ion-exchange membrane separation processes such as in electrodialysis the conservation of the electrical charge requires that fluxes of cations and anions multiplied with the chemical valence are identical but in opposite direction. Thus is:

$$\sum_a z_a J_a = \sum_c z_c J_c \qquad (2.156)$$

Here J is the flux and the subscripts a and c refer to anions and cations, respectively.

The concentration of a salt in a solution is related to that of the individual ions by:

$$C_a = \alpha\, \nu_a C_s \quad \text{and} \quad C_c = \alpha\, \nu_c C_s \qquad (2.157)$$

Here α is the dissociation constant of the electrolyte, ν is a stoichiometric coefficient, the subscript s refers to the electrolyte.

For a completely dissociated monovalent salt such as NaCl, is $\alpha = 1$ and $\nu_c = \nu_a = 1$. For a salt such as $CaCl_2$ is $\alpha = 1$ and $\nu_c = 1$ and $\nu_a = 2$.

4.1.6 Transport of Ions in an Aqueous Solution

The overall transport in ion-exchange membrane separation processes involves the transport of ions within the membranes and in the solutions between the ion-exchange membranes. To describe the ion transport in an electrolyte solution the solvent, i.e. generally water, is used as a frame of reference and its velocity is set to be 0. First, the most simple case, i.e., a solution of a single completely dissociated salt without an electrical current will be discussed. For this system the electroneutrality condition and the conservation of charges require:

$$|z_a| C_a = |z_c| C_c = C_{s,eq} \qquad (2.158)$$

and

$$|J_a| = |J_c| = J_s \qquad (2.159)$$

Equations 2.158 and 2.159 show that for a single monovalent salt the concentrations and absolute values of the fluxes of anions and cations are equal and identical with the salt concentration and flux. However, when an electrical potential is applied as driving force anions and cations move in opposite directions.

The fluxes are determined by the electrochemical potential gradients and by mobility coefficients of the ions in the solution. The electrochemical potential of a salt η_s is given by the sum of the electrochemical potential of the cation η_c and the anion η_a:

$$\eta_s = \eta_a + \eta_c \tag{2.160}$$

The flux of a component with the electrochemical potential gradient as driving force is given by Equation 2.133. The electrochemical potential gradient is a function of pressure, concentration, and electrical potential and is given in Equation 2.142. If there is no pressure difference in an ideal solution with an activity coefficient of 1 and if the kinetic coupling of components are neglected the flux of the individual ions can be expressed by the phenomenological Equation 2.133 in which the phenomenological coefficient L_{ii} has been replaced by the diffusion coefficient according to Equation 2.146. Combining Equations 2.133 and 2.146 with Equation 2.159 gives the fluxes of the salt and individual ions as a function of the electrochemical potential gradients of the individual ions and the electrolyte:

$$J_s = -D_s\left(\frac{C_s}{RT}\frac{d\eta_s}{dx}\right) = J_a = -D_a\left(\frac{C_a}{RT}\frac{d\eta_a}{dx}\right) = J_c = -D_c\left(\frac{C_c}{RT}\frac{d\eta_c}{dx}\right) \tag{2.161}$$

Here J is the flux, C is the concentration, F is the Faraday constant, η is the electrochemial potential, D is the diffusion coefficient, R is the gas constant, T the absolute temperature, and the subscripts s, a, and c refer to salt, anion, and cation, respectively.

Introducing Equation 2.160 into Equation 2.161 leads to:

$$D_s = 2\left(\frac{1}{D_a} + \frac{1}{D_c}\right)^{-1} \tag{2.162}$$

Combination of Equations 2.160 to 2.162 gives the salt and ion fluxes for a completely dissociated monovalent salt at constant temperature and pressure in an ideal solution, i.e. the activity coefficients are 1:

$$J_s = |J_a| = |J_c| = -2\left(\frac{D_a D_c}{D_a + D_c}\right)\left(\frac{dC_s}{dx} + \frac{C_s F}{RT}\frac{d\varphi}{dx}\right) \tag{2.163}$$

Here J is the flux, C is the concentration, F is the Faraday constant, φ is the electrical potential, D is the diffusion coefficient; the subscripts s, a, and c refer to salt, anion, and cation, respectively.

Equation 2.163 is identical to the Nernst-Planck equation and describes the flux of ions in an aqueous solution of a single monovalent salt as a function of a concentration, an electrical potential gradient, and an ion mobility term which is expressed by the diffusion coefficients of the individual ions. The electrical potential gradient is the result of different diffusivities of the cat- and anions, i.e. a diffusion potential which does not lead to a net electrical current. The system contains three individual components, i.e. anion, cation, and water which is used as frame of reference. Therefore, two identical ion fluxes are obtained which are coupled by the electroneutrality condition and the conservation of charges. A consequence of the electroneutrality condition for the transport of ions which have different diffusivities is that the faster moving ion will cause the build-up of a minute electrical potential gradient. This potential gradient will slow down the faster ion and speed up the slower ion so that at steady state both ions move with the same velocity.

Generally, an average diffusion coefficient is defined which is then used to describe the fluxes of the different ions which are coupled by the conservation of charges and the electroneutrality condition. The average diffusion coefficient $\overline{D}$ is given by:

$$\overline{D}_c = \overline{D}_a = D_s = 2\left(\frac{D_a D_c}{D_a + D_c}\right) \tag{2.164}$$

For a salt that consists of multivalent ions a similar relation can be derived and the average diffusion coefficient is given by:

$$\overline{D}_c = \overline{D}_a = D_s = \frac{D_a D_c (|z_a| + z_c)}{D_a |z_a| + D_c z_c} \tag{2.165}$$

If an external potential driving force is applied and an electrical current is obtained the fluxes of the individual ions are no longer restricted by the requirement of the conservation of electrical charges. As a consequence the diffusion coefficient of cations and anions in the solution can be quite different and conductivity of a solution is a function of the individual ion diffusivity and given by:

$$\kappa = \frac{F^2}{RT} \sum_i z_i^2 C_i D_i \tag{2.166}$$

4.1.7 Mass Transport in Ion-Exchange Membranes

In ion-exchange membrane separation processes the mass transport through the membranes is the key part of the process determining to a very large extent its efficiency. To describe the mass transport of a monovalent single salt in an aqueous solution through a membrane which is used as frame of reference 3 independent fluxes must be considered, i.e. the fluxes of the cations and anions, and the flux of the solvent. In ion-exchange membrane separation processes such as electrodialysis the transport of ions through a kinetic coupling with the flux of the solvent is much smaller than the ion transport due to a gradient in the electrochemical potential and can be neglected. The transport of the solvent through the membrane as a result of osmotic effects and a kinetic coupling with the fluxes of the ions, however, can under certain conditions be significant and must be taken into account.

Ion-exchange membranes carry positive or negative electrical charges fixed to a solid matrix and they are therefore permselective as far as the transport of ions is concerned, i.e. the cation-exchange membranes are preferentially permeable to cations and the anion-exchange membranes are preferentially permeable to anions.

The flux of an ion in an ion-exchange membrane under ideal conditions, i.e. no convection and an activity coefficient of 1, is given by:

$$J_i = -D_i^m \left(\frac{dC_i^m}{dx} + \frac{z_i F C_i^m}{RT} \frac{d\varphi}{dx} \right) \tag{2.167}$$

Here the superscript m refers to the membrane phase.

Equation 2.167 is identical to Equation 2.145 with the exception that all concentrations and transport parameters refer to the situation in the membrane and may thus be significantly different from those in the solution. In a strictly permselective anion-exchange membrane the concentration of the mobile cation is 0 and the current is transported by anions only, and their transport number T_a^m is 1. In a strictly permselective cation-exchange membrane the concentration of the mobile anion is 0 and the current is carried by cations only, and their transport number T_c^m is 1. In the solution between the membranes the current is carried by both ions according to their mobility, and their transport numbers are generally not very different.

In ion-exchange membrane separation processes such as electrodialysis the flux of cations through the cation-exchange membrane and the flux of the anions through the anion-exchange membrane are interdependent because of the electroneutrality requirement in the solution between the membranes, as can easily be seen when considering the electrodialysis of a monovalent single electrolyte solution such as NaCl. If strictly permselective membranes are assumed the fluxes of the cations through the cation-exchange membrane and the anions through the anion-exchange membrane are equal. Thus, in a solution containing a monovalent completely dissociated electrolyte such as NaCl and strictly permselective membranes is:

$$J_{Na^+}^{cm} = J_{Cl^-}^{am} = -2 \left(\frac{D_{Na^+}^{cm} D_{Cl^-}^{am}}{D_{Na^+}^{cm} + D_{Cl^-}^{am}} \right) \left(\frac{dC_i^{im}}{dx} + \frac{FC_i^{im}}{RT} \frac{d\varphi}{dx} \right) \tag{2.168}$$

Here $J_{Na^+}^{cm}$ and $J_{Cl^-}^{am}$ are the flux of Na^+-ions through the cation-exchange membrane and the flux of Cl^--ions through the anion-exchange membrane; $D_{Na^+}^{cm}$ and $D_{Cl^-}^{am}$ are the diffusion coefficients for the transport of Na^+- and Cl^--ions through the corresponding membrane; the subscript i refers to cation or anion, and the superscript im to cation or anion-exchange membrane.

If the membranes are not strictly semipermeable Equation 2.168 holds only if the coionfluxes in the membranes are identical. Generally the permselectivity of ion-exchange membranes used in separation processes such as electrodialysis is quite high, i.e. in the range of 0.97 to 0.98 and the transport numbers of the coions very similar, and Equation 2.168 can be applied to estimate the salt transport in electrodialysis.

4.1.8 Membrane Counterion Permselectivity

The transport number of counterions in an ion-exchange membrane compared to the coions is always quite high, but the transport rate of different counterions can be quite different. The transport rates of ions in a solution or through a membrane are proportional to their permeabilities in the solution and in the membrane. The permeability of ions is the product of their concentration and mobility. In most practical applications of ion-exchange membrane separation processes the concentration of the ions in the solution is a given process parameter and their mobility is determined by the ion radius according to the Stokes-Einstein relation. In an ion-exchange membrane the concentration of the coions is determined by the Donnan exclusion as expressed by Equation 2.102. The concentration of the counterions is always close to the concentration of the fixed charges of the membrane. The mobility of the ions in the membrane depends mainly on the radius of the hydrated ions and the membrane structure. The mobility of different ions in an aqueous solution do not differ very much from each other. An exception are the H^+- and OH^--ions. Their mobility is an order of magnitude higher than that of other ions as can be seen in Table 2.3 which shows ion mobility in an aqueous solution of infinite dilution and at 298°K. The table also shows that ions such as Li^+ or F^- have lower mobility than ions such as Ca^{2+} or SO_4^{2-} because of the smaller hydration shell and the higher charge of the latter ions [G. Kortüm, 1957].

Table. 2.3 Ion mobility in water at 298°K in the limit case of infinitely dilute solutions [P.W. Atkins 1990]

cations	u [10^{-8} m^2 s^{-1} V^{-1}]	anions	u [10^{-8} m^2 s^{-1} V^{-1}]
Li^+	4.01	F^-	5.70
Na^+	5.19	Cl^-	7.91
K^+	7.62	Br^-	8.09
NH_4^+	7.63	NO_3^-	7.40
Ca^{2+}	6.17	SO_4^{2-}	8.29
Cu^{2+}	5.56	CO_3^{2-}	7.46
H^+	36.23	OH^-	20.64

The exceptionally high mobility of the H^+-ion can be explained by the transport mechanism of protons. Because of the molecular interaction of water dipoles with electrical charges, protons form hydronium-ions with the water as indicated in Figure 2.16 a). In a hydronium-ion the three attached H-atoms are no longer distinguishable in terms of their electric charges. They form a tripod with three H-atoms equally apart carrying the positive charge together. The formation of the hydronium ions has consequences for the mobility of the protons in aqueous media. Common salt ions move with their hydrate shell through the solution. The proton, however, is transported mostly via a so-called tunnel mechanism from one hydronium-ion to the next water molecule as indicated in Figure 2.16 b).

This explains not only the extraordinary high mobility of protons but it is also one of the reasons for the high permeability of anion-exchange membranes for protons while these membranes generally have a very low permeability for salt cations. The same mechanism as the one described here for the transport of protons also holds true for the transport of hydroxide ions and thus the permeability of hydroxide ions in an aqueous solution and also in a cation-exchange membrane is much higher than that of other salt anions which generally is very low. Because protons and hydroxide ions are transported only to a small extent as individual ions surrounded by a hydration shell they contribute very little to the electroosmotic transport of water and their water transport number is always quite low as has been demonstrated in experimental studies.

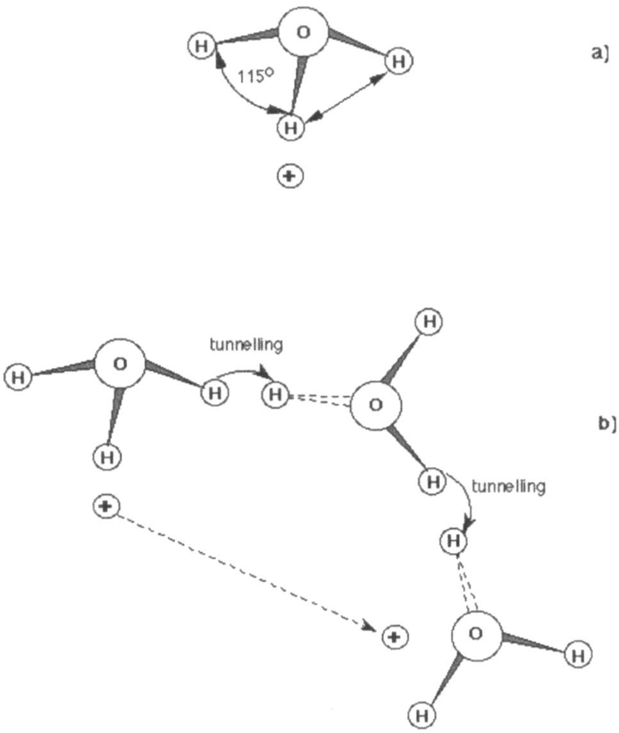

Fig. 2.16 Schematic drawing illustrating a) the structure of a hydronium-ion and b) the proton tunnel transport mechanism.

The permselectivity of an ion-exchange membrane for different counterions is determined by the concentration and the mobility of the different ions in the membrane as indicated earlier. The concentration of the different counterions in the membrane is determined mainly by electrostatic effects referred to as "electroselectivity" [F. Helfferich, 1962]. It can conveniently be expressed by an ion-exchange selectivity coefficient which for a solution containing H^+- and Na^+-ions by equation 2.85. For monovalent counterion the selectivity coefficient is:

$$K_{A^{+,-},B^{+,-}} = \frac{\left[{}^{m}A^{+,-}\right]\left[{}^{s}B^{+,-}\right]}{\left[{}^{s}A^{+,-}\right]\left[{}^{m}B^{+,-}\right]} \tag{2.169}$$

Here $K_{A^{+,-},B^{+,-}}$ is the equilibrium constant for the counterion distribution between the membrane and the solution, A and B refer to different counterions, the superscripts s and m refer to the solution and the membrane, and the superscript +,- refer to cation and anion.

A typical counterion exchange sequence of a cation-exchange membrane containing SO_3^--group as fixed charge is:

$$Ba^{2+} > Pb^{2+} > Sr^{2+} > Ca^{2+} > Mg^{2+} > Ag^{+} > K^{+} > NH_4^{+} > Na^{+} > Li^{+}$$

A similar counterion exchange sequence is obtained for anions in an anion-exchange membrane containing quaternary ammonium groups as fixed charges [R.A. Hann et al., 1983]:

$$I^- > NO_3^- > Br^- > Cl^- > SO_4^{2-} > F^-$$

The permselectivity is the product of ion-exchange selectivity and mobility selectivity. The mobility of different ions is determined mainly by steric effects, i.e. the size of the ions and the cross-linking density of the membrane. As a general rule the counterions with the higher valence and the smaller hydrated radius have a higher permeability in an ion-exchange membrane than ions with lower valence and larger hydrated radius

This is a very general rule which does not always apply. For some membranes and larger ions the "sieving action" overrules the electroselectivity and the permeability of larger ions is drastically reduced.

4.1.9 The Flux of Solvent Through an Ion-Exchange Membrane

The solvent transported through an ion-exchange membrane is the sum of three terms:

- the solvent flux due to a coupling with the electric current which passing through the membrane as the result of electrical potential gradient, which is referred to as the electroosmotic flux,
- the solvent flux coupled to the flux of ions in the hydration shell, and
- the flux of solvent due to a chemical potential gradient of the solvent in the membrane which can be expressed by the osmotic flux.

Each of the three terms may be dominant depending on the permselectivity of the ion-exchange membrane, the concentration gradient, and the current density. In a highly permselective membrane and moderate differences in the salt concentration in the two solutions separated by the membrane the electroosmotic flux and the solvent flux in the hydration shell of the ions are dominating and are much higher than the osmotic solvent flux. In most practical application of ion-exchange membrane separation processes the solvent is water and the water flux due to electroosmosis and due to the transport in the hydration shell of the ions can be expressed to a first approximation by a solvent transport number which gives the number of water molecules transported by one ion:

$$J_w = T_w^m \sum_i J_i \qquad (2.170)$$

Here T_w^m is the water transport number, J_w is the water flux, and J_i is the flux of ions through a given membrane.

The water transport number thus is:

$$T_w^m = \frac{J_w}{\sum_i J_i} \qquad (2.171)$$

The water transport number refers to the number of water molecules transferred by one ion through a given membrane. It depends on the membrane and on the electrolyte, i.e. on the size of the ions, their valence, and their concentration in the solution. In aqueous salt solutions and commercial ion-exchange membranes the water transport number is in the order of 4 to 8, i.e. one mole of ions transport ca. 4 to 8 moles of water through a typical commercial ion-exchange membrane.

Instead of the water transport number T_w^m often the water transference number t_w^m is used. The water transference number relates the transport of water to the current expressed in Faraday instead to the flow of ions and is given by:

$$t_w^m = \frac{J_w}{\sum_i z_i J_i} = \frac{F J_w}{i} \qquad (2.172)$$

Here t_w^m is the water transference number, J_w is the molar water flux, F is the Faraday constant, i is the current density, z is the valence, and the subscript i refers to different ions.

4.2 Membrane Permselectivity and the Donnan Exclusion

The membrane permselectivity is an important parameter for determining the performance of a membrane in a certain ion-exchange membrane process. It describes the degree to which a membrane passes an ion of one charge and retains an ion of the opposite charge. The permselectivity of cation- and anion-exchange membranes can be defined [K.S. Spiegler, 1956] by the following relations:

$$\Psi^{cm} = \frac{T_c^{cm} - T_c}{T_a} \qquad (2.173)$$

and

$$\Psi^{am} = \frac{T_a^{am} - T_a}{T_c} \qquad (2.174)$$

Here Ψ is the permselectivity of a membrane, T is the transport number, the superscripts cm and am refer to cation and anion exchange membranes, and the subscripts c and a refer to cation and anion, respectively.

Thus, the permselectivity of an ion-exchange membrane relates the transport of electric charges by a specific counterionto the total transport of electric charges through the membrane and the transport number of the ion in the solution. An ideal permselective cation-exchange membrane would transmit positively charged ions only, i.e. for a transport number of a counterionin an cation-exchange membrane is $T_c^{cm} = 1$ and the permselectivity of the membrane is $\Psi^{cm} = 1$. The permselectivity approaches zero when the transport number within the membrane is identical to that in the electrolyte solution, i.e. for $T_c^{cm} = T_c$ is $\Psi^{cm} = 0$. For the anion-exchange membrane holds the corresponding relation.

The transport number of a certain ion in the membrane is proportional to its concentration in the membrane which again is a function of its concentration in the solutions in equilibrium with the membrane phase, due to the Donnan exclusion. Thus the selectivity of ion-exchange membranes results from the exclusion of coions from the membrane phase.

The causes and consequences of the Donnan exclusion have been discussed earlier and are illustrated in the schematic drawing of Figure 2.10. The Donnan exclusion is the result of large differences in the concentration of ions between the ion-exchange membrane and the adjacent electrolyte solutions. Due to this concentration difference a strong driving force in form of a chemical potential gradient exists for the counterions to diffuse from the ion-exchange membrane into the electrolyte solution and for the coion to diffuse from the solution into the ion-exchange membrane. This diffusion of co- and counterions in opposite directions leads to an accumulation of charges in the solutions and in the membrane at the interface between the ion-exchange membrane and the electrolyte solution leading to a space charge, i.e. the Donnan potential which is counteracting the concentration driven diffusion process.

4.3 The Membrane Potential

If two solutions of different ion concentrations are separated by a permselective membrane a potential difference between the two solutions is established. This potential difference is referred to as membrane potential. Contrary to the Donnan potential which cannot be measured directly the membrane potential can be measured directly in an electrolytic cell using reversible electrodes [K.S. Spiegler, 1958]. The cell for measuring the membrane potential is described schematically in Figure 2.17.

The test is composed of two compartments separated by a membrane and filled with an electrolyte solution of different concentration. The voltage drop between the two solutions is measured with reversible electrodes, e.g. calomel electrodes.

The membrane potential is composed of two parts, i.e. the two Donnan potentials which are established between the membrane and the adjacent solution and the diffusion potential caused by differences in the transport number of different ions in the membrane. The membrane potential is given by:

$$\varphi_m = \varphi^{s2} - \varphi^{s1} = \varphi_{diff} + \varphi_{Don^2} - \varphi_{Don^1} \qquad (2.175)$$

Here φ_m, φ_{diff}, and φ_{Don} are the membrane potential, the diffusion potential, and the Donnan potential, respectively; φ^s is the potential at the surface, and 1 and 2 denote the two different sides of the membrane.

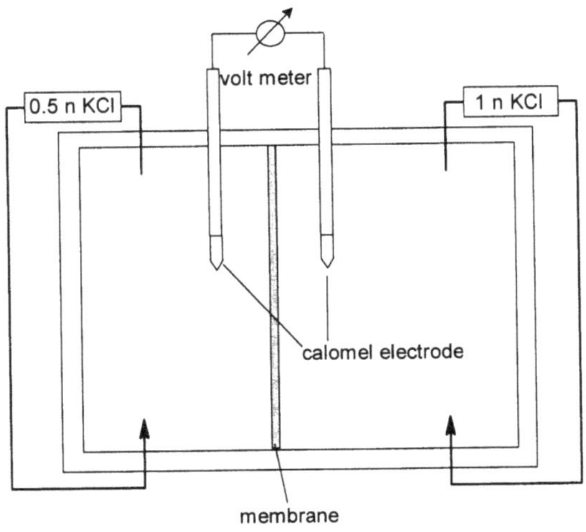

Fig. 2.17 Schematic drawing of a test cell to determine the membrane potential.

The concentration and potential gradients between the two solutions separated by a membrane are shown schematically in Figure 2.18.

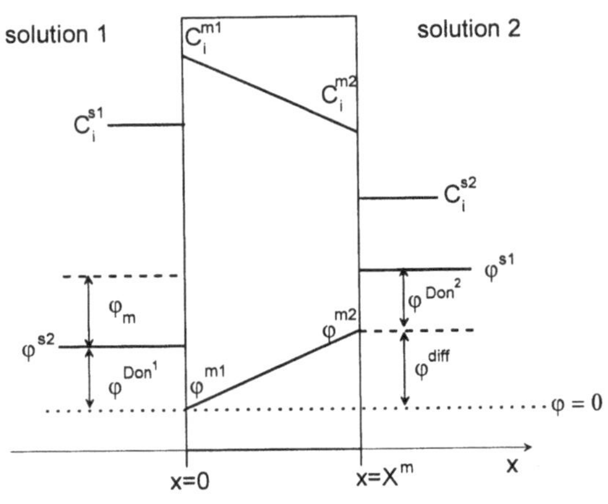

Fig 2.18 Schematic drawing illustrating the concentration and resulting potential differences in a solution separated by a permselective membrane. (φ_{Don}, φ_m, and φ_{diff} refer to the Donnan potential, the membrane potential, and the diffusion potential, respectively, C is the concentration, x is a directional coordinate, and X^m the membrane thickness; the subscript i refers to the ion, and the superscripts s, m, 1, and 2 refer to the solution, the membrane, and the solution 1 and 2, respectively).

For a completely permselective membrane the diffusion potential is 0 and the membrane potential is the sum of the two Donnan potentials.

The membrane potential is the result of the diffusive transport of ions through a membrane under constant temperature and pressure and without electrical current and the Donnan potentials. It can be derived from the general transport equation. For a monovalent salt, such as KCl the membrane potential φ_m is:

$$\varphi_m = -\frac{1}{F}\left(T^m_{cou}\,\Delta\mu_{cou} - T^m_{co}\Delta\mu_{co} + T^m_w\Delta\mu_w\right) \tag{2.176}$$

Here φ_m is the membrane potential, T is the transport number, μ is the chemical potential, Δ denotes the difference between the two solutions, the subscripts cou, co, and w refer to counterion, coion, and water, the superscript m refers to the membrane.

The chemical potential of the individual ions can be expressed by the chemical potential of the salt. For a monovalent salt is:

$$\nu_{cou}\,\Delta\mu_{cou} + \nu_{co}\,\Delta\mu_{co} = \Delta\mu_s \tag{2.177}$$

The chemical potential difference of the water between two solutions can be expressed by the chemical potential difference of the salt due to the Gibbs-Duhem relation [P.W. Atkins, 1990]:

$$0 = n_{cou}\,\Delta\mu_{cou} + n_{co}\,\Delta\mu_{co} + n_w\,\Delta\mu_w \tag{2.178}$$

Rearranging Equation 2.178 and introducing $n_w\,\overline{V}_w = V$ leads to:

$$\Delta\mu_w = \frac{n_{cou}\Delta\mu_{cou} + n_{co}\Delta\mu_{co}}{n_w} = \overline{V}_w\left(C_{cou}\Delta\mu_{cou} + C_{co}\Delta\mu_{co}\right) \tag{2.179}$$

and

$$\Delta\mu_w = C_s\overline{V}_w\Delta\mu_s \tag{2.180}$$

The chemical potential difference of water between two solutions for a monovalent salt is given by the salt concentration and the chemical potential difference of the salt.

Introducing Equation 2.180 into Equation 2.176 and rearranging leads to:

$$\varphi_m = -\left(T^m_{cou} + T^m_{co} - T^m_w C_s\overline{V}_w\right)\frac{\Delta\mu_s}{F} \tag{2.181}$$

Assuming that there is no hydrostatic pressure difference between the two solutions the chemical potential μ of the salt can be expressed by the activity. Thus, the membrane potential as a function of the salt concentrations of the solutions separated by a membrane is given by:

$$\varphi_m = \left(T^m_{cou} - T^m_{co} - T^m_w C_s \overline{V}_w\right)\frac{RT}{F}\ln\frac{a^{s1}_s}{a^{s2}_s} \qquad (2.182)$$

To eliminate the transport number of water generally an "apparent" transport number of ions $T^{m,app}$ is defined. For low water transport numbers the apparent ion transport number is very close to the real ion transport number. It is:

$$T^{m,app}_{cou} = T^m_{cou} - T^m_w C_s \overline{V}_w \cong T^m_{cou} \qquad (2.183)$$

To a first approximation the membrane potential is identical to the measured membrane potential and given by:

$$\varphi_m = \left(T^m_{cou} - T^m_{co}\right)\frac{RT}{F}\ln\frac{a^1_s}{a^2_s} = \left(2T^m_{cou} - 1\right)\frac{RT}{F}\ln\frac{a^{s1}_s}{a^{s2}_s} \qquad (2.184)$$

Here is T the transport number, a is the activity, the subscripts cou, and co refer to counterionand coion, the superscript m, 1, and 2 refer to the membrane and the two solutions, R is the gas constant, T is the absolute temperature, and F is the Faraday constant.

Equation 2.184 is important for the practical characterization of the permselectivity of ion-exchange membranes as will be shown later.

5 LIST OF SYMBOLS:

A	surface area	$[m^2]$
A	Helmholtz free energy	[J]
A	Debye-Hückel constant	$[mol\ kg^{-1}]^{0.5}$
a	activity	$[\ mol\ m^{-3}]$
a	shortest distance between ions	[m]
C	concentration	$[mol\ m^{-3},\ mol\ L^{-1}]$
C	capacitance	$[Farad,\ As\ V^{-1}]$
C	electrical charge	[Coulomb, As]
D	diffusion coefficient	$[m^2\ s^{-1}]$
Đ	Stefan-Maxwell diffusion coefficient	$[m^2\ s^{-1}]$
d	distance	[m]
E	energy	[A V s]
E	elementary charge	1.6019×10^{-19} [A s]
E	electric field	$[V\ m^{-1}]$
F	Faraday constant	96485 $[A\ s\ eq^{-1}]$
F_{fr}	friction force	$[kg\ m^{-1}\ s^{-2}]$
F_{el}	friction force	$[A\ U\ s\ m^{-3}]$
f	friction coefficient	$[N\ s\ m^{-1}\ mol^{-1}]$
G	Gibbs free energy	[J]
g	osmotic coefficient	[-]
H	enthalpy	[J]

I	current	[A]
i	current density	[A m^{-2}]
J_i	flux of moles	[mol m^{-2} s^{-1}]
J_v	convective flux	[m s^{-1}]
J_e	flux of electrical charges	[A s m^{-2} s^{-1}]
J	ionic strength	[eq^2 mol^{-1} L^{-1}]
K	equilibrium constant	[various]
K_w	water dissociation constant	1.008×10^{-14} at 25°C [mol^2 L^{-2}]
K_{diss}	dissociation constant	[mol L^{-1}]
K_a	acid dissociation constant	[mol L^{-1}]
K_b	base dissociation constant	[mol L^{-1}]
K	force	[N, U A s m^{-1}]
k	coefficient	[various]
L	phenomenological coefficient	[mol^2 N^{-1} m^{-2} s^{-1}]
l	distance	[m]
N_A	Avogadro number	6.0221×10^{23} [mol^{-1}]
n	mole number	[mol]
p	pressure	[Pa]
q	cross-section	[m^2]
q	electrical charge	[C, A s]
q	heat, energy	[J]
dq_{rev}	reversible heat exchange	[J]
dq_{irrev}	irreversible heat exchange	[J]
R	resistant coefficient	[N m^2 s mol^{-2}]
R	gas constant	[J mol^{-1} K^{-1}]
R	electrical resistance	[Ω]
r	area resistance	[Ω m^2]
r	distance between charges	[m]
r_D	Debye radius	[m]
S	entropy	[J K^{-1}]
$\overline{S}$	molar entropy	[J K^{-1} mol^{-1}]
S	conductivity	[Siemens, $Ω^{-1}$]
S	surface	[m^2]
T	temperature	[K]
T	transport number	[-]
t	transference number	[-]
t	time	[s, h]
U	electrical potential	[V]
u	ion mobility	[m^2 s^{-1} V^{-1}]
u	velocity	[m s^{-1}]
V	volume	[m^3]
v	velocity	[m s^{-1}]
$\overline{V}$	partial molar volume	[m^3 mol^{-1}]
W_e	electrical work	[A U s]
X	driving force	[N mol^{-1}]
X	mole fraction	[-]

x_D	Debye length	[m]
x	directional co-ordinate	[m]
y	directional co-ordinate	[m]
z	valence	[eq mol^{-1}]

Greek letters

α	dissociation coefficient	[-]
γ	activity coefficient	[m^3 mol^{-1}]
ε	permittivity	[C^2 J^{-1} m^{-1}]
ε_o	permittivity in vacuum	8.854×10^{-12} [C^2 J^{-1} m^{-1}]
ε_r	relative permittivity, dielectric constant	[-]
Δ	difference	[-]
η	electrochemical potential	[A V s mol^{-1}, J mol^{-1}]
η	dynamic viscosity	[kg m^{-1} s^{-1}]
κ	specific conductivity	[Ω^{-1} m^{-1}]
μ	chemical potential	[J mol^{-1}]
μ^o	standard chemical potential	[J mol^{-1}]
μ	dipole moment	[C m]
ν	stoichiometric coefficient	[-]
λ	ion equivalent conductivity	[m^2 Ω^{-1} eq^{-1}]
Λ_{eq}	electrolyte eqivalent conductivity	[m^2 Ω^{-1} eq^{-1}]
Λ_m	electrolyte molar conductivity	[m^2 Ω^{-1} mol^{-1}]
π	osmotic pressure	[Pa, N m^{-2}]
φ	electrical potential	[V]
φ_{emf}	electromotive force	[V]
θ	angle	[degree]
ρ	space charge	[C m^{-3}, As m^{-3}]
ρ	specific resistance	[Ω m]
σ	surface charge	[C m^{-2}]
Ψ	membrane permselectivity	[–]
Ψ	dissipation function	[N m s^{-1}]
ξ	current utilization	[-]
ξ	extent of reaction	[-]
ζ	zeta potential	[V]

Subscripts

i	component i
j	component j
k	component k
s	salt
a	anion
c	cation
diff	diffusion potential

Don Donnan potential or Donnan exclusion
co coion
cou counterion
fix fixed ion
r referring to a reaction
red referring to reduction
oxi referring to oxidation
max referring to a maximum value
w water
+ positive charge
- negative charge
± average value of cation and anion

Superscripts
o standard state
m membrane
cm cation-exchange membrane
am anion-exchange membrane
im anion- or cation-exchange membrane
s solution
s referring to membrane surface
mig migration
diff diffusion

LITERATURE:

Atkins, P.W., 1990. Physical Chemistry, fourth edition, Oxford University Press, Oxford.

Donnan, F.G., Guggenheim E.A., 1932. Exact thermodynamics of membrane equilibrium. Z. Phys. Chem. A , 162: 346-360.

Haase R., 1963. Thermodynamik der irreversiblen Prozesse, Steinkopff Verlag Darmstadt.

Hann, R.A., Eyres, R., Cottier, D., 1983. Separation of anions by electrodialysis, in: Ion-exchange membranes, edt. Flett D.S., Ellis Horwood Ltd. Chichester, UK.

Harned, H.S., Owen, B.B., 1958. The Physical Chemistry of Electrolyte solutions, 3. Edition, Reinhold, New York.

Helfferich, F., 1962. Ion-Exchange, McGraw-Hill, London.

Hunter, R.J., 1981. Zeta Potential in Colloid Science, Principle and Applications, Academic Press, London.

Katchalsky, A., Curran, P.F., 1965. Nonequilibrium Thermodynamics, Harvard University Press, Cambridge.

Kortüm, G., 1957. Lehrbuch der Elektrochemie, Verlag Chemie, Weinheim, Germany.

Kraaijeveld, G., 1994. The Maxwell-Stefan description of mass transfer in ion-exchange and electrodialysis, PhD.-thesis, University of Groningen, The Netherlands.

Lacey, R.E., 1972. Basis of electromembrane processes. In: Industrial Processing with Membranes, Edts.: R.E. Lacey and S. Loeb, John Wiley & Sons, New York.

Lyklema, J., 1995. Fundamentals of Interface and Colloid Science, Volume II: Solid-Fluid Interfaces, Academic Press, London.

Miller, D.G., 1966. Application of irreversible thermodynamics to electrolyte solutions: I. Determination of ionic transport coefficients l_{ij} for isothermal vector transport in binary electrolyte systems, J. Phys. Chem., 70: 2639-2659.

Narebska, A., Kujawski, W., Koter, S., 1984. Ion and water transport across charged Nafion membranes: Irreversible Thermodynanic approach, Desalination, 51: 3-17.

Onsager, L., 1931. Reversible relations in irreversible processes, Physic. Rev.,38: 2265-2279.

Robinson, R.A., Stokes, R.H., 1959. Electrolyte Solutions, Butterworths, London.

Schlögl, R., 1964. Stofftransport durch Membranen, D. Steinkopff Verlag, Darmstadt, Germany.

Shaffer, L.H., Mintz, M.S., 1966. Electrodialysis in: Principle of Desalination, edt. Spiegler K.S., Academic Press, New York.

Spiegler, K.S., 1957. Electrochemical operations, in: Ion-Exchange Technology, edts.: F.C. Nachod, J. Schubert, Academic Press, New York.

Spiegler, K.S., 1958. Transport process in ionic membranes, Trans. Faraday Soc. 54: 1408-1428.

Starzak,M. E., 1984. The Physical Chemistry of Membranes, pp. 44 –64, Academic Press New York.

Wagenen, R.A. Andrade, J.D., 1980. Flat plate streaming potetial investigations: Hydrodynamics and electrokinetic equivalency, J. Coll. Int. Sci., 76: 305-314.

Wesselingh, J.A., Krishna, R., 1990. Mass Tranfer, Ellis Horwood Ltd., London.

Chapter 3

Preparation and Characterization of Ion-Exchange Membranes

1 INTRODUCTION

Key components in ion-exchange membrane separation processes are membranes which are used in various applications. Their properties determine to a large extent the technical feasibility and the economics of the processes. A significant amount of work has been concentrated over the last 50 years in academic institutions as well as in industry to develop efficient membranes with high permeability, good permselectivity for certain ionic components and a long useful life under operating conditions. This work is documented in a very large number of publications and especially in patents which describe in great detail preparation procedures [R. Kunin, 1958], [F. Bergsma et al., 1961], [F. Helfferich, 1962], [N.P., Suryanarayana, et al., 1963] and [T. Sata,. 1986]. With the development of the low temperature polymer electrolyte fuel cell to be used for the methanol direct conversion in mobile applications the interest in new ion-exchange membranes with special properties such as low methanol permeability has drastically increased [J. Kerres et al., 2001].

Due to the importance of the membranes in electrodialysis and related processes it is not surprising that there are numerous detailed recipes described mainly in the patent literature for the manufacturing of ion-exchange membranes with special application adjusted properties. The preparation procedures of ion-exchange membranes are closely related to those of ion-exchange resins [E.B. Tooper et al., 1956]. As with resins, there are many possible types with different polymer matrixes and different functional groups to confer ion-exchange properties on the product. Although there are a number of inorganic ion-exchange materials [F. Helfferich, 1962] most of them are based on zeolites, zirconium phosphates, and bentonites. These materials are rather unimportant today in electrodialysis and related processes with the exception of the high temperature fuel cell which will not be discussed further.

Ion-exchange membranes can be classified by their function as a separation media or according to their structure and basic material. As far as their function is concerned ion-exchange membranes may be classified as:

- cation-exchange membranes which contain fixed negatively charged ions and which have a selective permeability for cations,
- anion-exchange membranes which contain fixed positively charged ions and which have a selective permeability for anions,
- amphoteric ion-exchange membranes in which both negatively and positively fixed ionic groups are randomly distributed,
- bipolar membranes which consist of a cation- and an anion-exchange membrane laminated together ,
- mosaic ion-exchange membranes which are composed of macroscopic domains of polymers with negatively fixed ions and those with positively fixed ions randomly distributed in a neutral polymer matrix.

The structure of an ion-exchange membrane based on a polymer matrix and fixed charge groups are illustrated in Figure 3.1.

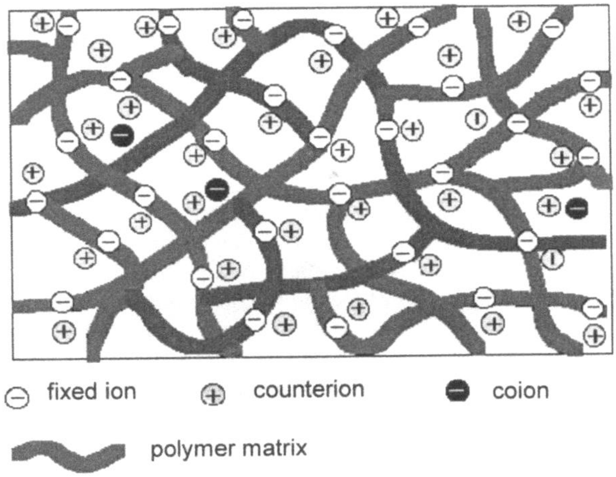

Fig. 3.1 Schematic drawing illustrating the structure of a cation-exchange membrane.

This figure shows schematically the matrix of a cation-exchange membrane with fixed anions and mobile cations, which are referred to as counterions. In contrast, the mobile anions, called coions, are more or less completely excluded from the polymer matrix because of their electrical charge which is identical to that of the fixed ions. Due to the exclusion of the co-ions, an ideal cation-exchange membrane permits transfer of cations only. Ideal anion-exchange membranes carry positive charges fixed on the polymer matrix. Therefore, they exclude all cations and are permeable to anions only.

The most desired properties for ion-exchange membranes are:

- high permselectivity - an ion-exchange membrane should be highly permeable for counterions, but should be impermeable to coions,
- low electrical resistance - the permeability of an ion-exchange membrane for the counterions under the driving force of an electrical potential gradient should be as high as possible,
- good mechanical and form stability - the membrane should be mechanically strong and should have a low degree of swelling or shrinking in transition from dilute to concentrated ionic solutions,
- high chemical stability - the membrane should be stable over the entire pH-range and in the presence of oxidizing agents.

It is often difficult to optimize the properties of ion-exchange membranes because the parameters determining the different properties often act contrary to each other. For instance, a high degree of cross-linking improves the mechanical strength of the membrane but also increases its electrical resistance. A high concentration of fixed ionic charges in the membrane matrix leads to a low electric resistance but, in general, causes a high degree of swelling combined with poor mechanical stability. The properties of ion-exchange membranes are determined by two parameters, namely the basic material they are made from and the type and concentration of the fixed ionic moiety. The basic material determines to a large extent the mechanical, chemical, and thermal stability of the membrane [P. Meares, 1983], [K.S. Spiegler, et al., 1980]. Based on the material of the ion-exchange membranes they can be classified as:

- membranes composed of hydrocarbon or partially halogenated hydrocarbon polymers,
- membranes composed of perfluorocarbon polymers,
- membranes made from inorganic materials,
- membranes composed of inorganic ion-exchange material and an organic polymer.

In most commercially available ion-exchange membranes the matrix consists of hydrophobic polymers such as polystyrene, polyethylene or polysulfone. Although these basic polymers are insoluble in water and show a low degree of swelling, they may become water soluble by the introduction of the ionic moieties. Therefore, the polymer matrix of ion-exchange membranes is very often cross-linked. The degree of cross-linking then determines to a large extent the degree of swelling and the chemical and thermal stability, but it also has a large effect on the electrical resistance and the permselectivity of the membrane.

The type and the concentration of the fixed ionic charges determine the permselectivity and the electrical resistance of the membrane, but they also have a significant effect on the mechanical properties of the membrane. The degree of swelling, especially, is effected by the concentration of the fixed charges.

The following moieties are used as fixed charges in cation-exchange membranes:

$-SO_3^-$, $-COO^-$, $-PO_3^{2-}$, $-PHO_2^-$, $-AsO_3^{2-}$, $-SeO_3^-$.

In anion-exchange membranes fixed charges may be:

$-\overset{+}{N}H_2R$, $-\overset{+}{N}HR_2$, $-\overset{+}{N}R_3$, $-\overset{+}{P}R_3$, $-\overset{+}{S}R_2$

These different ionic groups have significant effects on the selectivity and electrical resistance of the ion-exchange membrane. The sulfonic acid group, e.g. $-SO_3^-$ is completely dissociated over nearly the entire pH-range, while the carboxylic acid group $-COO^-$ is virtually undissociated in the pH-range < 3. The quaternary ammonium group $-\overset{+}{N}R_3$ again is completely dissociated over the entire pH-range, while the secondary ammonium group $-\overset{+}{N}H_2R$ is only weakly dissociated. Accordingly, ion-exchange membranes are referred to as being weakly or strongly acidic or basic in character. Most commercially available ion-exchange membranes have $-SO_3^-$ or $-COO^-$ groups, and most anion-exchange membranes contain $-R_3\overset{+}{N}$ groups.

Ion-exchange membranes can also be divided, according to their structure and preparation procedure, into two major categories: homogeneous membranes and heterogeneous membranes [F. Bergsma et al., 1961].

In homogeneous ion-exchange membranes the fixed charged groups are evenly distributed over the entire membrane polymer matrix. Homogeneous membranes can be produced, e.g. by polymerization or polycondensation of functional monomers such as phenolsulfonic acid with formaldehyde [F. Helfferich, 1962], or by functionalizing a polymer such as polysulfone dissolved in an appropriate solvent by sulfonation and cast into a film [P. Zschocke et al., 1985].

Heterogeneous ion-exchange membranes have distinct macroscopic domains of ion-exchange resins in the matrix of an uncharged polymer. They can be produced by melting and pressing of a dry ion-exchange resin with granulated polymer such as polyvinylchloride, or by dispersion of the ion-exchange resin in a polymer solution [G.E. Molau, 1981].

The completely homogeneous and the macroscopically heterogeneous ion-exchange membranes are extreme structures. Most ion-exchange membranes show a certain degree of heterogenity on the microscopic scale. The degree of heterogenity of ion-exchange membranes increases according to the following order [T. Sata, 1986]:

- homogeneous ion-exchange membranes,
- interpolymer membranes,
- microheterogeneous graft- and block-polymer membranes,
- snake-in-the-cage ion-exchange membranes, and
- heterogeneous ion-exchange membranes.

From the viewpoint of macromolecular chemistry all the intermediate forms may be considered as so-called polymer blends. As a consequence of the polymer/polymer incompatibility a certain phase separation of the different polymers and a specific aggregation of the hydrophilic and hydrophobic domains is obtained. A classification of the membrane morphology according to the type and the size of the microphase is possible. If membranes are translucent their structure is homogeneous or they contain inhomogenities that are smaller than the wavelength of visible light (400 nm). Thus, these membranes are called interpolymer or microheterogeneous membranes. Microheterogeneous membranes for example are produced by means of block-copolymerization of ionogenic and non-ionogenic monomers, or by graft-copolymerization of functional monomers [J. A. Kerres, 2000]. Interpolymer membranes are produced by dissolving compatible, functional polymers such as sulfonated polyetherketon and polysulfone in one solvent to form a homogeneous, macroscopically transparent solution, followed by the evaporation of the solvent [W. Cui et al., 1998].

2 PREPARATION OF ION-EXCHANGE MEMBRANES

As far as their chemical structure is concerned ion-exchange membranes are very similar to normal ion-exchange resins; from the chemical point of view, these resins would make excellent membranes of high selectivity and conductivity. The difference between membranes and resins arises largely from the mechanical requirements of the membrane process. Unfortunately, ion-exchange resins are mechanically weak, cation-exchange resins tend to be brittle and anion-exchange resins to be soft [F. Helfferich, 1962]. They are dimensionally unstable due to the variation in the amount of water imbibed into the gel under different circumstances. Changes in electrolyte concentration, in the ionic form or in temperature may cause major changes in the water uptake and hence in the volume of the resin. These changes can be tolerated in small spherical beads, but in large sheets which have been cut to fit an apparatus they are not acceptable. Thus, it is generally not possible to use sheets of material which has been prepared in the same way as a bead resin. The most common solution to this problem is the preparation of a membrane with a backing of a stable reinforcing material which gives the necessary strength and dimensional stability. Preparation procedures for making ion-exchange resins and membranes are described in great detail in the patent literature [A. A. Kasper, 1955], [W. Juda et al., 1950, 1953], [K. Haagen et al.,1959], [W.G. Grot, 1974, 1975], and [Y. Onoue et al., 1980], [S.R. Caplan et al., 1974], [D.S. Flett, 1983].

2.1 Preparation of Homogeneous and Microheterogeneous Membranes

The definition of homogeneous ion-exchange membranes used here includes micro-heterogeneous structures such as inter-polymers, graft-, block- and co-polymers. The different techniques of the preparation of ion-exchange membranes can be schematically illustrated as:

- Polymerization of monomers that contain a moiety that either is or can be made anionic or cationic, respectively:

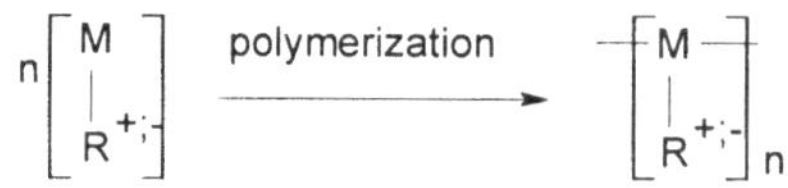

Membranes prepared by this technique often show a rather high charge density which is fixed by the molecular weight of the monomer and show excessive swelling.

- Polymerization of a monomer that contains an anionic or a cationic moiety with a neutral polymer:

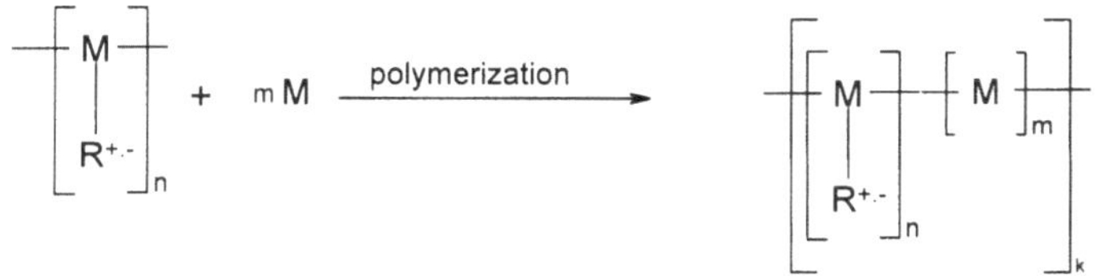

By co-polymerization of a monomer containing a positive or negative charge with a neutral monomer the fixed charge density can easily be controlled by the ratio of the two monomers.

- Preparation of block polymers or polymer mixtures of a neutral polymer with polymer containing anionic or cationic moieties:

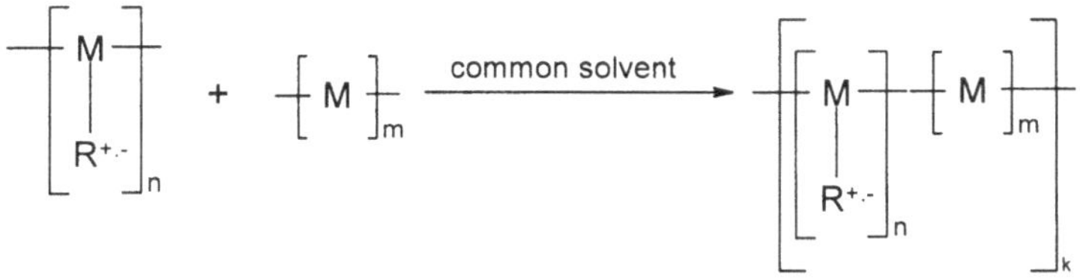

By preparing block polymers or mixtures of neutral and electric charges carrying polymers ion-exchange membranes with well controlled properties concerning the electrical charge density or hydrophilic and hydrophobic character can easily be prepared.

- Introduction of anionic or cationic moieties into a polymer dissolved in a solvent by a chemical reaction such as sulfonation with sulfuric acid or chlorosulfonic acid:

$$\text{polymer-Y} + \text{X-R}^{+,-} \xrightarrow{\text{chemical reaction}} \text{polymer-R}^{+,-} + \text{X-Y}$$

Here polymer-Y presents a polymer with a reactive end group Y while $X\text{-}R^{+,-}$ is a charge group containing reagent such as chlorosulfonic acid.
This is a very simple method to introduce electrical charges into a polymer matrix and to prepare an ion-exchange membrane by casting a film of the modified polymer and evaporating the solvent. However, the degree of functionalization is often difficult to control.

- Introducing functional groups into a preformed polymer film by grafting:

$$\text{polymer} + \text{M-R}^{+,-} \xrightarrow{\text{irradiation, plasma}} \text{polymer-M-R}^{+,-}$$

Here $M\text{-}R^{+,-}$ is a charge group containing monomer.
In this method of preparing ion-exchange membranes a polymer, that is not readily soluble in a common solvent such as a polytetrafluorethylene film, is exposed to a monomer carrying the desired functional group and then irradiated or plasma treated. While the actual process is simple the degree of functionalization of the polymer is difficult to control and often very heterogeneous structures are obtained.

2.1.1 Polymerization and Polycondensation of Monomers

The literature contains a multitude of detailed recipes for making ion-exchange resins and membranes by polymerization. Some of these preparation procedures shall briefly be described. One of the first membranes made by polymerization or polycondensation of monomers was prepared from phenol by polycondensation with formaldehyde. In the first step an ion-exchange resin is prepared which then is converted into a membrane. For the preparation of the resin phenol is treated with concentrated H_2SO_4 at 80°C for about 3 hours which leads to the phenolsulfonic acid in para position, a brown, crystalline material. This acid is reacted with a 38%-solution of formaldehyde in water initially at -5°C for about 30 minutes and then at 85°C for several hours. The solution is then cast into a film which solidifies when the solution is cooled down to room temperature. Excess monomer is removed by washing the film with water. Since the sulfone group is directly bound to the benzene ring a strong acid type ion-exchange resin is obtained which is completely dissociated over the entire pH-range [N. Ishibashi et al., 1954], [F. Helfferich, 1962], [W.A. McRae et al., 1960]. The reaction is shown in the following scheme:

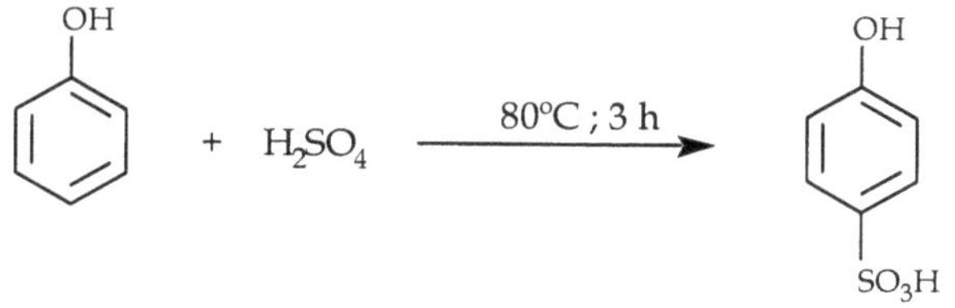

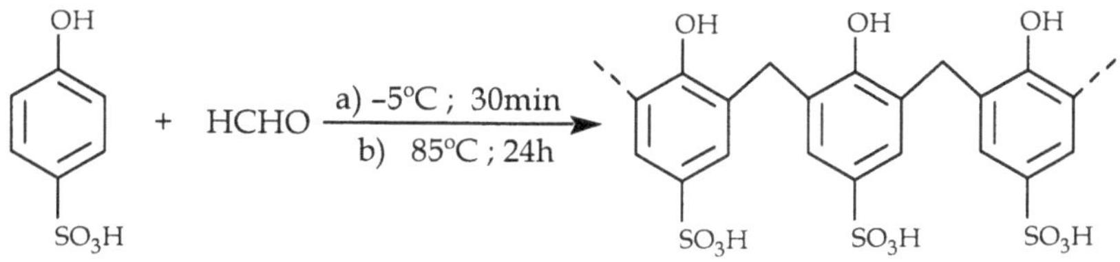

A weak ion-exchange membrane is obtained when instead of phenolsulfonic acid, e.g. salicylic acid is reacted with phenol and formaldehyde according to the following reaction scheme:

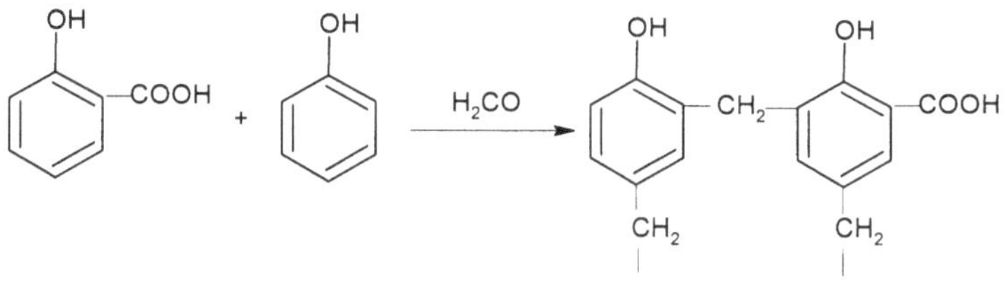

These films, however, are chemically and mechanically not very stable and are difficult to handle without a proper support material.

A method of preparing both cation- and anion-exchange membranes, which is widely used for the preparation of commercial membranes, is the polymerization of styrene and divinylbenzene and its subsequent sulfonation or amination.

The cation-exchange membrane is obtained according to the following reaction scheme [K. Kusomoto et al., 1976]:

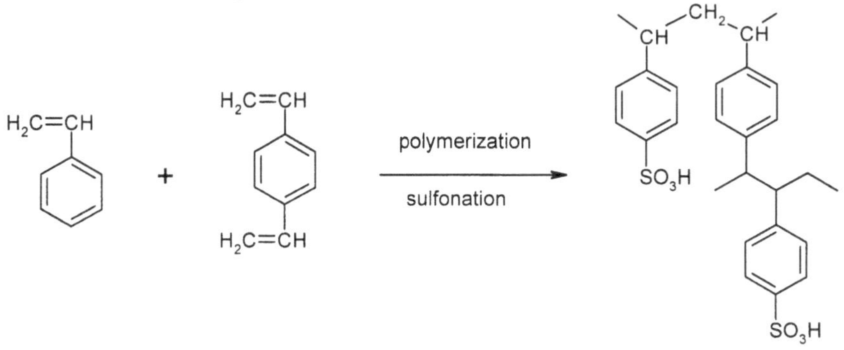

In a first step styrene is partially polymerized with divinylbenzene at about 60°C using benzoyl peroxide as an initiator for the polymerization. The polymer is partially cross-linked and obtained as a block which is cut into slices and in a second step sulfonated with concentrated sulfuric acid at room temperature. The obtained membranes show high ion-exchange capacity and low electrical resistance but their mechanical strength is insufficient to be used without a proper support material.

However, by addition of linear polymers such as polyvinylchloride, polyethylenestyrene-butadiene rubber, etc. to a mixture of styrene, divinylbenzene, and a plasticizer a pasty mixture is obtained. The mixture is then coated on a fabric backing such as a net and covered on both sides with a glass plate or a separating plastic sheet of polyvinylalcohol or polyethylene-terephtalate. By heating the composite layer vinyl monomers, divinylbenzene, peroxides, and additives are polymerized. The obtained film is then sulfonated with concentrated sulfuric acid as described earlier. The result is a reinforced ion-exchange membrane with excellent electrical and mechanical properties. Many of the commercial hydrocarbon-type ion-exchange membranes marketed by Tokuyama Soda and Asahi Glass are manufactured this way today [Asahi Glass, 1990].

A homogeneous anion-exchange membrane can be obtained by introducing a positively charged quaternary amine group into a preformed polymer by a chloromethylation procedure followed by an amination with a tertiary amine according to the following reaction scheme:

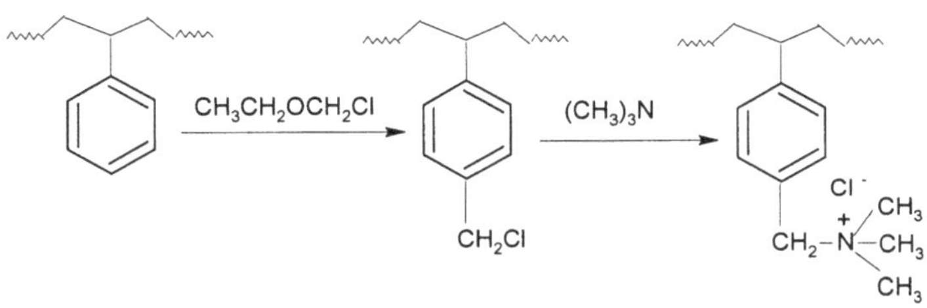

Membranes with good electrical and mechanical properties were obtained by additional polymerization with divinylbenzene or by first polymerizing the monochloromethylstyrene with divinylbenzene and then treating the resulting polymeric film with trimethylamine.

The membrane structures and their preparation described above are just two examples. There are many variations of the basic preparation procedure resulting in slightly different products. Instead of styrene often substituted styrenes such as methylstyrene or phenylacetate are used and instead of divinylbenzene monomers such as divinylacetylene or butadiene are used. Furthermore, instead of sulfonic acid phoposphoric or arsenic acid is introduced in the cross-linked polystyrene. However, most of these membranes have no, or only very little commercial relevance.

In most ion-exchange membranes based on polystyrene the ion-exchange moiety is introduced into the cross-linked polymer. This preparation technique has the disadvantage that the degree of sulfonation is difficult to control. In principle the monomers can be sulfonated prior to polymerization and cross-linking. By adjusting the ratio of sulfonated to non-sulfonated monomers the desired degree of sulfonation in the final membrane can be controlled. There are several detailed recipes described in the literature. A typical example is the preparation of a weak acidic ion-exchange membrane by cross-linking methacrylic acid with divinylbenzene according to the following reaction scheme:

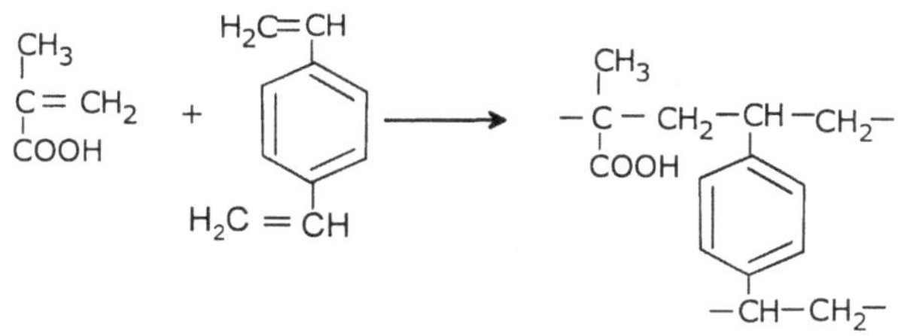

Numerous references exist in the literature for the preparation of ion-exchange membranes by polymerization. One of the technically and commercially most important cation-exchange membrane developed in recent years is based on perfluorocarbon polymers. Membranes of this type have extreme chemical and thermal stability and they are the key component in the chloralkali electrolysis and in most of today's fuel cells [A. Eisenberg et al., 1982]. They are prepared by copolymerization of tetrafluoroethylene with perfluorovinylether having a carboxylic or sulfonic acid group at the end of a side chain. There are several variations of a general basic structure commercially available today. The properties of these membranes have been optimized over the years with its special application in the chlorine-alkaline industry in mind. The various preparation techniques are described in detail in the patent literature [D.J. Connolly et al., 1966], [W.G. Grot, 1973, 1974] and [M.S. Seko et al., 1988].

Today's commercially available perfluorocarbon membranes have the following basic structure:

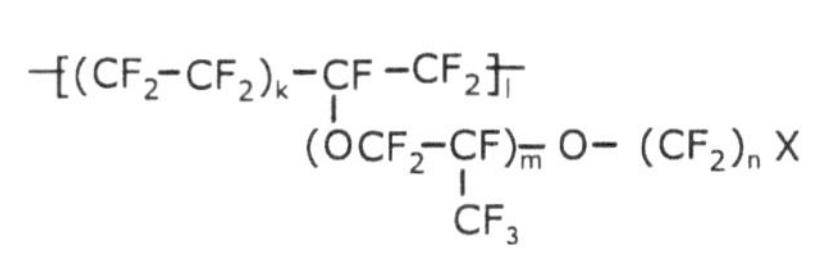

k = 5-8, l = 600-1200, m = 1-2, n = 1-4, X = SO_3^- , COO^-

The synthesis of the perfluorocarbon membranes is rather complex and requires a multistep process. Several companies have developed synthesis routes, however, the differences in the final structures are relatively small. One of the earliest publications describes a route developed by DuPont in the 70'ties [W.G. Grot, 1973, 1974]. It contains 4 mayor steps:

1. Synthesis of the sulfonated monomer

The first step is a reaction of tetrafluoroethylene with sulfur trioxide followed by a ring opening of cyclic product to form the fluorosulfonylacetylfluoride according to the following reaction scheme:

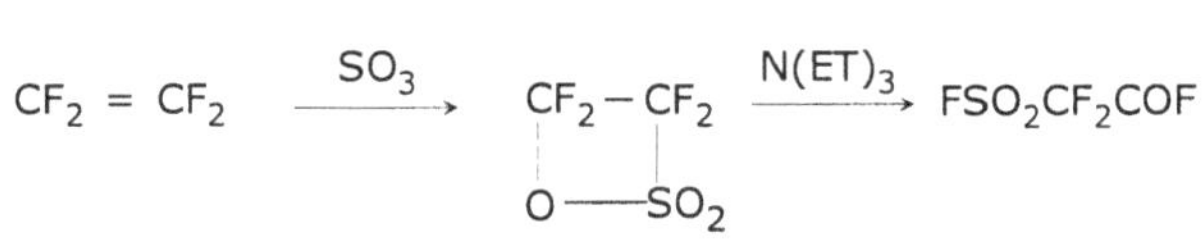

2. Synthesis of the polyvinylethersulfonylfluoride

In the second step the fluorosulfonylacetylfluoride is heated with sodium carbonate and the polyvinylethersulfonyl fluoride is formed according to the following reaction scheme.

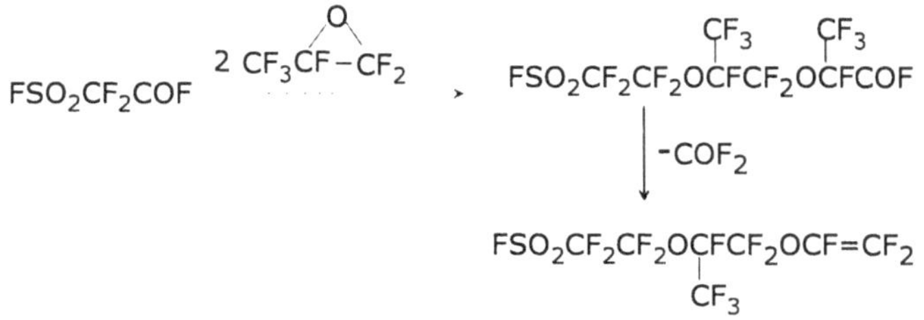

3. Formation of the fluorocarbonvinylether resin

In the third step the vinylether is copolymerized with tetrafluoroethylene to form a neutral thermoplastic polymer which can be extruded in sheet form. This polymer is generally referred to as XR-resin.

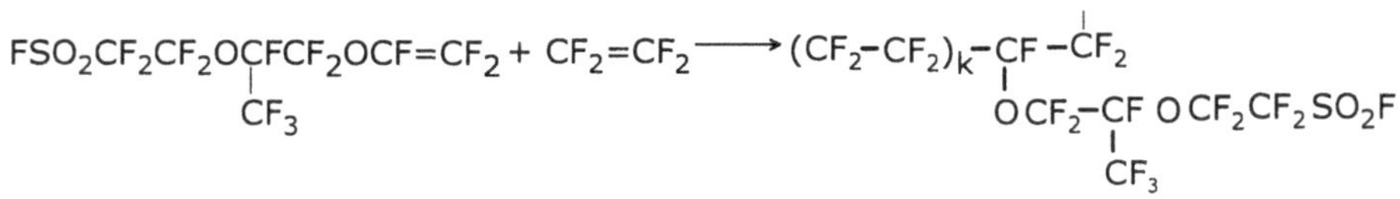

4. Conversion of the XR-resin to an ion-exchange membrane

The last step in the preparation of the final membrane is the conversion of the neutral XR-resin sheet into a negatively charged cation-exchange membrane. This is done by a treatment with hot sodium hydroxide according to the following reaction scheme:

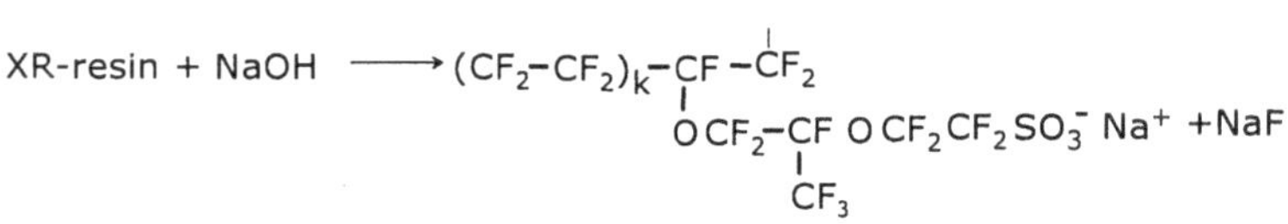

The sodium counter-ion in the membrane can readily be exchanged for other cations by immersing the film into the appropriate aqueous solution.

Perfluorinated membranes have excellent chemical and thermal stability. But because of the strong acidity and high hydrophilicity of the perfluorosulfonic group the membrane shows a high degree of swelling which results in a low electric resistance but also in relatively poor permselectivity, since excessive swelling can produce water filled channels which permit the passage of co-ions. The degree of swelling is also a function of the fixed ion concentration, i.e. the equivalent weight of the polymer, which is typically between 600 and 1500 grams of dry polymer per mol of ion-exchange site. Thus, the ion-exchange capacity of the membrane is between 0.5 to 2 milliequivalent per gram. The degree of swelling could be reduced by reducing the ion-exchange capacity of the membrane. This, however, would increase the electrical resistance and reduce the permselectivity in concentrated electrolyte solutions because of the decrease in the Donnan exclusion. The permselectivity of the membrane in concentrated electrolyte solutions is therefore often unsatisfactory [M.V. Coulter, 1980].

To improve the overall performance of the membrane or to adapt it to certain applications various modifications of the basic polymer have been introduced mainly to change the ion-exchange groups either of the entire membrane or only at its surface. In addition to the $-SO_3H$ group, carboxylic acid –COOH, sulfonamide $-SO_2(CH_2)NH_2$, or quaternized alcohol groups -COH are used in commercial membranes. Ion-exchange membranes with carboxcylic acid as charged group have significantly lower water sorption and, thus, higher permselectivity but also drastically increased electrical resistance. Because of the weak acidic character of the carboxylic acid group perfluorocarboxylate membranes may either be prepared from the appropriate monomers or by replacing the sulfonic acid group in the polymer by a carboxylic acid group. The second method is used mainly for the modification of the membrane surface. The $-CF_2SO_2F$ groups are converted to –COOH groups in various ways by the different manufacturers [Y.T. Onoue et al., 1980]. Tokuyama Soda Co. transforms the $-CF_2SO_3H$ groups to $-CF_2SO_2Cl$ with a solution of phosphorepentachloride which then is reduced to sulfinic acid by reaction with hydrazine and hence oxidizes to –COOH. Homogeneous perfluorocarboxylic acid membranes can be prepared by different routes. Asahi Chemical produces the carboxylic acid monomers from tetrafluoroethylene according to the following scheme:

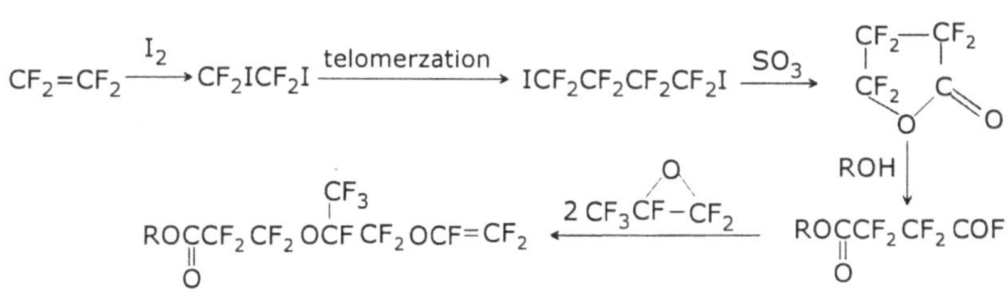

By addition of iodine to the tetrafluoroethylene, telomerization and oxidation by SO_3 a cyclic compound is obtained. Further treatment with alcohol leads to a ring opening and the formation of a bifunctional component which is then reacted with hexafluoropropylene epoxide to form the carboxalfluorodevinylether. The copolymerization of the monomer with tetrafluoroethylene leads to the thermoplastic polymer as described previously.

The synthesis of the bifunctional compounds $RCOO-(CF_2)_n-COF$ and $SO_2F-(CF_2)_n-COF$ is the key step in the entire membrane production process. Copolymerization of the functionalized monomers with hexafluoropropylene epoxide and tetrafluoroethylene are comparatively simple procedures.

The perfluorinated cation-exchange membrane is today by far the commercially most relevant membrane with applications not only in the chlorine-alkaline industry but also in fuel cells, electrolysis, and electrodialysis of certain industrial effluents. In recent years an entire series of modifications have been introduced in the original membrane manufacturing process to adapt the properties of the membrane to the anticipated practical application and today a number of perfluorinated membranes are commercially available which are quite different in their structure and electrochemical properties [H.L., Yeager, 1982].

In the electrolytic chlorine and alkaline production often perfluorinated membranes are used which contain a thin layer of a resin with the weakly acidic carboxylic acid group on a core resin containing the strongly acidic sulfonic acid group. These membranes are made either by lamination or by surface treatment according to the scheme of Figure 3.2.

For fuel cells used in the methanol direct conversion perfluorinated cation-exchange membranes having a low methanol permeability are being developed to avoid excessive methanol loss.

Asahi Chemical, Dupont and others are producing today both homogeneous and laminar perfluorocarboxylic acid membranes [R.L. Dotson et al., 1982], [T. Sata et al., 1983], [Tokuyama Soda Product Bulletin, 1988] and [Asahi Chemical Product Bulletin, 1988].

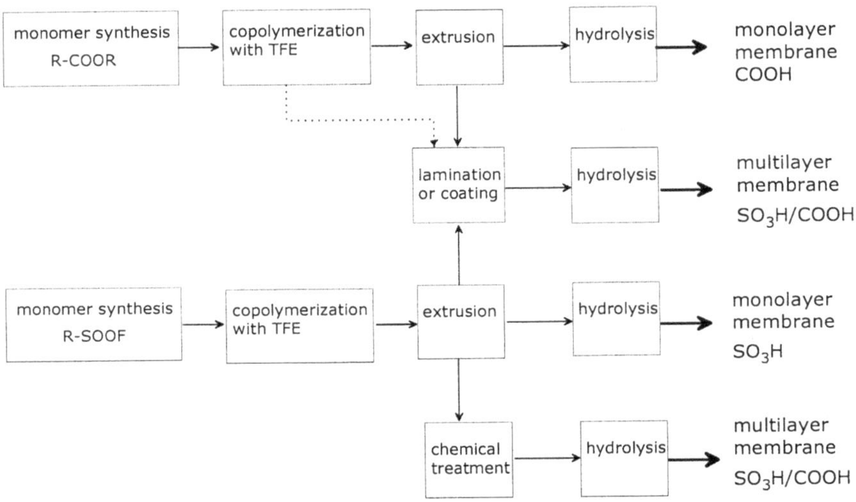

Fig. 3.2 Schematic diagram of the different pathways for the preparation of mono- and multi-layer fluorocarbon ion-exchange membranes.

In addition to the various perfluorinated cation-exchange membranes also an anion-exchange membrane can be prepared according to the following scheme [K.E. Matsui et al., 1986]:

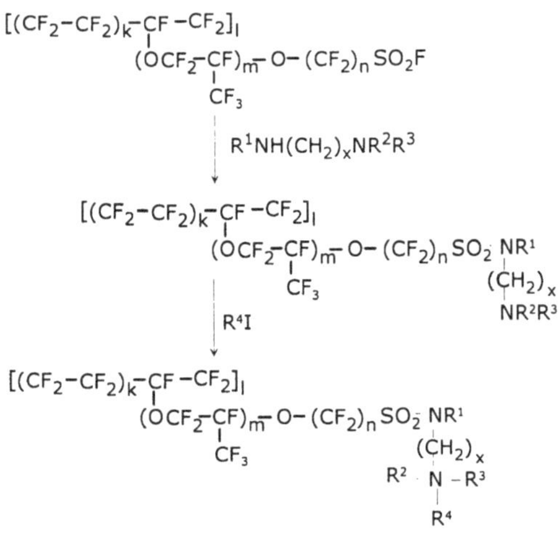

The preparation is based on the reaction of the sulfonylfluoride polymer with a diamine to a sulfonamide with a pendant amine group which then further reacts with an alkyl iodine to a quaternary amine. according to the following scheme [K.E. Matsui et al., 1986]:

The anion-exchange membrane is quite stable at elevated temperature up to 120°C in an oxidizing environment as well as in strong acids and bases. However, in spite of excellent chemical stability the perfluorinated anion-exchange membrane has not reached nearly the commercial relevance as the cation-exchange membrane.

2.1.2 Introduction of Ionic Groups into a Preformed Solid Polymer Film

Concerning the introduction of anionic or cationic moieties into a preformed film two techniques are applied:

1. A film of a commercial polymer such as polyethylene, polyvinylchloride or polytetrafluoroethylene is swollen with a monomer such as divinylbenzene. Then the monomers are polymerized with the film which also contains a radical initiator [I. Ishigaki et al., 1978]. Alternatively, a vinyl monomer may be grafted onto a polymer film by radiation techniques. The film is exposed to X-rays to form radicals in the polymer matrix which bind the vinyl monomer. Ion-exchange groups are then introduced by sulfonation, or chloromethylation, and amination. Starting with a film makes the membrane preparation rather easy. The starting material may be a hydrophilic polymer, such as cellophane or polyvinylalcohol. More often, however, a hydrophobic polymer such as polyethylene or polystyrene is used. In recent years a large number of papers have been published on the preparation of ion-exchange membranes by this method. However, the commercial success is rather limited.

2. A film of a commercial polymer such as polyethylene or polypropylene is directly sulfonated or aminated. A typical example is the preparation of ion-exchange membranes by sulfochlorination and amination.

a) Preparation of the cation-exchange membrane

Exposure of a polyethylene film to a mixture of SO_2 and Cl_2 gases under an UV-light source results in the sulfochloronation of the polymer [E. Korngold, 1984], [F. De Korocy et al., 1967]. The membrane is then reacted with sodiumhydroxide to hydrolyse the sulfonylchloride group to sodiumsulfonate resulting in a cation-exchange membrane according to the following reaction scheme:

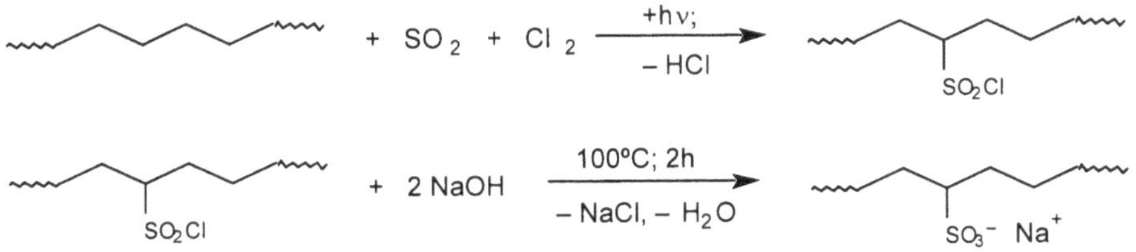

b) Preparation of the anion-exchange membrane

Alternative reaction with a diamine which contains both primary and tertiary amino groups results in an acid amide bonding. The pendant tertiary amine is quaternized with methylbromide resulting in an anion-exchange membrane.

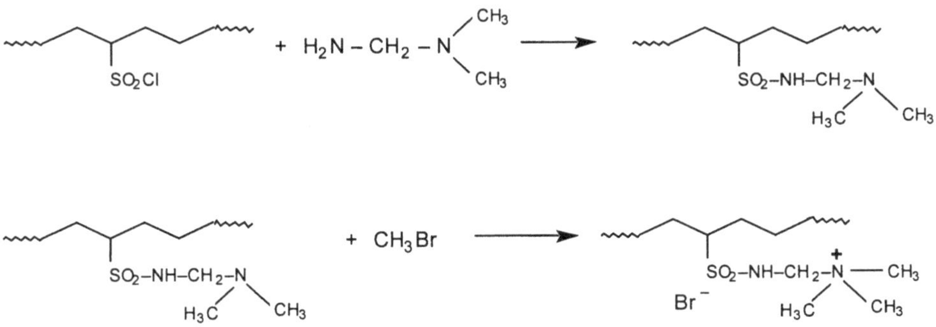

Ion-exchange membranes prepared by this method from polyethylene sheets have low electrical resistance combined with high permselectivity and excellent mechanical strength. However, in spite of their good properties these membranes are no longer commercially available.

2.1.3 Introduction of Ionic Groups into a Dissolved Polymer

Membranes can also be prepared by dissolving and casting a functionalized polymer, such as sulfonated polysulfone, sulfonated polyetheretherketone or polyphenyleneoxide into a film. As a typical example the preparation of a cation- and anion-exchange membrane based on polysulfone is described in the following reaction schemes.

Preparation of a cation-exchange membrane by sulfonation of polysulfone

The sulfonation is carried out with chlorosulfonic acid according to the following scheme:

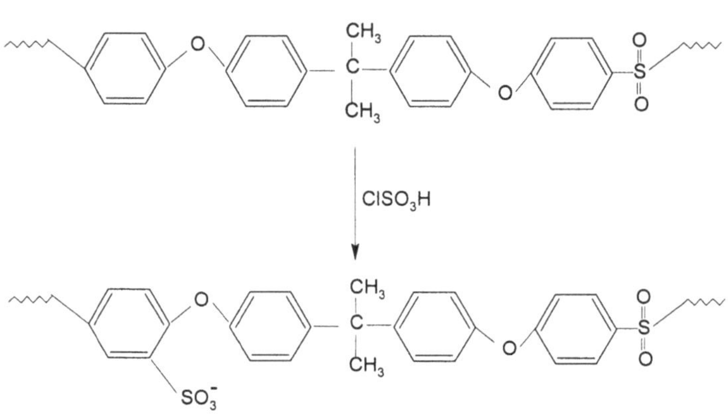

The sulfonated polysulfone can be cast as a film on a screen and precipitated after the evaporation of most of the solvent such as dichloroethane. This leads to a reinforced membrane with excellent chemical and mechanical stabilities and good electrochemical properties [P. Zschocke et al., 1985].

Preparation of the anion-exchange membrane based on polysulfon

The anion-exchange membrane based on polysulfone is prepared by halomethylation of the backbone polymer and subsequent reaction with a tertiary amine as follows:

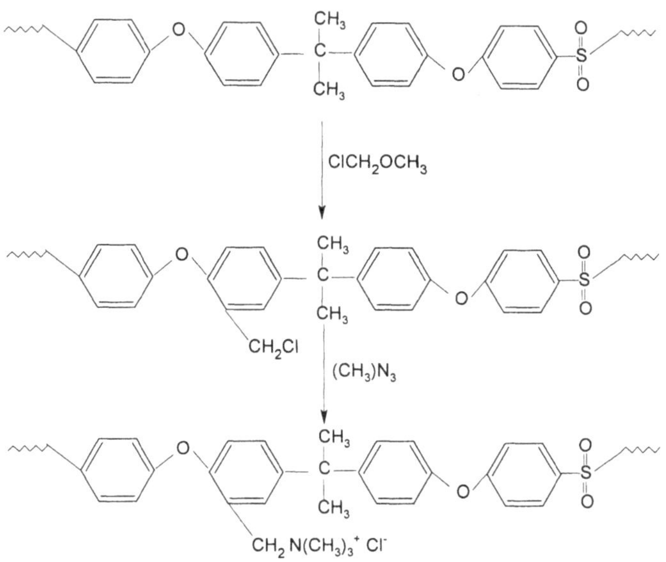

For the preparation of cation-exchange membranes also polyetheretherketone is used as the basic polymer. It can very easily be sulfonated with concentrated sulfuric acid according to the following scheme:

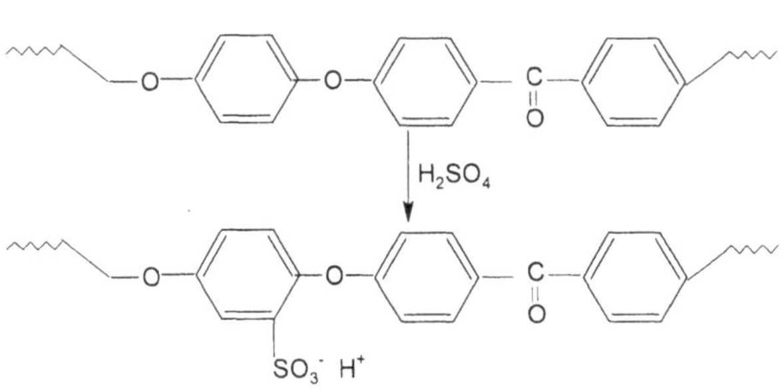

The sulfonation occurs on one polyetheretherketone block only and is thus very easy to control. To obtain membranes with different ion-exchange capacity the sulfonated polyetheretherketone can be mixed with polyethersulfone in a solvent such as N-methylpyrrolidone. By changing the ratio of the sulfonated polyetheretherketone to polyethersulfone the fixed charge density can easily be adjusted to a desired value [F.W. Wilhelm, 2001].

2.2 Preparation Procedure of Heterogeneous Ion-Exchange Membranes

Ion-exchange membranes with a heterogeneous structure consist of fine ion-exchange particles embedded in an inert binder such as polyethylene, phenolic resins, or polyvinylchloride. Heterogeneous ion-exchange membranes are characterized by the discontinuous phase of the ion-exchange material. The efficient transport of ions through a heterogeneous membrane requires either a contact between the ion-exchange particles or an ion conducting solution between the particles. Heterogeneous ion-exchange membranes can easily be prepared by different methods:

- An ion-exchange powder is mixed with a dry binder polymer and molded under the appropriate conditions of pressure and temperature into a sheet.
- Ion-exchange particles are mixed with a binder polymer which is brought into a semi-fluid state by adding a plasticizer or by heating and then extruded as a sheet.
- Ion-exchange particles are dispersed in a solution containing a dissolved film-forming binder, then the mixture is cast into a film and the solvent evaporated.
- Similarly, ion-exchange particles are dispersed in a partially polymerized binder polymer and cast into a film, then the polymerization is completed.

The size of the ion-exchange particles in the membranes made by the above described preparation procedures is generally below 50μm. Heterogeneous ion-exchange membranes with useful low electrical resistances contain more than 65% by weight of the ion-exchange particles. Since these ion-exchange particles swell when immersed in water, it is difficult to achieve adequate mechanical strength combined with low electrical resistance. Most heterogeneous membranes that possess adequate mechanical strength generally show poor electrochemical properties. On the other hand, a membrane that contains ion-exchange particles in a concentration high enough to show the desired electrochemical performance often exhibits poor mechanical strength. Furthermore, it has been shown that heterogeneous membranes with an ion-exchange resin content equivalent to that of a homogeneous ion-exchange membrane exhibit a greater degree of swelling when immersed in water than would be expected from the density of the wet ion-exchange material and the binder polymer. Apparently those membranes develop water filled interstices during the swelling process. This results not only in poor mechanical properties but also poor permselectivity.

However, there are significant differences between ion-exchange membranes and ion-exchange resins concerning the polymer structures which are primarily due to the differences in size and shape between ion-exchange beads and ion-exchange membrane sheets. In both cases the fixed ion-exchange groups lead to a swelling of the polymer when it is in contact with an aqueous solution. The degree of swelling depends on the ionic strength of the solution and on the degree of cross-linking of the polymer. Due to the spherical symmetry of granular ion-exchange resins and due to the fact that they are not physically constrained in use, there is generally no functionally important physical damage to the resins from drying and rewetting or from changes of the ambient ionic strength.

Dimensional changes which are tolerable during use in the case of granular ion-exchange resins, however, are not acceptable in ion-exchange membranes due to the large sizes of the latter and the fact that they are physically constrained in the electrodialysis stacks in which they are used. As a result, the ion-exchange capacities of ion-exchange membranes are about 1–3 equivalent per kilogram of dry membrane and thus significantly lower than that of a granular ion-exchanger resin which is 3–5 equivalent per kilogram dry resin, a fact that results in a reduced swelling of ion-exchange membranes compared to resins. In addition to covalent cross-linking of polymers other strategies are used to limit swelling by using, e.g. reinforcing fabrics, or by including polymer segments in the ion-exchange resin polymers which yield microcrystalline regions in the resins.

In general, heterogeneous ion-exchange membranes have relatively high electrical resistances. Homogeneous ion-exchange membranes have a more even distribution of fixed ions and often lower electrical resistances.

2.3 Special Property Ion-Exchange Membranes

Generally, cation-exchange membranes are permeable for cations and more or less impermeable for anions, while anion-exchange membranes are permeable to anions and more or less impermeable for cations. A separation of cations from anions is therefore quite efficient in electrodialysis. In other electromembrane processes such as diffusion dialysis and electrodialysis with bipolar membranes or specific application of electrodialysis, e.g. in the production of table salt from sea water or the removal of nitrate from potable water, additional properties such as the selectivity for ions of the same electrical charge may be required [Y. Kobuchi, et al., 1987], [G. Pourcelly et al., 1994]. Significant effort has also been concentrated on the development of anion-exchange membranes with low fouling, i.e. adsorption of high molecular weight charged materials, tendencies, and high chemical and thermal stability. In addition to excellent chemical stability and ion permselectivity bipolar membranes must exhibit high water dissociation capability and low electric resistance to be useful in practical applications. In the literature, there are numerous methods reported for the preparation of ion-exchange membranes with special properties to be used in the chemical process industry, for the treatment of industrial effluents, or as battery separators and ion-selective electrodes.

2.3.1 Monovalent Ion Permselective Membranes

Since 1972, in Japan table salt is produced by electrodialytic concentration of sea water. For the specific requirements of this process, ion-exchange membranes have been developed which can separate to a large extent monovalent ions from a solution mixed with multivalent ions. An increase in the cross-linking density of a cation-exchange membrane or the formation of a thin layer of a highly cross-linked resin on a low cross-linked cation-exchange membrane [Y. Onoue et al., 1961] results in a significant increase in permeability of monovalent over multivalent cations. The introduction of a small amount of anion-exchange groups in a cation-exchange membrane or the formation of a thin layer of a positively charged polyelectrolyte on the surface of a cation-exchange membrane also increases the membrane selectivity for monovalent ions [T.K. Sata, 1972]. Tokuyama Soda has commercialized monovalent cation selective membranes (Neosepta® CMS) prepared by forming a thin polyelectrolyte layer on the surface [T.K. Sata, 1986]. The permeability ratio of sodium to calcium is more than 20 for such a membrane.

Monovalent anion selective membranes have been developed mainly to prevent the permeation of sulfate which can cause scaling due to the precipitation of calcium sulfate in desalination of brackish water by electrodialysis. The permselectivity of anion-exchange membranes for monovalent anions can be increased by the formation of a highly cross-linked layer of the surface of an anion-exchange membrane [R. Yamane et al., 1964]. Also the formation of a negatively charged polyelectrolyte layer on the surface of an anion-exchange membrane increases its permselectivity for monovalent ions [K. Mihara et al., 1973]. The hydrophilicity of anion-exchange groups in the membrane also effects the permselectivity for specific anions. More hydrophobic anion-exchange groups have a higher permselectivity for less hydrated ions and vice versa [T.K. Sata et al., 1995]. Based on this concept membranes with high selectivity for nitrate and bromide over chloride have been developed. There are more preparation techniques to increase the permselectivity of anion-exchange membranes for certain anions described in the literature [T.K. Sata et al., 1997]. A monovalent anion selective membrane which shows a permeability ratio of sulfate to chloride of about 0.01 is the Neosepta® ACS which has a thin, highly cross-linked layer on the membrane surface.

2.3.2 Proton Permselective Cation-Exchange Membranes

Generally the permeation of protons is significantly higher, i.e. 4 to 5 times, than that of other cations. However, for certain applications such as the recovery of acids from a mixture with salts it is desirable to have a cation-exchange membrane with very high proton selectivity. Today two types of cation-exchange membranes with these properties are available. One is an amphoteric type ion-exchange membrane in which sulfonic acid and ammonium groups exist at a certain ratio [R. Yamane et al., 1965]. The second type is a cation-exchange membrane having a thin positively charged layer on the surface [T.K. Sata et al., 1979]. Both membranes show a proton permeability of more than 40 times that of sodium.

2.3.3 Anion-Exchange Membranes of High Proton Retention

Generally protons easily permeate through a conventional anion-exchange membrane because of the different transport mechanism of protons in an aqueous environment. Thus, it is difficult to obtain high acid concentrations in the electrodialysis of acid solutions or in the electrodialytic production of acids and bases with bipolar membranes without a significant loss in current utilization due to proton leakage through the anion-exchange membranes. Thus, a substantial effort has been concentrated on the development of anion-exchange membranes with decreased proton leakage. To reduce the proton permeability of anion-exchange membranes the fixed charge density of the membrane should be as high as possible. Furthermore, membranes containing weakly dissociated anion-exchange groups that are less hydrated also reduce the proton permeability [G. Pourcelly et al., 1994]. Commercially available membranes with improved proton retention contain 4-vinylpyridine as a key component. Thin layers of polypyrrole on conventional anion-exchange membranes show a remarkable decrease in proton permeability. The membranes are prepared by equilibrating a conventional anion-exchange membrane with an aqueous pyrrole solution. The membrane is then immersed in a ferric chloride solution to polymerize the pyrrole adsorbed by the membrane. The electrical resistance of the membrane hardly increases in spite of the decreased proton permeability because the polypyrrole layer at the surface of the anion-exchange membrane is extremely thin [T.K. Sata et al., 1993]. The Neosepta® ACM resembles this type of membrane. It exhibits low proton permeability and enables an efficient acid concentration.

2.3.4 Anti-Fouling Anion-Exchange Membranes

Fouling of membranes is a common problem in all membrane separation processes. Ion-exchange membranes are fouled by ionic components of medium molecular weight such as ionic surface active agents or proteins having charges opposite to the fixed charges of the membrane. Generally, the anion-exchange membrane is more sensitive to fouling than cation-exchange membranes [E. Korngold, 1970]. However, significant improvements have been made in the development of anion-exchange membranes with low fouling tendencies. Fouling of anion-exchange membranes often occurs when the anion is small enough to penetrate into the membrane structure, but its electro-mobility is so poor that the membrane is virtually blocked. To overcome this problem, different companies developed membranes, such as the Neosepta® AMX which is characterized by a high permeability for large organic anions. In general, the permselectivity of these membranes is lower than that of regular membranes. Since the pore diameter of ion-exchange membranes is in the range of 1 nm, polyelectrolytes of high molecular weight, i.e. molecular weights > 1000 Dalton are not harmful to ion-exchange membranes. However, ionic compounds with molecular weight of several hundred Dalton can cause membrane fouling. According to some earlier studies concerning the permeability of commercial anion-exchange membranes, the upper molecular weight limit for practical electrodialytical separations is in the range of 150 Dalton [H. Itoh et al., 1986], [R. Dohno et al., 1975]. For instance, a molecular weight of 350 Dalton is to be considered as a maximum size for any charged component to effectively being transported through the Ionac MA-3475 membrane.

Ionics Inc. produces a macroreticular membrane which is less sensitive to traces of detergents [R.B. Hodgdon et al., 1973]. It is produced by dissolving an organic compound in the membrane forming system. When the material diffuses from the membrane after the polymerization, large pores are left behind. Alternatively, certain salts such as potassium iodide are added to the solvent of the binder polymer, which mostly is dimethylformamide. Through these pores, large anionic molecules can penetrate, thus preventing a steep increase in the electrical resistance.

Another type of anti-fouling anion-exchange membrane is produced by Tokuyama Soda. The membrane is coated with a thin layer of cation-exchange groups causing electrostatic repulsion of organic molecules. In practice the coating is done by weak sulfonation of the membrane surface, followed by the ordinary chloromethylation and quaternization steps.

One method to improve the permeability of anion-exchange membranes for organic acids having molecular weights in excess of 300 Dalton is based on the adjustment of the degree of cross-linkage and the chain length of the cross-linker in the polymer network, as indicated by the following structure [W.G. Gudernatsch et al.,1990]:

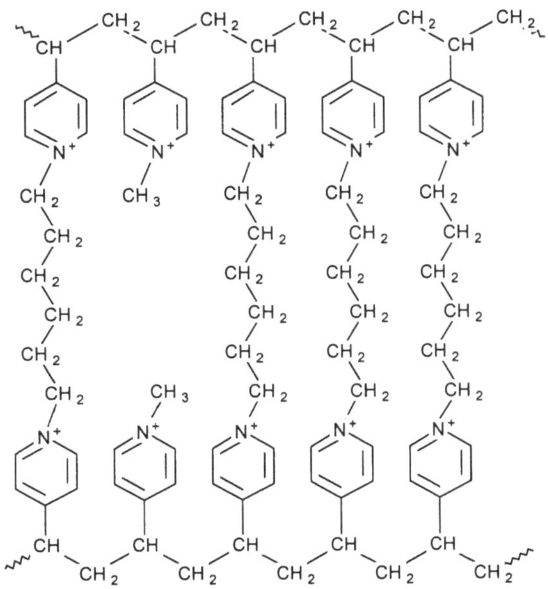

During the membrane formation, poly-4-vinylpyridine is quaternized with methyliodide and simultaneously cross-linked together with different dibromoalkanes.

Ion-exchange membranes based on aliphatic polymers show reduced organic fouling in natural waters when compared to membranes based on aromatic polymers [R.B. Hodgdon, 1992] . The membranes prepared from aliphatic polymers also allow operations with solutions containing 0.5 ppm chlorine and for shock chlorination up to 20 ppm free chlorine.

2.3.5 Alkaline Stable Anion-Exchange Membranes

In several technical interesting applications, the economics of the process are affected by the limited stability of currently available anion-exchange membranes in strong alkaline solutions. In the case of cation-exchange membranes the chemical stability could be improved by perfluorination of the polymer backbone, resulting in membranes such as the Nafion® or Flemion®. Comparable attempts with anion-exchange membranes to overcome the poor alkaline stability have so far failed. Since the fluorcarbon-type anion-exchange membranes cannot overcome the problem of instability under strong basic conditions, it is assume that the alkaline stability of an anion-exchange membrane is determined by the stability of the positively charged groups. By determining the disintegration rate of quaternized amines in alkaline solutions an anion-exchange membrane with considerable improved alkaline stability was developed [B. Bauer et al., 1990]. It was shown, that due to their higher acidity the cross-linked bis-quaternary structures have a much higher alkaline stability than the mono-quaternary ammonium groups. The highest alkaline stability was obtained with 4-diazabicyclo-[2.2.2]-octane (DABCO). Chloromethylated polystyrene can react with DABCO according to the following scheme:

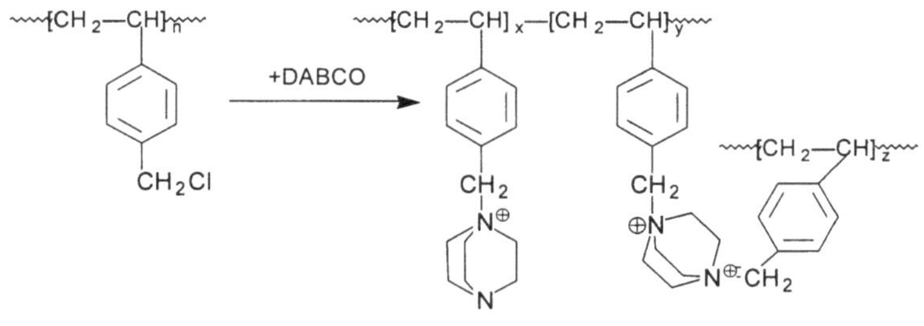

An equimolar ratio of chloromethyl groups to the DABCO yields a membrane which is not cross-linked. By reducing the ratio of the chloromethyl groups to the DABCO monomer to >1 the membrane is partially cross-linked. The cross-linking reduces the swelling of the membrane drastically and increases its chemical and mechanical stability but it also increases the electrical resistance significantly. This is due to the fact that the bis-quaternary DABCO structure converts the positive charges into a piperazine system. However, the chemical stability of the membrane is not effected.

Instead of polystyrene other chloromethylated polymers such as polysulfone or polyethersulfone can be used.

2.3.6 Bipolar Membranes

Bipolar membranes have recently gained increasing attention as an efficient tool for the production of acids and bases from their corresponding salts by electrically enforced accelerated water dissociation. The process which has been known for many years is economically very attractive and has a multitude of interesting technical applications [K.J. Liu et al., 1977]. So far, however, large-scale technical use of bipolar membranes has been rather limited by the availability of efficient membranes.

The principal structure of a bipolar membrane and its function is illustrated in Figure 3.3.

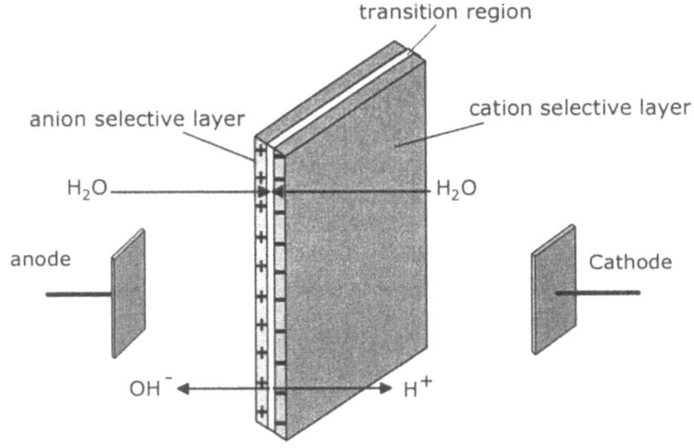

Fig. 3.3 Schematic drawing illustrating the structure of a bipolar membrane.

This figure shows an anion- and a cation-exchange membrane arranged in parallel between two electrodes similar to conventional electrodialysis. If there are no salt ions in the transition region between the two membranes and an electrical potential gradient is applied the transport of the electrical charges through the membranes is accomplished exclusively by protons and hydroxyl ions which are available even in pure water in a concentration of 10^{-7} mole per liter due to the dissociation equilibrium of water. Thus, an alkaline solution is formed on the anion-exchange side and an acid solution on the cation-exchange side of the bipolar membrane. The dissociated water which is transferred from the bipolar membrane to the outside phases as protons and hydroxide ions are continuously replenished from the outer phases by diffusion.

The most important properties of bipolar membranes are:

- low electric resistance,
- high selectivity for the counter-ions, i.e. complete exclusion of co-ions,
- excellent stability in solution with extremely low and high pH-values,
- high water dissociation capability,
- high water permeability of the monopolar layers.

Low electrical resistance, high permselectivity, and good chemical stability are properties required in all ion-exchange membranes. In addition bipolar membranes must be able to split water into protons and hydroxyl ions at a very fast rate[K.N. Mani, 1991]. The mechanism of water dissociation in bipolar membranes will be discussed later. Although the actual mechanism of water dissociation in the bipolar membrane is still subject of a rather controversial discussion it is generally assumed that the water dissociation in bipolar membranes is based on a proton transfer reaction that is catalyzed by weak acids and bases. It is assumed that the best catalytic effect can be achieved if the pK_a- or pK_b-value of the interphase material is close to 7 [H.J. Rapp, 1995]. Some weak ion-exchange groups such as tertiary amines, phosphoric acid, carboxylic acids or pyridine show the required pK_a-values. Certain heavy metal ion complexes, such as chromium(III)- or iron(III)-complexes provide the required catalytic water dissociation effect. In principle there are many more suitable metal ions available. But there is an additional requirement for the catalytic material to be effective for a long term period, i.e. it must stay in the interphase for the anticipated lifetime of the membrane. Some metal hydroxides such as chromium (III) and iron(III) hydroxide have an extremely low solubility constant of $K = 6.7\times10^{-31}$ at 25°C and $3{,}8\times10^{-31}$ at 18°C and they are thus especially suited to be deposited as catalyst in the interphase between the cation- and anion-exchange layers of a bipolar membrane. Other catalytic components used in the preparation of bipolar membranes such as tertiary amines or phosphoric acids are generally directly bonded to the polymer matrix of the membrane. The most commonly used catalytic components in bipolar membranes are listed in Table 3.1.

For the preparation of bipolar membranes several techniques are used. The most important ones are [F.G. Wilhelm et al., 2000]:

- The lamination by pressure at room temperature or at elevated temperature of solid cation- or anion-exchange resin sheets one of which containing the catalytic active component.
- The lamination by pressure at room temperature or at elevated temperature of solid cation- or anion-exchange resin sheets with an intermediate layer containing the catalytic active material.

- Casting a film from a solution containing an anion-exchange resin with the catalytic active component on a solid sheet of the cation-exchange resin.
- Casting a film from a solution containing a cation-exchange resin with the catalytic active component on a solid sheet of the anion-exchange resin.
- Co-extrusion of a cation- and anion-exchange resin one of which is containing the catalytically active component.

There are certainly many more techniques for the preparation of bipolar membranes described in the literature. However, very few are utilized commercially.

Table 3.1 Components used as catalytic material for the preparation of bipolar membranes.

material	form of application	pK_a-value	reference
$Cr(OH)_3$	as salt in the cation-exchange layer or the interphase		R. Simons, 1993
$Fe(OH)_3$	as salt in the cation-exchange layer or the interphase		F. Hanada et al., 1991
$-NR_2$	bond to the matrix of the anion-exchange membrane	~ 9	B. Bauer et al., 1988
$R-PO_3H$	bond to the matrix of the cation-exchange membrane	~7	N.V. Sheldeshov et al., 1986

The first bipolar membranes were prepared by simply laminating conventional cation- and anion-exchange membranes back to back. These membranes showed relatively high electric resistance. The high resistance of the membranes was caused by the interphase between the cation- and anion-exchange layers which contains only completely de-ionized water. To provide an area resistance which is in the same order of magnitude as that from the ion-exchange layers the interphase must be very thin, i.e. between 1 and 5 nm. Since the surface roughness of conventional ion-exchange membranes is in the order of several μm the contact between the laminated membranes is rather poor as indicated in Figure 3.4 which shows a schematic drawing of the a cross-section of two bipolar membranes.

The Figure 3.4 a) shows a cation-exchange and an anion-exchange resin layers which are in direct contact only at a limited number of points. In between these contact points there are areas where the distance between the two layers is comparatively large. Since the gap in these areas is filled with pure water its electric resistance is very high. The current through such a bipolar membrane passes mainly through contact points where very high current densities are obtained which often result in so-called "hot spots" where very high temperatures can be reached leading to a destruction of the membrane. If bipolar membranes are prepared by simply laminating conventional cation- and anion-exchange membranes back to back more satisfactory properties can be obtained by introducing a special surface treatment [R. Simons, 1986, 1993] or by introducing a thin layer containing the catalytically active groups [B. Bauer, 1993]. The electrical resistance of a bipolar membrane depends to a very large extent on the resistance of the area between the two ion-exchange membranes.

Membranes with a more uniform and thin interphase are obtained when a cation- or anion-exchange resin is cast on a solid film carrying the opposite charge [F.P. Chlanda et al., 1976],

[R.B. Hodgdon, et al., 1989]. Even when the solid film has a certain surface roughness there will be a close and uniform contact over the entire surface as indicated in Figure 3.4 b).

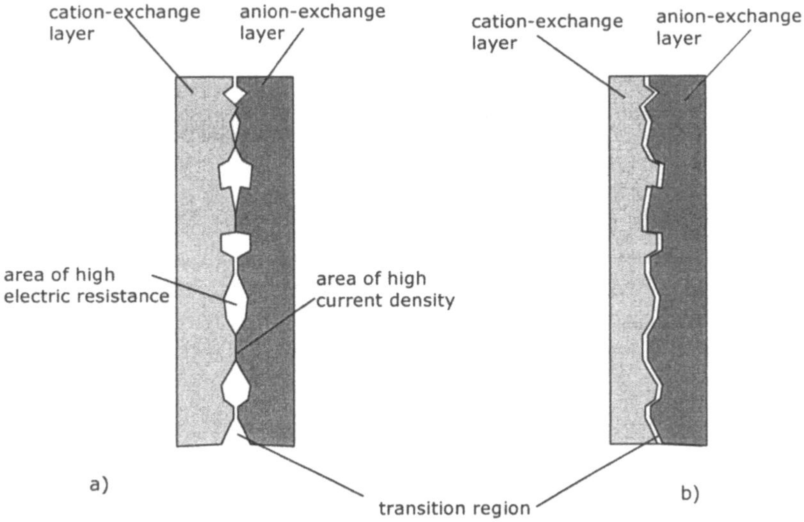

Fig. 3.4 Schematic diagram illustrating the transition region in a bipolar membrane prepared a) by laminating two conventional ion-exchange membranes back to back by pressing and b) by casting liquid anion-exchange resin on a solid ion-exchange membrane.

Today, most bipolar membranes are composed of a three layer structure, i.e. a cation-exchange layer, an anion-exchange layer and an intermediate layer containing a weak acid or base catalyst. This intermediate layer is a transition region between the two ion-exchange membranes where the water dissociation takes place. To have the required water dissociation capability it must contain the appropriate concentration of catalytic components. It must also be very thin to exhibit acceptable ion conductivity. Weakly dissociated ionic groups such as tertiary amines or certain metal hydroxides fulfill the requirements for catalytic components. However, their specific electrical resistance is relatively high. To minimize the electrical area resistance the thickness of the interphase or transition region between the oppositely charged layers should be as thin as possible, preferably less than 5 nm. In some commercial bipolar membranes heavy metal hydroxides are deposited in the interphase to catalyze the water dissociation [F. Hanada et al., 1993]. Tertiary ammonium groups as fixed charges at the surface of an anion-exchange membrane have the same effect and are therefore also used [B. Bauer et al., 1988].

A bipolar membrane with satisfactory properties, i.e. low electrical resistance at high current density, high water dissociation rates, low coion transport rate, high ionselectivities, good chemical and thermal stability in strong acids and bases can be prepared, e.g. as a laminate of anion- and cation-exchange layers.

An anion-selective layer with the required properties can be obtained by reaction of chloromethylated polysulfone dissolved in N-methyl-2- pyrrolidone with the mono-quaternary salt of 4,4'-diazabicyclo-[2.2.2]-octane (DABCO) to the following structure [H. Strathmann et al., 1993].

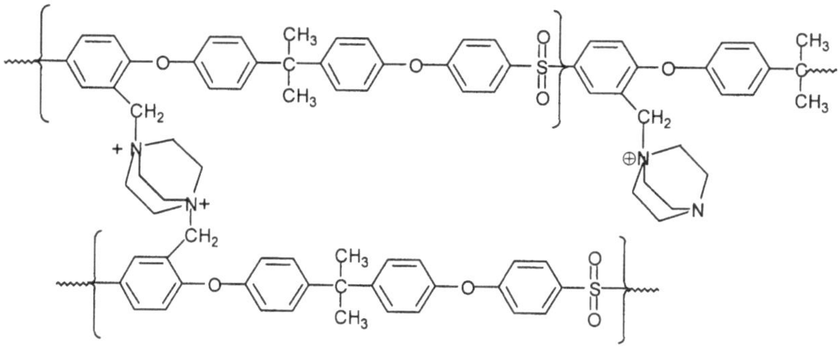

The cross-linking density can be adjusted by the ratio of DABCO to the chloromethylated polysulfone. The morphology of the anion-exchange layer is of a homogeneous type structure.

The cation-selective layer can be prepared by introducing sulfonic acid groups as fixed charges into a polyetheretherketone matrix using chlorosulfonic acid. The coion transport and the swelling behavior can be decreased by partial cross-linking which is done during the membrane formation step as indicated in the following scheme:

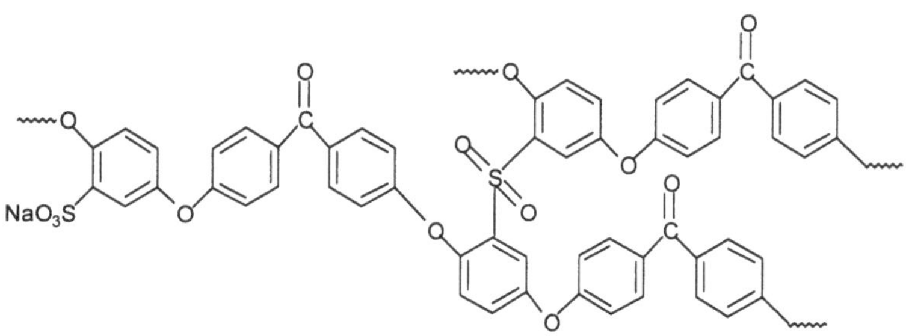

The above described bipolar membrane is composed of:

- a highly permselective anion-exchange layer which shows excellent alkaline stability,
- an equally highly permselective cation-exchange layer which exhibits excellent acid stability, and
- a transition region with a limited number of catalytically active tertiary ammonium groups.

This bipolar membrane shows most of the properties requested in practical applications where the cation-selective layer of the bipolar membrane is generally in contact with an acid solution and should thus be stable in strong acids. The anion-exchange layer is in contact with an alkaline solution and should thus be stable in strong bases.

The electrochemical properties of the ion-exchange layers prepared by the above described procedures are listed in Table 3.2. Properties of commercial bipolar membrane do not vary significantly from the data given in the table [R.B. Hogdon et al. 1989], [R. Simon, 1986, 1993], [F. Hanada et al., 1993]. However, there are also single sheet membranes prepared by plasma induced polymer grafting [Y. Yokoyama et al., 1989], [H. Hurwitz et al.,1998].

Table 3.2 Electrochemical properties of the cation- and anion-selective layers of a bipolar membrane prepared by the above described technique

membrane properties	anion-exchange layer	cation-exchange layer
ion-exchange capacity [meq g^{-1}]	1.2	1
membrane thickness [μm]	60	60
area resistance [Ωcm^2]	1.05	1.31
permselectivity [%]	97.5	98.5
swelling [%]	8	12.5

2.3.7 Ion-Exchange Membranes for Diffusion Dialysis

Since anion-exchange membranes are in general quite permeable for protons acids can be recovered from a mixture with salts by diffusion dialysis. The principle of the process was described earlier. The driving force for the transport of protons is the concentration of the acid in a feed solution and a stripping solution. For economical reasons high acid recovery rates are required. Thus, the concentration difference between the stripping and the feed solution is relatively small and the transport rates are low. To obtain acceptable fluxes the anion-exchange membranes should be as thin as possible and have very low salt ion leakage. Thin and mechanically strong anion-exchange membranes are difficult to prepare from copolymers cross-linked by divinylbenzene. Thin anion-exchange membranes with satisfactory mechanical strength and high proton fluxes can be prepared from block type aromatic condensation polymers.

Accordingly, hydroxide ions easily permeate through cation-exchange membranes. Thus, bases such as sodium hydroxide can be recovered from a mixture with salts by diffusion dialysis. The cation-exchange membranes developed for this process are relatively thin and they are stable in a strong alkaline environment. The main application of diffusion dialysis is the recovery of sodium hydroxide from aluminum etching baths. Since the aluminate ion is relatively bulky, leakage through the membrane is relatively small even in a loosely structured anion-exchange membrane.

There are many more special property membranes developed with a very specific application in mind.

2.3.8 Other Special Property Ion-Exchange Membranes

Large effort is concentrated to develop ion-exchange membranes as separators for fuel cells and especially the methanol direct conversion fuel cell [J. Kerres, 2000]. In addition to low electrical resistance membranes used in fuel cells must have good chemical stability at elevated temperatures up to 150°C and well controlled water and eventual methanol permeability. Today mainly fluorocarbon cation-exchange membranes of the Nafion® or Flemion® type are used. But new structures with better temperature stability and lower methanol leakage are under development [J. Kerres et al., 2001]. Also, certain electro-organic syntheses require special property membranes with satisfactory electrical properties and good chemical stability in the reaction environment.

Mosaic membranes which are composed of macroscopic domains of cation- and anion-exchange resins separated by a neutral polymer were developed for the concentration of salt solutions by a process referred to as piezodialysis. In this process a hydrostatic pressure is applied to transport salt solutions through a mosaic membrane. Since both cations and anions are simultaneously adsorbed by the cation- and anion-exchange domains of the membrane an electric circuit is formed which allows the salt to pass the membrane. As long as the fixed charge density of the membrane is higher than the ion concentration in the feed solution salt will preferentially permeate the membrane under a hydrostatic pressure driving force and will thus be concentrated in the permeate. Mosaic membranes can be prepared by distributing cation- and anion-exchange particles of sub-micron size in a neutral hydrophilic polymer film. Alternatively, mosaic membranes can be prepared by a film casting method using a block polymer composed of a part in which cation-exchange groups and a part where anion-exchange groups can be introduced [Y. Miyamoto et al., 1991]. However, today mosaic membranes seem to be of little technical and commercial relevance.

3 CHARACTERIZATION OF ION-EXCHANGE MEMBRANES

Characterization of ion-exchange membranes is closely related to their preparation. The most interesting properties of ion-exchange membranes are [K.S. Spiegler, 1958]:

- the electrical resistance in different electrolyte solutions,
- the type and density of fixed charges and their distribution in the membrane matrix,
- the permselectivity of the membrane for different ions of the opposite and same polarity,
- the transport rate of neutral components, especially water, under operating conditions, i. e. under an applied current,
- the mechanical stability at ambient and elevated temperatures and the swelling behavior in different electrolyte solutions,
- the chemical stability in the presence of oxidizing agents and at extreme pH-values.

A series of procedures and techniques are applied to determine the different membrane properties. Microscopic examination yields information on heterogeneity of the membrane structure and in case of reinforced membranes the type of reinforcement used. The electrical charge of an ion-exchange membrane can be determined qualitatively by using indicator solutions. A drop of a 0.05% solution of methyleneblue and methylorange on a membrane sample stains a yellow spot on top of an anion-exchange membrane and a deep blue spot on top of a cation-exchange membrane, respectively. For a quantitative characterization more complex procedures are applied such as the determination of the ion-exchange capacity of the membrane, its electrical resistance and the membrane transport properties for different ions and neutral components under various conditions which are close to the operating conditions in a specific application. Additional information can be obtained by impedance spectroscopy and transition time measurements as well as from the determination of osmotic and electro-osmotic water transfer. Bipolar membranes are further characterized in terms of the water dissociation capability and salt leakage under operating conditions [H. Strathmann, 1995].

3.1 The Structure of Ion-Exchange Membranes and its Characterization

The properties of ion-exchange membranes are closely related to their structures on a microscopic scale. Most hydrocarbon polymer membranes have a rather regular distribution of cross-linkings and ionic groups attached to a back bone of aromatic rings. As a result these membranes are quite homogeneous in their structure even on a microscopic scale. Membranes based on linear fluorocarbon polymers have a backbone of a highly crystallized polymer with ionic groups fixed on a side chain. These membranes are quite heterogeneous on a microscopic scale with distinct crystalline areas formed by the linear fluorocarbon polymer and separated by amorphous, highly hydrophilic domains formed by the side chains with the ionic groups. The structure of the fluorocarbon polymers based membrane have been studied in great detail using transmission electron microscopy, wide and small angle X-ray diffraction [T.H. Gierke et al., 1982], differential scanning calorimetry, infrared and nuclear magnetic resonance spectroscopy [M. Falk, 1982] and [R.A. Komoroski et al., 1982]. From the experimental evidence it can be concluded that the fluorocarbon polymer membranes have a heterogeneous two phase structure composed of crystalline polytetrafluorethylene and amorphous vinylether with fixed charges as depicted in Figure 3.5.

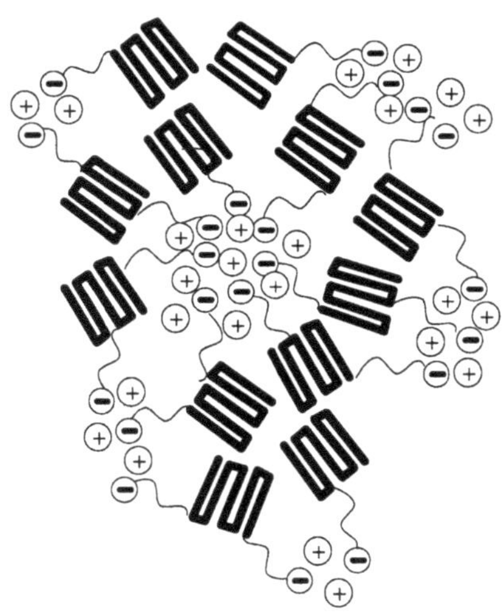

Fig. 3.5 Schematic drawing illustrating the structure of the fluorocarbon polymers based membranes showing the crystalline polytetrafluorethylene region and the negatively fixed charges on the vinylether linkages.

The amorphous region containing a cluster of fixed charges, counterions, and water is extremely hydrophilic and swells in water or aqueous solutions while the crystalline region acts as cross-linkage and restricts the swelling of the membrane. In the literature the hydrophilic clusters are pictured as spherical regions with a diameter of ca. 4 to 6 nm connected by "bottle necks" between crystalline regions of ca. 1 nm in diameter as depicted in Figure 3.6. The mobility of the counter-ions within the hydrophilic cluster is very high. The transport of the counter-ions from one cluster to the next, however, is restricted by the "bottle neck". The size of the hydrophilic cluster is a function of the swelling of the membrane and depends on the concentration of the electrolyte solution in equilibrium with the membrane.

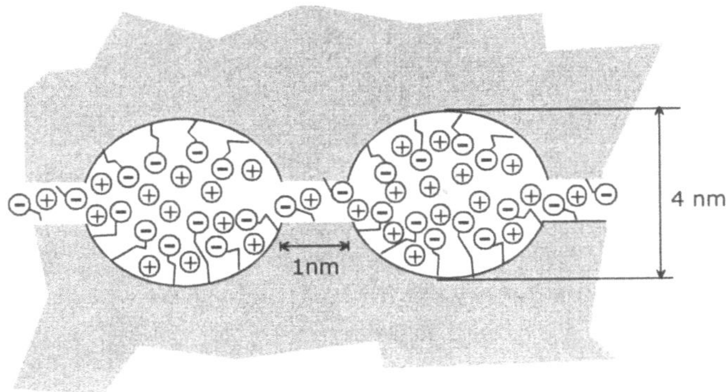

Fig. 3.6 Schematic picture illustrating the cluster network of a fluorocarbon polymer based ion-exchange membrane.

3.2 Mechanical Stability, Swelling, and Permeability of Membranes

A detailed mechanical characterization of ion-exchange membranes involves the determination of thickness, swelling, and dimensional stability in water or different test solutions, tensile strength and hydraulic permeability measurements. All mechanical characterization tests should be carried out with pretreated and well equilibrated membranes in a controlled environment. Relevant information related to the storage and handling requirements of the membranes is obtained from the determination of the dimensional changes between wet and dry states of the membrane and by tensile strength measurements. The tensile strength and information concerning the plastic or elastic deformation of a membrane is obtained from a stress versus strain diagram as shown schematically in Figure 3.7.

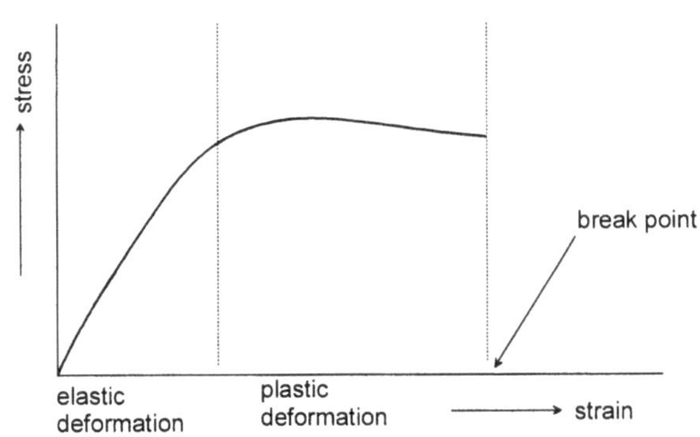

Fig. 3.7 Stress versus strain diagram of an ion-exchange membrane sample indicating the elastic and plastic deformation and the breakpoint.

The strain versus stress curves of ion-exchange membranes generally show three distinct areas. At relatively low strain the membranes show elastic deformation, with increasing strain the membranes show plastic deformation, and at a certain point they break. The strain versus stress curves can be measured with reinforced or non reinforced membranes. However, in reinforced membranes the reinforcing material is generally determining the tensile strength and very little information is obtained about the mechanical properties of the ion-exchange membrane material. Since the mechanical properties generally change drastically with the water content of the membranes they must be determined with dry membranes and with membranes equilibrated in water or different solutions similar to those used in practical applications.

3.2.1 The Membrane Water Content

The gel water content, i.e. swelling of a membrane not only determines its mechanical properties and its dimensional stability, it also effects its ion permselectivity, its electrical resistance, and its hydraulic permeability.

The swelling of a membrane depends on a number of different parameters such as the nature of the basic membrane polymer, the nature of the ion-exchange groups and their concentration in the membrane, the counter ions, the cross-linking density and the homogeneity of the membrane. The water content of an ion-exchange membrane depends not only on the properties of the membrane but also on the composition of the solution with which the membrane is in contact. Especially the concentration of the solution has a significant effect on the water content of the membrane because of osmotic effects that are directly related to the chemical potential difference of the water in the membrane and in the solution as has been shown earlier in the discussion of the electrochemical and chemical equilibrium of a membrane with its surrounding. The state of the water in the membrane can also be rather different. Part of the water is so-called "free" water. Another part of the water is strongly bound within the hydration shell of the counterions and the fixed charges of the membrane, while again another part is more loosely bound to the basic polymer matrix. The different water structures in ion-exchange membranes is extensively studied applying differential scanning calorimetry, infrared spectroscopy, and nuclear magnetic resonance. Most of these studies are carried out on perfluorinated ionomer membranes [M. Escoubes et al., 1982], [M. Falk, 1982], and [R.A. Komoroski et al., 1982]. These membranes are rather heterogeneous in structure with distinct, highly hydrophobic crystalline domains and pores filled with the electrolyte solution while apparently the water in these membranes shows very different structures. In sulfonated polystyrene or polysulfone the absorbed water is more homogeneously distributed in the membrane matrix and more homogeneous in its structure.

Studies of the water structure within the membrane and its interaction with the membrane polymer, fixed charges, and counter-ions require more sophisticated spectroscopic measurements. The total water uptake of the membrane in equilibrium with an electrolyte solution can be determined by measuring the weight difference between a membrane in the wet and dry state. To determine the water content of a membrane, a sample is equilibrated in deionized water or any other test solution. After removing the surface water from the sample, the wet weight of the swollen membrane is determined. The sample is then dried at elevated temperature over phosphorous pentoxide under reduced pressure until a constant weight is obtained. The water content of a membrane is obtained in weight percent by:

$$\text{wt\% swelling} = \frac{W_{wet} - W_{dry}}{W_{wet}} \times 100 \qquad (3.1)$$

Here W_{wet} and W_{dry} are the weight of a membrane sample in the wet and the dry state.

Determining the membrane water sorption by mass balance seems quite simple. However, some precautions must be taken to achieve satisfactory measurement accuracy. For example, surface water on a wet membrane must be removed without loss of water from the membrane bulk phase. This can be difficult with very thin membranes. During the drying procedure all water must be removed. This often requires drying for an extended time period at elevated temperature in vacuum which may lead to some irreversible structural changes of the membrane. In comparing commercial membranes from literature references it is mandatory to know the exact experimental procedure used in determining the membrane water sorption. The equilibrium water sorption of various commercial ion-exchange membranes are listed in Table 3.3.

Table 3.3 Water sorption of various commercial membranes in equilibrium with NaCl- solutions of different concentrations

membrane	solution [NaCl mol/L]	water sorption [g H_2O/ g dry membrane]	water sorption [mol /fixed ions]	reference
Nafion 117 [1)]	0.2	0.14	11.9	H.L. Yaeger, 1982
Flemion AR 1.25 [2)]	1.0	0.23	10.6	M. Tasaka, 1992
Neosepta CL 25T [2)]	0.5	0.34	10.5	T. Sata, 1986
Neosepta C 66 5T [2)]	0.5	0.42	9.5	T. Sata, 1986
Neosepta AM 1 [2)]	0.5	0.30	8.3	T. A. Davis, 1997
Neosepta ANR 1 [2)]	0.5	0.23	6.1	T. A. Davis, 1997
Ionics CR 67 [3)]	1.0	0.46	11.7	Ionics brochure

1) produced by Dupont, 2) produced by Tokuyama, 3) produced by Ionics

The fixed charges of the membranes listed in Table 3.3 are strong acidic ($-SO_3^-$-groups) in the cation-exchange membranes and strong basic ($-NR_3^+$-groups) in the anion-exchange membranes. Membranes containing weak acidic or weak basic groups generally show significantly lower water sorption. The water sorption of perfluorinated ion-exchange membranes with carboxyl groups as fixed charges is about half of that of membranes with sulfonic acid groups.

The water sorption depends also on the type and concentration of the counterions and is generally decreasing with the size of the hydrated counterion. Thus, for a cation-exchange membrane the water sorption is decreasing along the series $Li^+ > Na^+ > K^+ > Mg^{++} > Ca^{++}$.

3.2.2 Hydraulic Permeability of Ion-Exchange Membranes

Hydraulic permeability measurements provide information on the diffusive or convective transport of components through a membrane under a hydrostatic pressure driving force. In most practical applications of electrodialysis and related processes the hydraulic permeability plays a minor role for the overall performance of the membrane in a given application since the processes are generally operated without pressure gradients across the membrane. However, for the determination of pinholes in ion-exchange membranes which will not only obscure the membrane permeability for different components but will also invalidate most applications where hydrostatic pressure differences across the membrane are obtained hydraulic permeability tests are important as a quality control step. Larger pinholes can also be detected quickly by placing a wet membrane sheet on a sheet of white absorbent paper. Then a 0.2% solution of methylene blue for an anion-exchange membrane or a 0.2% solution of erythrocein-B for a cation-exchange membrane is spread over the entire surface of the membrane. If no spots of the dye can be observed on the paper, the membrane is free of large pinholes. The hydraulic permeability of the membrane is determined at room temperature using deionized water and a hydrostatic pressure driving force in a conventional filtration cell as used in reverse osmosis or ultrafitration experiments. The permeability can then be calculated from the volumetric flow rate.

3.2.3 Long-Term Chemical Stability of Membranes

The economics of processes using ion-exchange membranes in different applications is determined to a large extent by the chemical stability of the membranes under process conditions. The structure, the mechanical and electrochemical properties of ion-exchange membranes can be effected by oxidizing agents, high or low pH-values, extreme temperatures, loss of water, etc.. The structure and the mechanical properties of the membranes can be effected by degradation of the basic polymer or re-crystallization and aging due to loss of water or plasticizing agents. The deterioration of the electrical chemical properties can be caused by the loss of fixed charges due to the splitting-off functional ion-exchange groups. Also changes in the membrane structure and the irreversible loss of water effect the membrane ion transport properties and selectivity. Generally, the changes in membrane properties are time and temperature dependent. Reliable information about the long term stability of the membranes is of great importance for a process design and cost calculation. For a fast evaluation membrane samples are exposed to oxidizing agents or acids and bases and other chemical components in much higher concentration and at higher temperature than expected in practical use. Structural mechanical changes are determined by visual or microscopic investigation and standard mechanical and electrochemical tests comparing exposed membrane samples with new unexposed samples. Since the useful life of membranes under process conditions should be as long as possible and is generally in the range of a couple of years membrane stability tests are very time consuming. However, some information concerning chemical and mechanical stability under operating condition may be obtained in short term tests carried out at elevated temperatures that are well above the anticipated operating temperature. Nevertheless, in many cases long-term pilot plant tests are mandatory to get a reliable estimation of the life of the membranes in practical applications.

3.3 Determination of Electrochemical Properties of Membranes

The performance of ion-exchange membranes in the various electromembrane processes is determined to a very large extent by their electrochemical properties. Therefore, a major task in characterizing ion-exchange membranes is the determination of their electrochemical properties such as the fixed charge density, the electrical resistance, the ion permselectivity, and the transport of non-ionic components such as water or of other neutral molecules.

3.3.1 The Ion-Exchange Capacity of a Membrane

The ion-exchange capacity of charged membranes is a crucial parameter which effects almost all other membrane properties. Its determination is therefore an important step in characterizing an ion-exchange membrane. The ion-exchange capacity is a measure of the number of fixed charges per unit weight of dry polymer. It is usually expressed in milli-moles per gram which in almost all ion-exchange membranes is identical with milli-equivalents per gram dry membrane since in ion-exchange membranes the fixed charges are mono-valent. Experimentally the ion-exchange capacity of a membrane is readily determined by titration of the fixed ions with 1N NaOH or HCl, respectively. For these tests, cation- and anion-exchange membranes are equilibrated for about 24 h in 1N HCl or 1N NaOH, respectively, and then rinsed free from chloride or sodium for 24 h with deionized water. The ion-exchange capacity of the samples is then determined by back titration with 1N NaOH or 1N HCl, respectively. Weak base anion-exchange membranes are characterized by equilibration in 1N sodium chloride and titration with standardized 0.1 N silver nitrate solution. The samples are then dried, and the ion-exchange capacity is calculated for the dry membrane. The accuracy of the measurement depends on the complete exchange of ions in the membrane which can take some time. In most commercial membranes complete ion-exchange is achieved within a couple of minutes. The concentration of the fixed ion in commercial membranes is in the range between 1 and 3 milli-equivalent per gram. The experimentally measured ion-exchange capacity of a membrane is an average value assuming that the ions are homogeneously distributed over the entire polymer matrix. In many membranes such as most of the perfluorenated cation-exchange membranes the fixed ions are clustered and the local fixed ion concentration may be significantly higher than the average value.

3.3.2 Electrical Resistance of Ion-Exchange Membranes

The electrical resistance of ion-exchange membranes is one of the factors which determine the energy requirements of electrodialysis processes. However, in most practical cases the membrane resistance is considerably lower than the resistance of the dilute solutions, since the ion concentration in the membrane is relatively high. The specific membrane resistance is usually reported as Ω cm or Ω m. From the engineering point of view, the membrane area resistance in units of Ω cm^2 or Ω m^2 is more useful and generally given in the literature describing commercial products. The electrical resistance of a membrane is determined by the ion-exchange capacity and the mobility of the ion within the membrane matrix. The ion mobility in the membrane depends strongly on the nature of the mobile ion species, i.e. its valence, its size, and its extent of hydration, and on the water content of the membrane. It is furthermore effected by the temperature and by the nature of the fixed ions, i.e. the interaction between the mobile ions with the fixed ions. The area resistance of ion-exchange membranes can be determined by direct current (DC) measurements or by alternating current (AC) measurements. In DC measurements the membrane is installed in a cell which consists of two chambers containing the test solution separated by the test membrane as indicated in Figure 3.8.

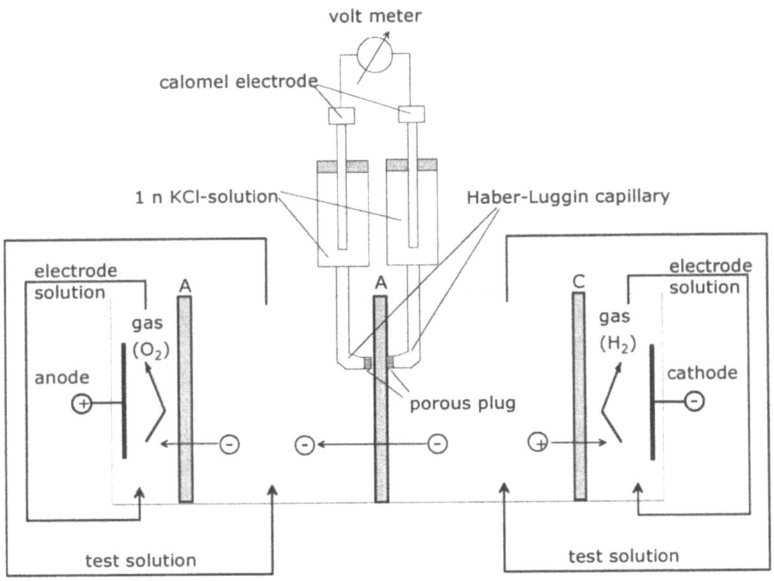

Fig. 3.8 Schematic drawing of a test cell for determining the resistance of ion-exchange membranes with direct current.

The two electrodes used to provide the electrical potential driving force to obtain a certain current are separated by membranes from the test solution to avoid gas bubbles to disturb the measurement. As test solution a 1 or 0.5 mol/L Na_2SO_4 solution is used. The same solution is also used to rinse the electrode chambers and remove the gases produced at the electrodes. The actual potential drop in DC measurements across the membrane is determined with calomel electrodes attached to Haber-Luggin capillaries placed with their tips close to the membrane surface. The potential drop between the Haber-Luggin capillaries is measured with and without the membrane in the test cell as a function of the current density passing through the cell. The resistance is given by the slope of the current versus the voltage drop curve. To obtain the membrane resistance the resistance of the cell without the membrane is subtracted from the resistance of the cell with the membrane.

The area resistance is given by:

$$r_{m+s} = R\,A_m = \frac{U}{i} \tag{3.2}$$

The membrane resistance is:

$$r_m = r_{m+s} - r_s \tag{3.3}$$

Here R is the resistance, A_m is the area of the membrane, U is the voltage drop measured between the Haber-Luggin capillaries and I is the current density, r_{m+s} and r_s are the area resistances of the cell with and without the membrane between the Haber-Luggin capillaries, and r_m is the area resistance of the membrane.

Typical test results obtained in a test cell of the above described design are shown in Figures 3.9 a) and b).

The experiments were carried out at 25 °C with a 1mol/L Na_2SO_4 test solution and a none reinforced prototype anion-exchange membrane based on functionalized polysulfone. Figure a) shows the test results obtained by measuring the area resistance between the Haber-Luggin capillaries with and without the test membrane as a function of the current density. Figure b) shows the area resistance of the membrane determined according to Equation 3.6.

a)

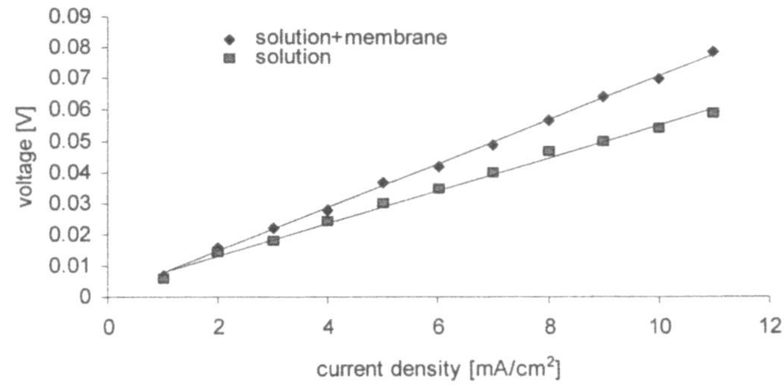

b)

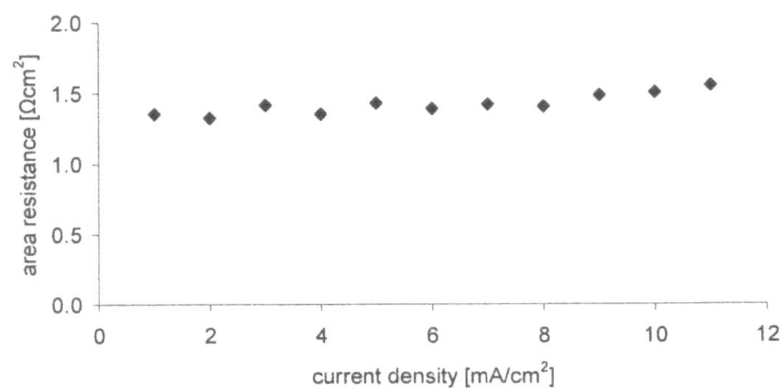

Fig. 3.9 Experimental results obtained with a prototype of non-reinforced anion-exchange membrane a) showing the voltage drop of the cell with a 1mol/L Na_2SO_4 solution at 25°C as function of the current density with a membrane and without a membrane, b) showing the area resistance as function of the current density calculated by Equation 3.3 from the experimental results.

The determination of the membrane resistance by direct current measurements is quite accurate as long as the position of the Haber-Luggin capillaries is as close as possible to the membrane and kept identical in the test with and without the membrane. To avoid boundary layer effects the bulk solution must be well mixed. But even if the boundary layer effects are well controlled the direct current measurements are always connected with a transport of ions resulting in a concentration change in the electrolyte solution at the membrane surface which can effect the accuracy of the measurement.

Concentration changes are avoided in resistance measurements with alternating current. Therefore, membranes have also been characterized in terms of their resistance by alternating current measurements. A typical test cell used for these tests is shown in Figure 3.10. Also in the case of alternating current the membrane resistance is determined from resistance measurements in a cell with and without the membrane. The resistance of the membrane can be calculated from the following relation:

The area resistance r_m is related to the specific resistance by:

$$r_m = \rho_{m+s}(d_m + d) - \rho_s d \tag{3.4}$$

Here r_m, are the area resistances of the membrane, ρ_{m+s} and ρ_s are the specific resistances of the cell with and without the membrane, d_m and d are the thickness of the membrane and the distance between the electrodes.

The specific resistance ρ is:

$$\rho = R\frac{A_m}{d} \tag{3.5}$$

Here A_m is the cross-sectional area of the cell as indicated in Figure 3.10 and R is the resistance measured between the electrodes.

For an accurate determination of the membrane resistance it is important that the difference between the measurement of the cell resistance with and without a membrane is as large as possible. Since the specific resistances of the membrane and the electrolyte solution are in the same order of magnitude and the membrane thickness d_m is small, the distance d has to be as small as possible. In reliable tests d should be smaller than 20 times d_m.

To compare the direct and alternating current membrane characterization a membrane which was first characterized with the alternating current was then characterized using direct current. The measurements were carried out in a 1 mol/L Na_2SO_4 solution in a cell with a distance d of 0.2 cm. The thickness of the membrane was 0.015 cm. The specific resistance that was measured in the cell with the membrane at 25°C was 6.62 Ωcm and that without the membrane was 5.75 Ω cm. The area resistance according to Equation 3.4 is then 0.273 Ωcm^2. The area resistance of the same membrane measured with direct current was 1.44 Ωcm^2. There is obviously a significant difference between the results obtained in the direct current and the results obtained in the alternating current membrane resistance measurements. The alternating current measurements give a much lower membrane resistance than the direct current measurements. This is a general finding. However, the magnitude of the difference in the results depends on the membrane structure. In very heterogeneous membranes the difference is larger than in more homogeneous membrane structures. The reason for the different results is that in direct current measurements the ions are physically transported through the membrane matrix, whereas in alternating current measurements the ions are only oscillating around their locations. Assuming the structure of an ion-exchange membrane consists on a microscopic scale of a solid neutral polymer phase and electrolyte filled pores, then the mobility of the ions in the pore liquid is high. But their transport through the membrane is hindered by the solid polymer structure which acts as a bottle neck between pores. This type of structure has been depicted in Figure 3.5 which shows a schematic drawing of a typical fluorocarbon membrane. In alternating current measurements the transport of the ions on a very small scale, i.e. in a pore is measured. In direct current measurements the transport of the ions is measured on a macroscopic scale which includes also the passage through a bottle neck between electrolyte filled pores. The direct current and alternating current difference is treated in detail in the literature [V.I. Zabolotsky et al., 1993].

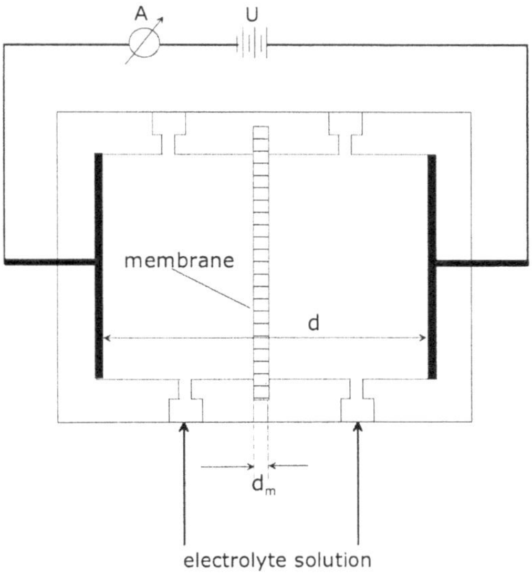

Fig. 3.10 Schematic drawing of a test cell used for membrane resistance measurements using alternating current.

In recent years, a rather elegant and reliable method of measuring resistances with alternating current is widely used. This method is based on impedance spectroscopy. The difference between the alternating current resistance measurements and the impedance spectroscopy is, that in the first case the frequency of the alternating current is kept constant while in impedance spectroscopy the frequency of the alternating current is changed and the response to the changing frequency is determined by a spectrometer.

Ion-exchange membranes display a variety of electrical properties that suggest that they can be modeled as a combination of electrical analog components. For example, an applied potential can produce a current flow through the membrane and this is indication that the membrane has a finite resistance. In addition, the charged groups at the membrane/solution interface provides sites for the redistribution of charges at the membrane surface when the transmembrane potential is changed. This charge redistribution is similar to that observed for a capacitor in which a potential change produces a change in the net charges. The electrolyte solutions and the electrodes on both sides of the membrane also have a certain resistance. The properties of the ion-exchange membrane, the solution, and the electrodes can be described electrically by an equivalent circuit as indicated in Figure 3.11, where C_m and R_m are the membrane capacitance and resistance, respectively, and R_s is the resistance of electrodes and the electrolyte solution.

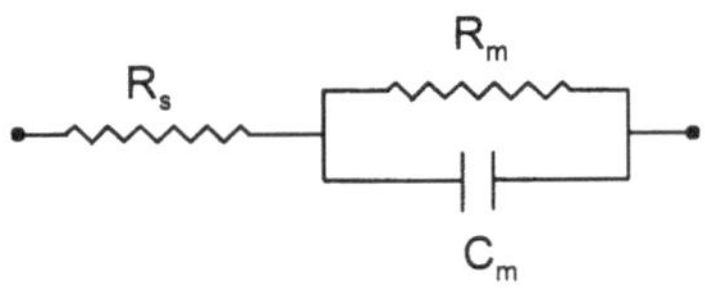

Fig. 3.11 Equivalent resistance-capacitance circuit for a simple membrane in an electrolyte solution neglecting electrode resistances and capacities.

To determine membrane resistances by impedance spectroscopy the entire system, i.e. the membrane, the electrolyte, and the electrodes are treated as a "black box". An alternating sinusoidal voltage of a given frequency and amplitude is applied to the system. The resulting current is measured, and then the phase shift compared to the input signal is determined. The procedure is repeated at different frequencies. An experimental set-up using the black box approach for determining the electrical resistance of ion-exchange membranes is illustrated in Figure 3.12.

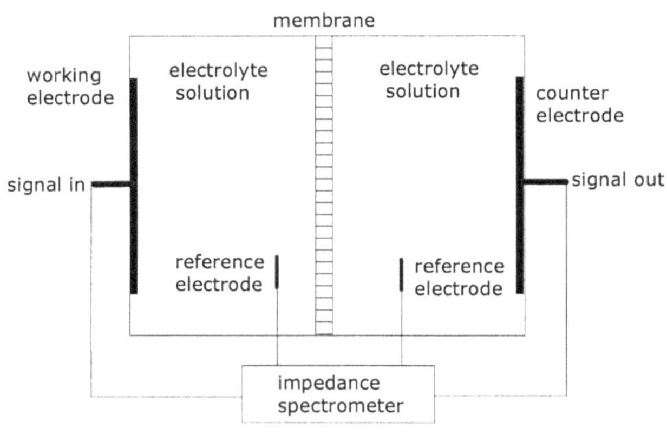

Fig. 3.12 Experimental setup using the "black box" approach for measuring the impedance of a system composed of a membrane, an electrolyte, and electrodes.

The figure indicates the signal input through the working electrodes and the output via the counter electrodes. In certain cases it is advisable to use reference electrodes to exclude electrode reactions in the "black box" measurement system.

The impedance is defined in analogy to Ohm's law to:

$$Z_{(\omega)} = \frac{U_{(\omega)}}{I_{(\omega)}} \tag{3.6}$$

with:

$$U_{(\omega)} = U_o \cos \omega t = U_o\, e^{i\omega t} \tag{3.7}$$

and

$$I_{(\omega)} = I_o \sin \omega t + \varphi = I_o\, e^{i(\omega t + \varphi)} \tag{3.8}$$

Here Z is the impedance, U is the voltage and ω is the circular velocity which is also referred to as circular frequency of the alternating current, t is the time, φ is the phase shift between voltage and current, and the subscript o refers to the amplitude of voltage and current, and (ω) refers to a certain circular velocity. The circular velocity ω is a function of the frequency ν, and is given by:

$$\omega = 2\,\pi\nu \tag{3.9}$$

Furthermore is:

$$Z_{total} = Z_1 + Z_2 + \ldots.. Z_n \tag{3.10}$$

for impedances in series,

and

$$\frac{1}{Z_{total}} = \frac{1}{Z_1} + \frac{1}{Z_2} + \ldots\ldots \frac{1}{Z_n} \tag{3.11}$$

for impedances in parallel.

The impedance is given by the relation between the voltage and current:

$$Z_{(\omega)} = \frac{U_{(\omega)}}{I_{(\omega)}} = Z\ e^{-i\varphi} = Z\ \cos\varphi - i\,Z\ \sin\varphi \tag{3.12}$$

Equation 3.12 indicates that the impedance is composed of two parts, i.e. the real part given by Z cos φ and the imaginary part given by iZ sin φ.

The impedance related to an electric resistance and to a capacitance are different. For an electric resistance the imaginary part of the impedance is zero since the current and voltage are in phase and the real part is frequency independent.

$$Z = \frac{U_o}{I_o} = R \tag{3.13}$$

For a capacitance the impedance is given by:

$$Z = \frac{1}{i\,\omega\, C} \tag{3.14}$$

A capacitor is characterized by the phase shift between current and voltage with a phase angle of $\pi/2$. Thus the impedance of a capacitance is frequency dependent and imaginary.

To obtain information about the dynamic behavior of a system two different graphic illustrations are used. In one graph the two different parts of the impedance, i.e. the real and the imaginary part are shown as a function of the frequency. This graph is referred to as the complex plain plot, or the Argand-diagram. In the second graph, the so-called Bode-diagram, the impedance and the phase angle are shown as a function of the frequency.

A typical complex plain plot obtained for a system containing resistances and capacitance in series and in parallel are shown in Figures 3.13 a) and b).

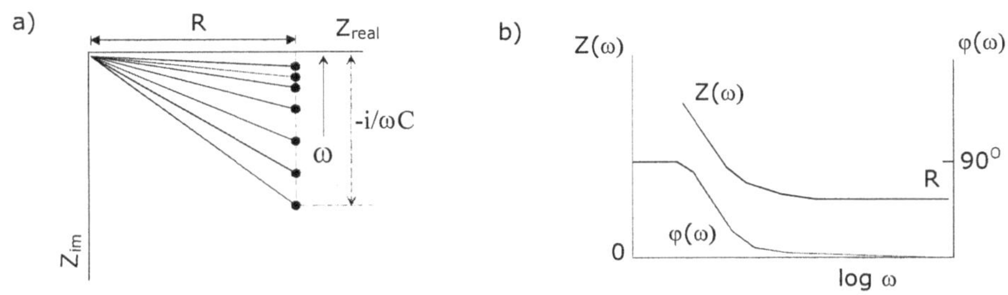

Fig. 3.13 Results of the impedance measurements of a system with resistance and capacitance in series shown schematically a) in a complex plain plot and b) in the Bode diagram.

The total impedance for a resistance and a capacitance in series is:

$$Z_{(\omega)} = R + \frac{1}{i\,\omega\,C} \qquad (3.15)$$

According to Equation 3.15 and the graphs a) and b) in Figure 3.13 the imaginary part will disappear at very high frequencies if resistance and capacitance are in series and the impedance is identical to the ohmic resistance. At very low frequencies the impedance of the capacitance increases with decreasing frequency and becomes infinitely high in direct current.

For resistance and capacitance in parallel the total impedance is given by:

$$Z_{(\omega)} = \frac{R}{1+\omega^2R^2C^2} - i\frac{\omega\,R^2C}{1+\omega^2R^2C^2} \qquad (3.16)$$

The results are shown in Figure 3.14 a) and b).

According to Equation 3.13 and the graphs in Figure 3.14 a) and b) the imaginary part of the impedance disappears at very low and very high frequencies. At very low frequency all current is passing through the resistance and the impedance is identical to the electric resistance. At very high frequencies both the imaginary and the real part of the impedance approach zero. With increasing frequency the phase shift between current and voltage is increasing and with the phase shift the imaginary part of the impedance is increasing and reaches a maximum at a frequency of $\omega=1/RC$. A further increase in frequency will increase the phase shift but will decrease the imaginary part of the impedance. This is indicated by the semi-circular curve in the diagram of Figure 3.14 a).

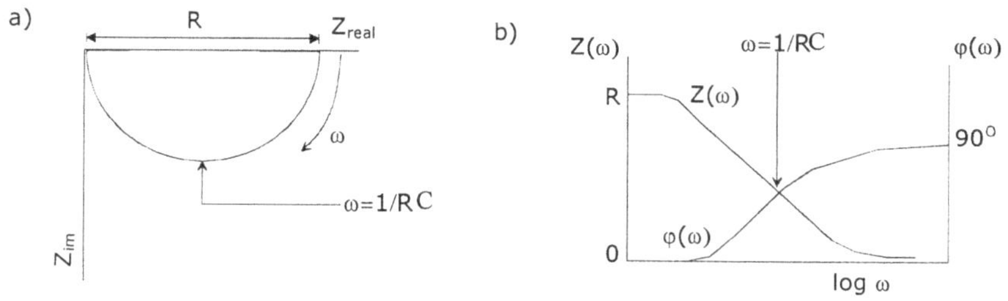

Fig. 3.14 Results of the impedance measurements of a system with resistance and capacitance in parallel shown schematically a) in the complex plain plot and b) in the Bode diagram.

In a system composed of an ion-exchange membrane, an electrolyte, and two electrodes there may be both resistances and capacitances in series and in parallel and the impedance can be rather complex and it is not always easy to determine the membrane resistance from the obtained diagram, and mathematical models may be used to obtain reliable data for the membrane resistance.

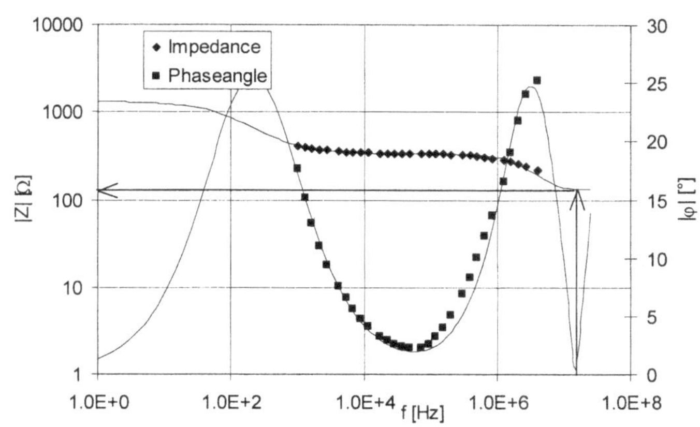

Fig. 3.15 Bode diagram of an impedance spectrum determined in a 0.5 mol/L sodium acetate solution with a CM-1 cation-exchange membrane manufactured by Tokuyama Soda.

Figure 3.15 shows the Bode plot of an impedance spectrum determined in a 0.5 mol/L sodium acetate solution with a CM-1 cation-exchange membrane manufactured by Tokuyama Soda. At a frequency of 815 kHz the phase angel approaches zero and the resistance is 118 Ω.

The practical determination of the impedance spectra can be carried out in the test cell shown in Figure 3.16 following the same procedure as in measurements with alternating current described earlier.

Generally, the measurements are carried out with and without a membrane in the test cell. The membrane resistance is calculated from the difference of the impedance determined with and without the membrane. Often, the specific resistance of the membrane is of the same order as that of the electrolyte solution. Therefore, the difference in the resistance of the cell with and without membrane is small compared to the overall resistance. Thus, for the relatively thin membrane the resistance must be determined from a small difference between two large numbers. This effects the accuracy of the calculated membrane resistance. To obtain more reliable data a stack of membranes is placed in the cell and the resistance of a single membrane is determined by dividing the resistance of the membrane stack by the number of membranes. This method results in more reliable data for membranes with smooth surfaces. It is not suited for membranes with very rough surfaces such as membranes reinforced by a screen. To avoid the problems related to determining the membrane resistance as a small difference between to large numbers from measurements in electrolyte solutions the test membrane is placed between two contact membranes and the working electrodes which also serve as reference electrodes placed directly on the surface of the membrane assembly as in Figure 3.16. The figure shows the cross-section of a test cell. The contact membranes should be homogeneous and unsupported and should have a relatively soft surface to provide good contact between the electrodes and the test membrane. The test membrane and the two contact membranes must have the same electrical charges and must be equilibrated in the test solution. The impedance measurement is carried out with the test membrane and without the test membrane and the resistance of the test membrane is obtained as the difference of the resistance measurements.

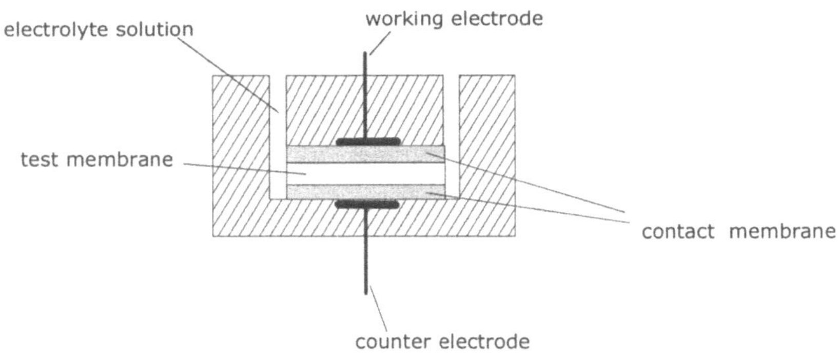

Fig. 3.16 Schematic drawing of the test cell cross-section used for the impedance measurements.

The determination of the membrane resistance by impedance measurements is a rather reliable and fast method which is especially suited for routine testing of homogeneous membranes having a smooth surface. In contrast to direct current measurements boundary layer effects caused by differences in the transport rate of ions in the membrane and in the solution at the membrane surface are excluded and thus do not effect the measurement. However, the membrane resistances determined with impedance measurements are generally significantly lower for heterogeneous and reinforced membranes than those measured in tests with direct current for reasons that have been explained earlier. As data base for the design of electrodialysis stacks the resistances determined in direct current measurements are better suited than those determined by impedance measurements.

3.3.3 Permselectivity of Ion-Exchange Membranes

The permselectivity of a membrane is determined by the ratio of the flux of specific components to the total mass flux through the membrane under a given driving force. In ion-exchange membranes the permselectivity is generally related to the transport of electric charges by the counterions to the total electrical current through the membrane. However, the transport rate of different components of the same charge in ion-exchange membranes can also be quite different because of kinetic parameters such as the size of the components and the structure of the membrane. The permselectivity of an ion-exchange membrane, i.e. its charge selectivity is determined by the concentrations of counter- and coions in the membrane, which again depends mainly on the ion-exchange capacity of the membrane and the ion concentration in the outside solutions because of the Donnan exclusion as discussed earlier. The permselectivity can be calculated from the transport or transference number of the counter- and coions in the membrane and the outside solutions.

The permselectivity of a membrane is given by Equation 2.174 to:

$$\Psi^m = \frac{T_{cou}^m - T_{cou}}{T_{co}} \tag{3.17}$$

The transport numbers are defined by Equation 2.102 to:

$$T_i = \frac{z_i\, J_i}{\sum_i z_i\, J_i} \tag{3.18}$$

Here Ψ is the permselectivity, T is the transport number, z is the valence, and J is the flux; the subscript i refers to cation or anion, the subscripts cou and co refer to counter- and coions and the superscript m refers to ion-exchange membrane. In a cation-exchange membrane the cation is the counterion and the anion the coion while in an anion-exchange membrane the anion is the counterion and the cation the coion.

An ideal permselective cation-exchange membrane, for instance, would be permeable for positively charged cations (counterions) only. The permselectivity of a membrane approaches zero when the transport numbers of the ions within the membrane are identical to those in the electrolytic solution.

To determine the transport number of ions in the membrane and the membrane permselectivity, respectively, two methods can be applied. The first method is based on a measurement of the increase in concentration of certain ions in the concentrate and the decrease in the diluate solution during electrodialysis of a test solution and by measuring the amount of current passing through the unit. From the current utilization the permselectivity of the membrane and the transport number of the counterion can be calculated. With this "dynamic" measurement water transport due to electroosmosis and osmosis is taken into account and the "true" membrane permselectivity is obtained. The dynamic method, however, is rather time consuming and effected by concentration polarization at the membrane surfaces facing the diluate and concentrate solutions.

A faster "static" method for the determination of membrane permselectivities, which is not effected by boundary layer transport phenomena, is based on the measurement of the potential gradient across a membrane which separates two electrolyte solutions of different concentrations. The static method, however, does not take the water transport through the membrane into account and is referred to as "apparent" permselectivity.

The potential between two electrolyte solutions of different concentrations, i.e. the membrane potential has been discussed earlier and it is illustrated in Figure 2.18 and consists of the two Donnan potentials between the membrane and the adjacent solutions and the diffusion potential across the membrane.

$$\varphi_m = \varphi_{dif} + \varphi^1_{Don} - \varphi^2_{Don} \tag{3.19}$$

Here φ is the potential, the subscripts m, dif, and Don refer to membrane, diffusion, and Donnan potential, and the superscripts 1 and 2 refer to the different solutions.

The diffusion potential is caused by the transport of co-ions, i.e. salt diffusion through the membrane. It is the result of the different permeability of the cation and anion in an ion-exchange membrane. For a completely permselective membrane the salt diffusion and thus, the diffusion potential, is zero. The membrane potential is then given by the difference in the two Donnan potentials. For a not strictly permselective membrane the diffusion potential must be added to the Donnan potential difference to determine the membrane potential. For a single, monovalent salt such as KCl as an electrolyte the membrane potential between two solutions of the same electrolyte but different concentration can be derived from the general flux equation described earlier with the boundary conditions of $\Delta P = 0$ and $I = 0$. By introducing several approximations, such as negligible osmotic flow between the two solutions, constant ion mobility, and small concentration gradients across the membrane phase etc., the potential difference between solutions can be expressed by:

$$\varphi_m = \left(T^m_{cou} - T^m_{co}\right)\frac{RT}{F}\ln\frac{a^1_s}{a^2_s} = \left(2T^m_{cou} - 1\right)\frac{RT}{F}\ln\frac{a^1_s}{a^2_s} \tag{3.20}$$

For a strictly permselective membrane the transport number of the counter-ion is 1, i.e. there is no salt diffusion through the membrane, and the membrane potential is given by:

$$\varphi_{m,sp} = \frac{RT}{F}\ln\frac{a^1_s}{a^2_s} \tag{3.21}$$

Here are φ_m and $\varphi_{m,sp}$ the membrane potentials of a real and a strictly permselective ion-exchange membrane, T^m_{cou} and T^m_{co} are the transport numbers of the counter- and coions in the membrane, a_s is the activity of the salt solution, R is the gas constant, T is the temperature, F is the Faraday constant and the superscripts 1 and 2 refer to the two solutions separated by the membrane.

The counterion transport number in a not strictly permselective membrane is obtained as follows:

$$\frac{\varphi_m}{\varphi_{m,sp}} = 2T^m_{cou} - 1 \tag{3.22}$$

and

$$T^m_{cou} = \frac{\dfrac{\varphi_m}{\varphi_{m,sp}} + 1}{2} \tag{3.23}$$

Introducing Equation 3.23 into Equation 3.17 and rearranging gives the permselectivity of a cation-exchange membrane as a function of the membrane potential:

$$\Psi^m = \frac{\dfrac{\varphi_m}{\varphi_{m,sp}} + 1 - 2T_{cou}}{2T_{co}} \tag{3.24}$$

Here Ψ^m is the apparent permselectivity of an ion-exchange membrane.

For a salt such as KCl the transport numbers of cation and anion in the solution are nearly identical, i.e. $T_{cou} \cong T_{co} \cong 0.5$ and the permselectivity can be expressed to a first approximation by:

$$\Psi^{cm} \cong \frac{\varphi_{m,real}}{\varphi_{m,sp}} \tag{3.25}$$

The membrane potential is measured with reference electrodes in a test cell consisting of two chambers separated by a membrane as illustrated in Figure 3.17.

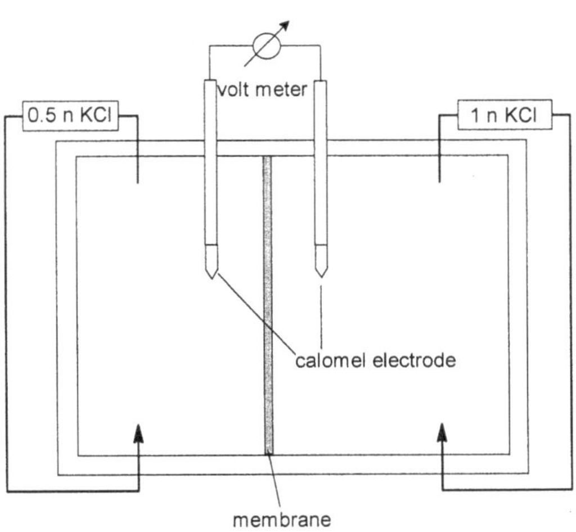

Fig.3.17 Schematic drawing illustrating the test cell for determining the membrane potential.

Both chambers are filled with the same electrolyte solution, e.g. KCl but with different concentrations, i.e. 0.5 and 1 mol L^{-1}. To avoid a concentration change due to diffusion the two solutions are well mixed. Then the membrane potential is measured and divided by the potential of a strictly permselective membrane at 25°C which is calculated by Equation 3.21 for the above given electrolyte concentration to:

$$\varphi_{m,sp} = \frac{RT}{F}\ln\frac{a_i^1}{a_i^2} = \frac{RT}{F}\ln\frac{C_i^1}{C_i^2}\frac{\gamma_i^1}{\gamma_i^2} = 0.02569\ln\frac{1}{0.5}\frac{0.6}{0.65} = 0.0158\,[\text{Volt}] \qquad (3.26)$$

Here γ_i^1 and γ_i^2 are the activity coefficients of KCl in the two solutions separated by the membrane which are obtained from the literature. The permselectivity of a membrane is according to Equation 3.25:

$$\Psi^m \cong \frac{\varphi_{m,real}}{0.0158} \qquad (3.27)$$

This is the apparent permselectivity of a membrane which is always larger than the real permselectivity since it does not take into account the water transport due to osmosis and electroosmosis. However, the difference between the apparent and real permselectivity is generally very small. As an advantage, the determination of the potential between two solutions of different concentrations is not effected by concentration polarization at the membrane surface. For characterizing membrane samples in manufacturing processes the static method is generally applied.

3.3.4 Membrane Permeation Selectivity for Different Counter-Ions

The permeation of counter- and co-ions in ion-exchange membranes is generally quite different and the counter-ion permselectivity is close to 1. It is determined mainly by the Donnan exclusion of the co-ions. However, certain ion-exchange membranes also show a permeation selectivity for different counter-ions. This selectivity is determined mainly by the mobility of the different ions in the membrane, i.e. by the ion size and the structure of the membrane. The permeation selectivity of different counterions is determined by measuring the fluxes of counterions in a mixture of electrolytes having different counterions but identical coions with reversible electrodes, or in a conventional electrodialysis test. The permeation selectivity of a membrane for different counter-ions is given by the ratio of their fluxes through the membrane which is identical to the ratio of their transport numbers in the membrane.

3.3.5 Diffusion Coefficients of Electrolytes in Ion-Exchange Membranes

When ion-exchange membranes are used in electrodialysis or diffusion dialysis one side of the membrane is in contact with a concentrated solution and the other side with a diluted solution. In practical applications often large concentration gradients across the membrane are obtained which results in a diffusion of low molecular weight charged and non-charged components through the membrane. The diffusion of charged components effects the current efficiency and leads eventually to a contamination of the product or to a loss of valuable components when certain biological materials are treated. The determination of the diffusion coefficients can be carried out by standard procedures such as the time-lag method [J. Crank et al., 1968]. In the time-lag experiment the membrane separates two solutions. One solution contains the test component in a certain concentration while the other solution and the membrane are free of this component. Then the flux through the membrane is measured as a function of time. In a first stage of the measurement, the so-called transition state, the flux is increasing exponentially with time. After a certain time the flux increase with time becomes linear and the system has reached a steady state. At steady state the mass flux is given by:

$$M_t = \frac{DC_o}{l}\left(t - \frac{l^2}{6D}\right) \qquad (3.28)$$

Here is M_t the steady state mass flux of the test component through the membrane, C_o is its original concentration in the feed solution, l is the thickness of the membrane, and D is the diffusion coefficient of the test component in the membrane.

If the linear part of the plot of $\frac{M_t}{lC_o}$ versus time is extrapolated to the time axis of the diagram as shown schematically in Figure 3.18 the resulting intercept L is referred to as time-lag L which is given by:

$$L = \frac{l^2}{6D} \qquad (3.29)$$

For a given membrane thickness l the diffusion coefficient can be determined from Equation 3.29:

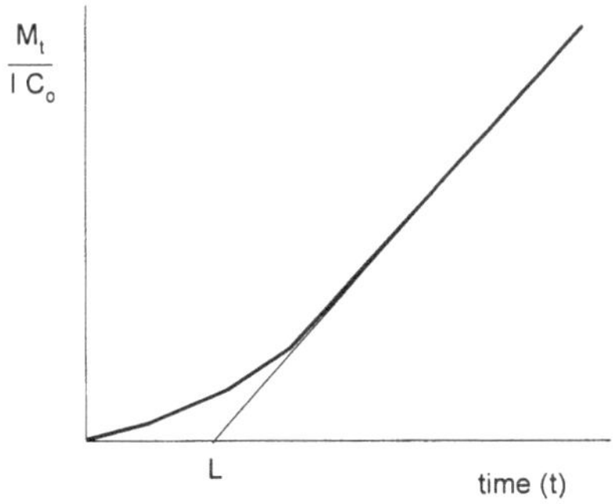

Fig. 3.18 Schematic drawing illustrating the determination of diffusion coefficients by the time-lag method.

The time-lag method is very convenient for the determination of diffusion coefficients. However, the results can be obscured by osmotic water transfer due to large concentration gradients across the membrane. To eliminate such effects radioactive tracer techniques may be used.

3.3.6 Water Transport in Ion-Exchange Membranes

The water transport through ion-exchange membranes is caused by osmosis and by electro-osmosis. The osmotic water transport is the result of concentration differences between the two solutions separated by the membrane. Electroosmosis refers to the water transported in the hydration shell of the ions migrating through the membrane due to an electrical potential gradient. The osmotic water transport is measured using a two compartment cell containing pure water in one compartment separated from the second compartment containing a concentrated solution. The volume change in the compartments is determined as a function of time in two horizontal capillaries connected with the compartments as indicated in the schematic drawing of Figures 3.19 and 3.20 which illustrate the experimental set-up to determine the water transport in osmosis and electroosmosis.

In osmotic water transport measurements one cell is filled with water and the other with a salt solution. The water flux from the cell filled with water into that filled with the solution is measured in the horizontally arranged capillaries as indicated in Figure 3.19. Since the osmotic pressure is a linear function of the concentration of a solution the osmotic water transport increases linearly with the salt concentration.

The electroosmotic water transport is expressed by a transport number in analogy to the transport number of ions and is given by:

$$T_w^m = \frac{F J_w}{i} \tag{3.30}$$

Here is T_w^m the transport number of water through a membrane, F is the Faraday constant, i is the current density, and J_w is the water flux expressed in moles per unit area and time.

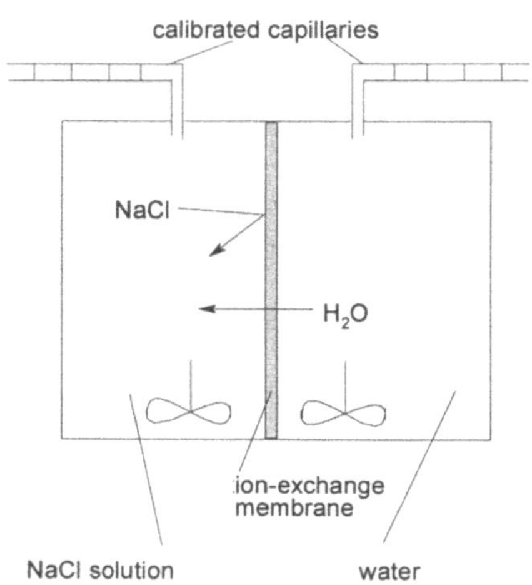

Fig. 3.19 Schematic drawing illustrating the test cell for determining the osmotic water flux through an ion-exchange membrane.

In the electroosmotic water transport measurement a direct current is applied using reversible electrodes, i.e. Ag/AgCl-electrodes. Both compartments of the test cell are filled with a test solution of identical concentrations. The electroosmotic water transport of a cation-exchange membrane is a function of the migration of the cation and is determined by water connected to the ion in its hydration shell. The electroosmotic water transport through a cation-exchange membrane due to the migration of a Na^+-ion can be determined by measuring the volume change in the two compartments of a test cell filled with a NaCl-solution as indicated schematically in Figure 3.20.

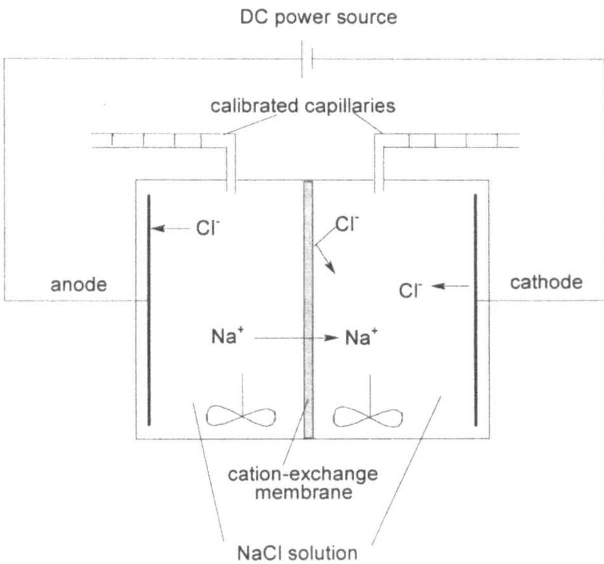

Fig. 3.20 Schematic drawing illustrating the test cell to determine the electroosmotic water transport.

In the test it is assumed that the membrane is strictly semipermeable, i.e. it is only permeable for Na^+-ions. However, the volume in the two compartments of the cell is not only the result of the water transport. The volumes of the cells are also changed by the electrode reactions. At the anode Cl^--ions disappear and at the cathode Cl^- -ions are generated according to the following reaction scheme:

$$\text{anode: } Ag + Cl^- \rightarrow AgCl + e$$

and

$$\text{cathode: } AgCl + e \rightarrow Ag + Cl^-.$$

The electroosmotic water transport is calculated from the volume change in the two compartments:

$$\Delta V_F = \Delta V_{tot} + \overline{V}_{AgCl} - \overline{V}_{Ag} - \overline{V}_{Cl^-} = T^m_{Na^+} n_w^{Na^+} + \overline{V}_{Na^+} \qquad (3.31)$$

Here ΔV_{tot} and ΔV_F are the total volume change and the volume change due to electro-osmosis in the cell when electrical charges of 1 Faraday are transported, $\overline{V}$ is the partial molar volume of AgCl and Ag, and T_{Na}^{m} is the transport number of Na^+-ions in the membrane, and $n_w^{Na^+}$ is number of water molecules carried by on Na^+-ion.

The electroosmotic water flux is:

$$\Delta V_F = T_w^m \overline{V}_w \tag{3.32}$$

Here is T_w^m the water transport number, ΔV_F is the electroosmotic water flux, and $\overline{V}_w$ is the partial molar volume of water.

Different cations show rather different electroosmotic water transport numbers. Ions with a relatively large hydration shell such as the Na^+-ion have a water transport number of 6 to 8. Protons have a very low water transport number of less than 3 because of the different transport mechanism which has been discussed earlier. The water transport number is decreasing with increasing salt concentration. The main reason for the decreasing transport number with increasing salt concentration is the decrease of the membrane permselectivity with increasing salt concentration which leads to a considerable cointransport in the opposite direction.

3.3.7 Characterization of Special Property Membranes

In many applications of electrodialysis and related processes special property membranes are required. For instance for the pre-concentration of sea water for the production of table salt membranes with a low permeability for divalent ions are required. In diffusion dialysis anion-exchange membranes with high proton permeability or cation-exchange membranes with high hydroxide ion permeability are needed, and in bipolar membrane applications generally acid and base blocking mono-polar membranes as well as efficient bipolar membranes with a low electric resistance and good water dissociation capability are needed [G. Pourcelly et al., 1993, 1994]. Most mono-polar membranes can be characterized by applying the procedures described earlier. In addition to the electrical resistance and the permselectivity the water dissociation capability is for bipolar membranes of prime importance. Therefore, membranes are generally characterized by a so-called current versus voltage plot. Determination of the current through a bipolar membrane as a function of the applied voltage provides information about the water dissociation, the electrical resistance, and the salt leakage of the membrane. A typical current versus voltage curve of a commercial bipolar membrane is shown in Figure 3.21 a) and b) [J.J. Krol et al., 1998].

The Figure 3.21 a) shows the current density as a function of the voltage applied across the commercial bipolar membrane in a 1mol L^{-1} Na_2SO_4 solution. At low voltage between 0 and 0.1 V the current increases linearly with the applied voltage to about 1 to 2 mA cm^{-2}. A further increase of the voltage across the membrane does not lead to a significant current increase until a voltage drop of 0.6 to 0.8 Volt is reached. Then the current increases sharply to about 200 to 400 mA cm^{-2} with increasing applied voltage until a voltage of ca. 1 V is reached. A further increase in voltage does not increase the current significantly and generally leads to a destruction of the membrane.

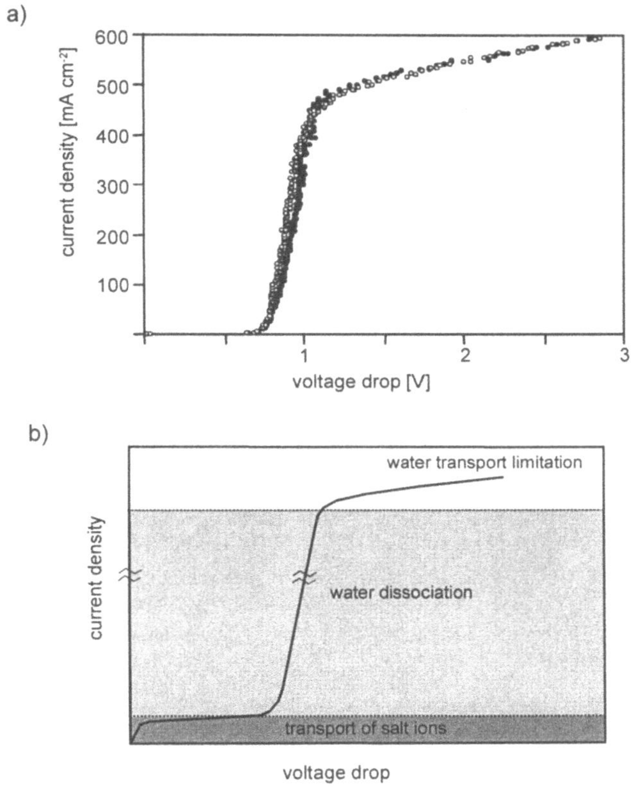

Fig. 3.21 Current density versus applied voltage curve a) measured with a commercial bipolar membrane in a 1 mol L^{-1} Na_2SO_4 solution and b) schematic drawing indicating 3 distinct areas where the current is determined by the salt transport, the water dissociation, and by the water diffusion into the reaction zone of the membrane.

As illustrated in the schematic drawing of Figure 3.21 b) three distinct areas in the current versus voltage curve can be identified. The first area is characterized by a linear increase in current with the applied voltage until a limiting value of ca. 1 mA cm^{-2} is reached at ca 0.1 V. A further increase in the applied voltage does not lead to an increase in the current. The current in this region is the result of salt transport from the transition region of the bipolar membrane into the outside solutions due to the driving force of the applied potential. Up to an applied voltage of ca. 0.1 V this transport is compensated by salt diffusion from the outside solutions into the bipolar membrane due to a concentration gradient. The salt transport out of the membrane can not exceed the diffusive salt transport into the membrane independent of the applied voltage. Thus, a steady state is reached and the current is limited by the salt diffusion into the membrane until the voltage drop across the membrane reaches 0.6 to 0.8 V. The drastic current increases when the applied voltage is increased further, is the result of the transport of H^+- and OH^--ions that are produced within the bipolar membrane. In the region between 0.5 to 1.2 V applied voltage and a current density of up to 500 mA cm^{-2} the current is exclusively carried by H^+- and OH^--ions which are provided by the water dissociated. The water dissociation is a very fast reaction and does not limit the transport of the ions from the transition region of the bipolar membrane into the outer phases. If the applied voltage has reached ca. 1.2 V, a further increase does not lead to a significant increase in current. This limitation of the current density is the result of a decrease in the H^+- and OH^--ion concentration in the transition region of the bipolar membrane which is determined by the water dissociation rate which again is limited by the water diffusion rate from the outside phases into the membrane. Operation of the membrane in the third, i.e. the water diffusion limited region, generally leads to a destruction of the membrane [J.J. Krol et al., 1998].

In addition to determining the current density as a function of the applied voltage more complex testing methods such as impedance spectroscopy [V.I. Zabolotsky et al., 1978], [H. Holdik et al., 1998] or chronopotentiometric measurements [F. Wilhelm, 2001] are used to obtain more detailed information about the mechanism of the water dissociation and the resistance of the bipolar membrane. The mechanism of the water dissociation in bipolar membranes will be discussed in more detail later. For a quick evaluation the current density versus the applied voltage plot provides reliable information about the electrical resistance, the water dissociation capability, and the salt leakage of a bipolar membrane.

Properties of selected commercially available membranes are listed in the Table.of the Appendix I This table, however, is by far not complete. There are many more special property membranes on the market today.

4 LIST OF SYMBOLS

A_m	membrane surface area	$[m^2]$
a	activity	$[mol\ m^{-3}]$
AC	alternating current	[A]
C	concentration	$[mol\ m^{-3}]$
D	diffusion coefficient	$[m^2\ s^{-1}]$
DC	direct current	[A]
d	distance	[m]
d_m	membrane thickness	[m]
i	current density	$[A\ m^{-2}]$
J	flux	$[mol\ m^{-2}\ s^{-1}]$
L	time lag	[s]
l	distance	[m]
M_t	steady state mass flux	$[mol\ m^{-2}\ s^{-1}]$
N	number of moles	[mol]
R	resistance	[Ω]
R	gas constant	$[J\ mol^{-1}\ {}^oK]$
T	temperature	$[{}^oK]$
T	transport number	[-]
t	time	[s]
V	volume	$[m^3]$
$\overline{V}$	partial molar volume	$[m^3\ mol^{-1}]$
U	voltage drop	[V]

r	area resistance	[Ω cm^2]
W	weight	[kg]
z	valence	[eq mol^{-1}]
Z	impedance	[Ω]

Greek letters

ρ	specific resistance	[Ω m]
Ψ	permselectivity	[-]
ω	circular velocity	[s^{-1}]
ν	frequency	[s^{-1}]
φ	phase shift angle	[degree]

subscripts

i	refers to component
cou	counter-ion
co	coion
m	membrane
s	solution

LITERATURE

Asahi Chemical 1990. Product Bulletin, Tokyo, Japan.

Asahi Glass 1990. Product Bulletin, Tokyo, Japan.

Bauer, B., 1993. Bipolar membrane and process of its preparation, European Patent 056851

Bauer, B., Effenberger, F., Strathmann, H.. 1990. Anion-exchange membranes with improved alkaline stability, Desalination, 79:125-144.

Bauer, B., Gerner, F.-J., Strathmann, H., 1988. Development of bipolar membranes, Desalination 68: 279-292.

Bergsma, F., Kruissink, CH.A., 1961. Ion-exchange membranes, Fortschr. Hochpolym.-Forsch., 21: 307-362.

Caplan, S.R., Sollner, K., 1974. Influence of the characteristics of the activating polyelectrolyte in the preparation and on the properties of interpolymer ion-exchange membranes: rational principles of membrane preparation and their experimental test, J. Colloid Interface Sci., 40: 46-66.

Chlanda, F.P., Lee,L.T.C., Liu, K.J., 1976. Bipolar membranes and method of making the same, U.S. Patent 4,116,889.

Connolly, D.J., Gresham, W.F., 1966. Fluorocarbon vinylether polymers, U.S. Patent 3,282,875.

Coulter, M.V., 1980. Modern chlor-alkali technology, Ellis Horwood Ltd., Chichester, U.K.

Crank, J., Park, G.S., 1968. Diffusion in polymers, Academic Press, London.

Cui, W., Kerres, J. Eigenberger, G., 1998. Development and characterization of ion-exchange polymer blend membranes, Separation & Purification Technology, 14: 145-154.

Davis, T.A., Genders, J.D., Pletcher, D.,1997. Ion permeable membranes, Alresford Press Ltd. Alresford, Hants, England.

De Korocy, F., Zeigerson, E., 1967 Investigation by electron microscopy and x-ray diffraction of the microstructure of permselective membranes made of polyethylene, J. Appl. Polymer Sci., 11: 909-919.

Dohno, R., Azumi, T., Takashima, 1975. Permeability of monocarboxylate ions across an anion-exchange membrane, Desalination, 16: 55-64.

Dotson, R.L., Woodard, K.E., 1982. Electrosynthesis with perfluorinated ionomer membranes in chlor-alkali cells, in: Perfluorinated ionomer membranes, edts.: Eisenberg, A., Yeager, H.L.. ASC-Symposium Series 180, American Chemical Society, Washington DC.

DuPont, 1988. Product Bulletin.

Eisenberg, A., Yeager, H.L., 1982. Perfluorinated ionomer membranes, in: Perfluorinated ionomer membranes, edts.: Eisenberg, A., Yeager, H.L., ASC-Symposium Series 180, American Chemical Society, Washington DC.

Escoubes, M., Pineri, M., 1982. Thermodynamic studies of the water-perfluorosulfonated polymer interactions, in: Perfluorinated ionomer membranes, edts.: Eisenberg, A., Yeager, H.L.. ASC-Symposium Series 180, American Chemical Society, Washington DC.

Falk, M., 1982. Infrared spectra of perfluorosulfonated polymer and of water in perfluorosulfonated polymer, in: Perfluorinated ionomer membranes, edts.: Eisenberg, A., Yeager, H.L.. ASC-Symposium Series 180, American Chemical Society, Washington DC.

Flett, D.S., 1983. Ion-exchange membranes, E. Horwood Ltd., Chichester, U.K.

Grot, W.G., 1973. Laminates of support material and fluorinated polymer containing pendant side chains containing sulfonyl groups, U.S. Patent 3, 770,567.

Grot, W.G., 1974. Films of fluorinated polymer containing sulfonyl groups with one surface in the sulfonamide or sulfonamide salt form and a process for preparing such, U.S .Patent, 3, 784, 399.

Grot, W.G., 1975. Perfluorinated cation exchange polymers, Chem.-Ing.-Tech., 47: 617.

Gudernatsch, W., Krumbholz, CH., Strathmann, H., 1990. Development of an anion-exchange membrane with increased permeability for organic acids of high molecular weight. Desalination 79: 249-260.

Haagen, K., Helfferich, F., 1959. Ionenaustauscher Membranen, German Patent 971729.

Hanada, F., Hiraya, K., Ohmura, N., Tanaka, S., 1991. Bipolar membrane and methode for its production, European Patent, 0459820.

Hanada, F., Hiraya, K., Ohmura, N., Tanaka, S., 1993. Bipolar membrane and methode for its production, US Patent 5, 221, 455.

Helfferich, F., 1962. Ion-exchange, McGraw-Hill, London.

Hodgdon, R.B., 1992. Aliphatic anion-exchange membranes having improved resistance to fouling, US Patent 5,137,925.

Hodgdon, R.B., Alexander, S.S., 1989. Novel bipolar membranes and process of manufacture, US-Patent 4 851 100,

Hodgdon, R.B., Witt, E., Alexander, S.S., 1973. Synthesis and transport properties of macrorecticular anion-exchange membranes for electrodialysis in the presence of surface water foulants, Desalination, 13: 105-127.

Holdik, H., Alcaraz, A., Ramirez, P., Mafe, S, 1998. Electric field enhanced water dissociation at bipolar membrane junction from impedance spectroscopy, J. Electroanal. Chem., 442: 13-18.

Hurwitz, H., El Moussaoui, R., 1998. Single film membrane, process for obtaining it and use therefore, US Patent, 5 840 192.

Ionics Inc., 1990. Product Bulletin CR, Watertown, MA 02172.

Ishibashi, N., Seiyama, T., Sakai, W., 1954. Electrochemical studies on ion-exchangers, preparation of a cation-exchange membrane and its properties, J. Electrochemical Soc. Japan, 22: 684-687.

Ishigaki, I., Kamiya, N., Sugo, T., Machi, S.,1978. Synthesis of an ion-exchange membrane by radiation-induced grafting of acrilyc acid onto poly(tetrafluorethylene), Polym. J., 10: 513-519.

Itoh, H., Yoshizumi, T., Saeki, M., 1986. Sieving effect in electrodialysis with an ion-exchange membrane, J. Membrane Sci., 27:155-163.

Juda, W., McRae, W.A., 1950. Coherent ion-exchange gels and membranes, J. Amer. Chem. Soc. 72: 1044.

Juda, W., McRae, W.A., 1953. U.S. Patent 2, 636, 851.

Kasper, A. A., 1955. Electrolytically conductive membranes, US Patent 2,702,272.

Kerres, J.A., 2000. Development of ionomer membranes for fuel cells, J. Membrane Sci., 185: 3–27.

Kerres, J.A., Ullrich, A., Hein, M., 2001. Preparation and characterization of novel basic polysulfone polymers, J. Polymer Sci., Part A, Polymer chemistry, 39: 2874-2888.

Kobuchi, Y., Matsunaga, Y., Noma, Y., 1987. Application of ion-exchange membranes to the recovery of acids by diffusion dialysis and electrodialysis, in: Synthetic polymer membranes, edts.: B. Sedlaceek, J. Kahovec, W. de Gruyter & Co. Berlin.

Komoroski, R.A., Mauritz, K.A., 1982. Perfluorosulfonated ionomer membranes, edts: Eisenberg, A., Yeager, H.L., ASC Symposium Series 180, American Chemical Society, Washington, D.C.

Korngold, E., 1984. Electrodialysis: Membranes and mass transport, in: Synthetic membrane processes, edt.: Belfort, G., pp. 192-219, Academic Press, New York.

Korngold, E., De Körösy, F., Rahav, R., Taboch, M.F., 1970. Fouling of anion selective membranes in electrodialysis, Desalination, 8: 195-220.

Krol, J.J., Jansink, J.M., Wessling, M., Strathmann, H., 1998. Behaviour of bipolar membranes at high current density: Water diffusion limitation, Separation & Purification Technology, 14: 41-52.

Kunin, R., 1958. Ion-exchange resins, Wiley-Interscience, New York.

Kusomoto, K., Sata, T., and Mizutani, Y., 1976. Modification of anion-exchange membranes with polystyrene sulfonic acid, Polym. J., 8:225-226.

Liu, K.J., Chlanda, F.P., Nagasubramanian, K.J., 1977. Use of bipolar membranes for generation of acid and base: an engineering and economic analysis, J. Membrane Sci., 2: 109-124.

Mani, K.N., 1991. Electrodialysis water-splitting technology, J. Membrane Sci., 58: 117-138.

Matsui, K.,Tobita, E., Sugimoto, K., Kondo, K., Seita, T., Akimoto, 1986. Novel anion-exchange membranes having fluorocarbon backbone: preparation and stability, J. Appl. Polym. Sci., 32: 4137-4143.

McRae, W.A., Alexander, S.S., 1960. Sulfonating reagent and its use in preparing cation-exchange membranes, U.S. Patent 2,962,454.

Meares, P., 1983. Trends in ion-exchange membrane science and technology, in: Ion-Exchange Membranes, edts.: Flett, D.S., E. Horwood Ltd., Chichester, UK.

Mihara, K., Misumi, T., Miyauchi, H., Ishida, Y., 1973. Anion-exchange membranes with excellent permselectivity between anions with the same charge, Japan Patent 48-34676.

Miyamoto, Y., 1991. Fabrication of charge-mosaic membranes and their performance, Maku, 16: 233-238.

Molau, G.E., 1981. Heterogeneous ion-exchange membranes, J. Membr. Sci., 8: 309-330.

Onoue, Y., Mizutani, Y., Yamane, R., Takasaki, Y., 1961. Selectivity of cation-exchange membrane, for NaCl-$CaCl_2$ system, J. Electrochem. Soc. Japan, 29: 187-191.

Onoue, Y., Sata, T., Nakahara, A., Itoh, J., 1980. Process for preparing fluorine-containing polymers having carboxyl groups, U.S. Patent 4,200,711.

Pourcelly, G., Tugas, I., Gavach, C., 1993. Electrotransport of HCl in anion exchange membranes for the recovery of acids, J. Membrane Sci., 85: 195-204.

Pourcelly, G., Tugas, I., Gavach, C., 1994. Electrotransport of sulphuric acid in special anion-exchange membranes for the recovery of acids, J. Membrane Sci., 97: 99-107.

Rapp, H.-J., 1995. Die Elektrolyse mit bipolaren Membranen, Theorie und Anwendung. Universität Stuttgart, FRG.

Sata, T., 1972. Monovalent cation permselective exchange membrane, Kolloid-Z. u. Z. Polymere, 250: 980-982.

Sata, T., 1986. Recent trends in ion-exchange research, Pure & Appl. Chemistry 58: 1613-1626.

Sata, T., Yamane, R., Mizutani, Y., 1979. Modification of properties of ion-exchange membranes, J. Polymer Sci. Polymer Chem., 17: 2071-2085.

Sata, T., Motani, K., Ohaski, Y., 1983. Perfluorinated exchange membrane "Neosepta-F" and its properties. In: Ion Exchange Membranes, edt. Flett, D.S., E. Horwood Ltd., Chichester, U.K.

Sata, T., Ogura, S., Kishimoto, F., 1993. Properties of composite membranes from ion-exchange membranes and conducting polymers, J. Membrane Sci., 84: 259-269.

Sata, T., Yamaguchi, T., Matsusaki, K., 1995. Effect of hydrophilicity of ion-exchange groups of anion-exchange membranes on permselectivity between anions, J. Phys. Chem., 99: 12875-12882.

Sata, T., Yamaguchi, T. Kawamura, K., Matsusaki, K., 1997. Transport numbers of various anions relative to chloride ions in modified anion-exchange membranes during electrodialysis, J. Chem. Soc. Faraday Trans. 93: 457-462.

Seko, M., Ogawa, S., Kimoto, K., 1988. Perfluorocarboxylic acid membrane and membrane chlor-alkali process developed by Asahi Chemical Industry. In: "Perfluorinated ionomer membranes", edts.: Eisenberg, A., Yeager, H.L.. ASC-Symposium Series 180, American Chemical Society, Washington, D.C.

Shel'deshov, N.V., Zabolotsky, V.I., Pis'menskaya, N.D. Gnusin, N.P., 1986. Catalysis of water dissociation by the phosphoric acid groups of an MB-3 bipolar membrane, Elektrokhimiya, 22: 791-795.

Simons, R. 1993. Preparation of a high performance bipolar membrane, J. Membrane Sci., 78: 13-23.

Simons, R., 1984. Electric field effects on proton transfer between ionizable groups and water in ion-exchange membranes, Electrochemica Acta, 29:151-158.

Simons, R., 1986. A novel method for preparing bipolar membranes, Electrochimica Acta, 31: 1175-1176.

Spiegler, K.S., 1958. Transport process in ionic membranes, Trans. Faraday Soc., 54: 1408-1428.

Spiegler, K.S., Laird, A.D.K., edts., 1980. Principles of desalination, 2nd edition, Academic Press, New York.

Strathmann, H., 1995. Electrodialysis and related processes, in: Membrane Separation Technology, edts.: Nobel, R.D., Stern, S.A. pp 213-281, Elsevier, Amsterdam.

Strathmann, H., Bauer, B., Rapp, H.J., 1993. Better bipolar membranes, CHEMTECH, June 1993: 17-24.

Suryanarayana, N.P., Krishnaswamy, 1963. Ion-exchange membranes based on rubber, J. Polym. Sci., Part B. 1: 491-495.

Gierke, T.D., Munn, G.E., Wilson, F.C., 1982. Morphology of perfluorosulfonated membrane products, in: Perfluorinated ionomer membranes, edts.: Eisenberg, A., Yeager, H.L.. ASC-Symposium Series 180, American Chemical Society, Washington DC.

Tokuyama Soda, 1987, 1988. Product Bulletin, Tokyo, Japan.

Tooper, E.B., Wirth, L.F., 1956. Ion-exchange resins, in: Ion-exchange technology, edts.: Nachod, F.C., Schubert, J., Academic Press, New York.

Wilhelm, F.W., 2000. Bipolar membrane preparation, in: Bipolar membrane technology, edt.: Kemperman, A.J.B., Twente University Press, Enschede, The Netherlands, ISBN 9036515203.

Wilhelm, F.W., 2001. Bipolar membrane electrodialysis, PhD.-thesis Twente University, Enschede, The Netherlands, ISBN 9036515270.

Yamane, R., Mizutani, Y., Motomura, H., Izuo, R., 1964. Preparation of SO42- nonpermeable anion-exchange membrane, J. Electrochem. Soc. Japan, 32: 277-288.

Yamane, R., Izuo, R., Mizutani, Y., 1965. Permeability of the amphoteric ion-exchange membranes, J. Electrochem. Soc. Japan, 33: 589-593.

Yeager, H.L., 1982. Cation-exchange selectivity of perflurosulfonate polymer, in: Perfluorinated ionomer membranes, edts.: Eisenberg, A., Yeager, H.L.. ASC-Symposium Series 180, American Chemical Society, Washington DC.

Yokoyama, Y., Tanioka, A., Miyasaka, K., 1989. Preparation of a single bipolar membrane by plasma-induced graft polymerization, J. Membrane Sci., 43: 165-175.

Zabolotsky, V.I., Nikonenko, V.V., 1993. Effect of structural membrane inhomogeniety on transport properties, J. Membrane Sci., 79: 181-198.

Zabolotsky, V.I., Shel′deshov, N.V., Gnusin, N.P., 1978. Impedance of MB-1 bipolar membranes, Elektrokhimiya, 15: 1488-1493.

Zschocke, P., Quellmalz, D., 1985. Novel ion-exchange membranes based on an aromatic polyethersulfone, J. Membrane Sci., 22: 325-332.

Chapter 4

Operating Principle of Electrodialysis and Related Processes

1 INTRODUCTION

All ion-exchange membrane separation processes are based on the selective transport of ions from one solution to another through an ion-exchange membrane under the driving force of an electrochemical potential gradient. However, the various processes such as electrodialysis, diffusion or Donnan dialysis, and electrodeionization vary significantly in their basic process design, in the required properties of the membranes and other hardware components as well as in their practical application. In conventional electrodialysis, for instance, cation- and anion-exchange membranes with low electric resistance, high permselectivity, and good chemical and mechanical stabilities are required. In electrodialysis with bipolar membranes the bipolar membranes must have high water dissociation capability in addition to the properties required in conventional electrodialysis, and cation- or anion-exchange membranes used in diffusion dialysis must have a high permeability for hydroxide ions and protons, respectively. The current density applied in conventional electrodialysis is significantly lower than that used in electrodialysis with bipolar membranes, and in diffusion dialysis no electrical potential but only concentration gradient is applied as driving force for the transport of the different ions. The performance of ion-exchange membrane separation processes in the different practical applications depends not only on membrane properties, or the feed solution composition and product quality requirements, it is also determined by several process and equipment design parameters, such as stack construction, i.e. the cell geometry and the spacer configuration, the feed flow velocities, and the mode of operation, i.e. continuous or batch operation with co- or counter-current flow streams. These parameters effect the overall costs of the process by determining the required energy for the driving force necessary to transport the ions from a feed stream into a product stream or waste stream and by determining the required membrane area for a given capacity plant. In electrodialysis the various process and design parameters effect the voltage drop, the hydraulic pressure losses, and the current utilization. In diffusion dialysis the mass transfer through the membrane and thus, the required membrane area for a given plant capacity is strongly related to flow velocities and cell design. Therefore, not only the basic principles of conventional electrodialysis and electrodialysis with bipolar membranes or diffusion dialysis are very different but also many plant components and the complete system design can vary significantly for a given process depending on the specific application.

2 THE PRINCIPLE OF CONVENTIONAL ELECTRODIALYSIS

The basic concept of electrodialysis has been discussed earlier and is shown in Figure 4.1. This figure depicts the transport of ionic species in a cell arrangement consisting of cation- and anion-exchange membranes in alternating series forming an array of individual cells between two electrodes. If an electrical potential difference between the electrodes is established the cations will migrate towards the cathode, i.e. the negatively charged electrode while the anions migrate to the positively charged anode. The positively charged cations permeate easily the oppositely charged cation-exchange membrane but will be retained at the anion-exchange membrane. The negatively charged anions will permeate the anion-exchange membrane but will be retained by the negatively charged cation-exchange membrane. The overall result of the ion migration is an increase in the ion concentration in alternating compartments formed by cation- and anion-exchange membranes while the concentration in the other compartment is correspondingly decreased. The solution which is concentrated during the electrodialytic process is generally referred to as concentrate or brine. The solution with decreased concentration is referred to as diluate. A unit composed of a cation-exchange membrane, a compartment with the dilute, an anion-exchange membrane, and the concentrate compartment is referred to as cell pair.

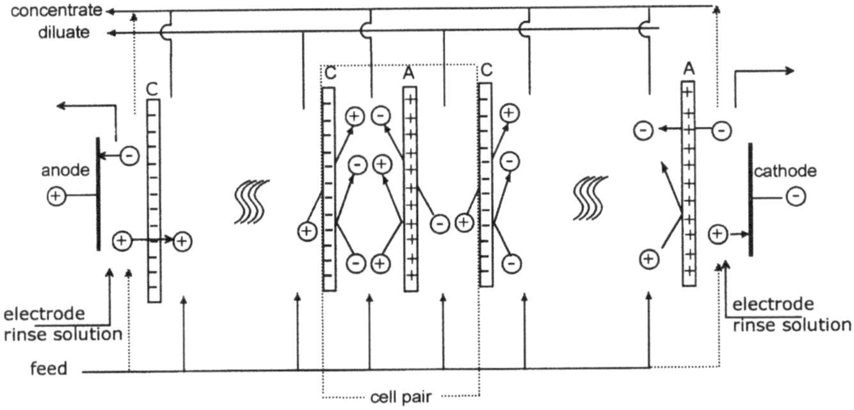

Fig. 4.1 Schematic diagram illustrating the principle of an electrodialysis stack design.

A cell pair may be arranged between two electrodes in different ways [J.R. Wilson, 1960]. The most common one is the so-called membrane stack. In an elctrodialysis stack a multitude of cell pairs, typically up to 200, are "stacked" between two electrodes. A cell pair is composed of a diluate and a concentrate cell and the cation- and anion-exchange membranes which separate these cells. A cell pair is a repeating unit in a stack. The cells containing the electrodes are sometimes rinsed with a separate solution which contains no Cl^--ions to avoid the formation of chlorine. In most large scale commercial electrodialysis stacks used for the desalination of brackish water the electrodes are rinsed with the feed solution and the effluent is disposed with the brine.

There are other membrane arrangements described in the literature such as a spiral-wound cell design. However, the system design with the stacked membranes has advantages as far as the investment costs and the energy consumption are concerned. In any ion-exchange membrane separation process design the total number of ions transferred from a feed to a concentrate solution is proportional to the installed cell pair area. In the electrodialysis stack the overall ion transport is proportional to the current passing through a cell pair multiplied by the number of cell pairs in the stack which is identical to the current between the two electrodes installed on both sides of the stack as indicated in Figure 4.1. That means that for a large membrane area a relatively small electrode area is needed and the costs of the expensive electrodes as well as the energy consumption for the electrode reactions are distributed over a large cell pair area. The total energy used for the actual ion transfer from a feed to a concentrate is the same for a system in which the required membrane area is installed in one cell pair between two electrodes as for a system in which the membrane area is installed in a multitude of cell pairs since the energy consumption is the product of current and voltage drop. If the cell pair area required for a certain plant capacity is installed a large number of cell pairs in parallel in a stack between two electrodes then a large voltage drop between the electrodes is obtained. However, the total current that is passing through the stack is low. The total current required for a certain degree of desalination is given by the area of one cell pair multiplied by the current density. The total voltage drop across the stack is that of one cell pair multiplied by the number of cell pairs. If the total required cell pair area would be installed in one cell pair between two electrodes the voltage drop would be low, but the current would be very high.

The properties of the ion-exchange membrane and their arrangement between the electrodes are the most important design parameters determining the technical feasibility and the economics of the process. However, there are many other components necessary for the proper and efficient operation of an electrodialysis plant such as the electrical power supply, pumps, and control and monitoring devices. The stack is a key component in an electrodialysis unit. For designing and operating an electrodialysis stack certain process parameters have to be taken into account and controlled such as the feed solution concentration and the desired product and brine concentrations. Various modes of operation are possible such as batch type or continuous. In many cases the desired desalination or concentration of a feed solution cannot be obtained in a single pass through the electrodialysis cells. To achieve a higher degree of desalination or concentration two or more stacks are placed in series. When high product recovery rates are required part of the brine may be recycled. The flow streams through a stack can be counter- or co-current. The velocities of the flow streams will effect the concentration polarization and thus the limiting current density which determines the required membrane area for a given capacity desalination plant. The flow velocities, on the other hand, determine the degree of desalination and concentration that can be achieved in one pass in a stack of a given geometry and the pressure loss which has to be overcome by the flow stream circulation pumps. Certain process parameters are fixed by the feed and product solution properties. Other parameters may be varied in a certain range and can thus be utilized to optimize the process, such as the current density, the applied voltage, the flow stream velocities etc. Many parameters are interdependent and often counteracting in terms of overall process efficiency. They all must be taken into account in constructing the various hardware components of an electrodialysis unit as well as designing the process. But before describing the different hardware components and the process schemes of an electrodialysis plant in detail some basic relations concerning the construction of an electrodialysis stack and the mass transport in the stack including concentration polarization effects will be discussed.

2.1 Mass Transport in an Electrodialysis Stack

An electrodialysis stack is composed of a multitude of identical cell pairs in parallel. The mass transport in an electrodialysis cell pair which is composed of two compartments bounded on each side by an anion- and cation-exchange membrane is determined by a mass balance, which postulates that all components that are removed from a feed solution will be transferred to the concentrate flow stream. There are different stack designs used in the various applications of electrodialysis. For water desalination the sheet flow and the tortuous-path stack design are mainly used [L.H. Schaffer et al., 1966]. The mass transfer in a electrodialysis cell pair of a sheet flow stack is illustrated in Figure 4.2 which shows two compartments bounded by ion-exchange membranes. For simplicity reason it is assumed that the individual cells are of identical geometry and the flow streams are co-current under equal hydrodynamic conditions. This is the case in most commercial electrodialysis plants. However, in special applications the flow streams in a stack can also be counter-current and have different flow velocities in the concentrate and diluate cell. The main reason for having stacks with diluate and concentrate cells of identical geometries and co-current flow with identical velocity is to obtain identical pressure losses in the cells and to avoid the build-up of pressure differences between the concentrate and diluate solution. Pressure differences between the flow streams can cause hydraulic fluxes through defects in the membranes and the spacer gaskets. A flux from the concentrate solution to the diluate is particularly disastrous. It reduces the current utilization of the process drastically when the concentration difference between the two solutions is high.

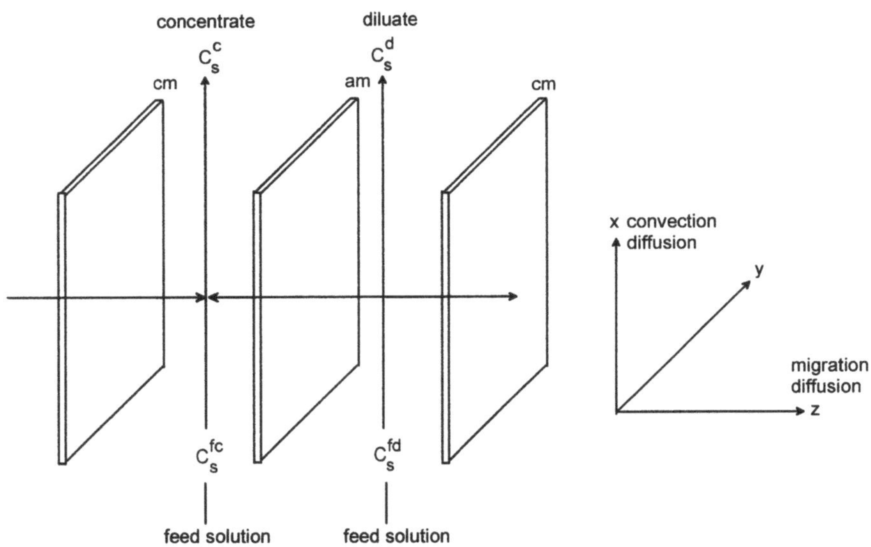

Fig. 4.2 Schematic diagram illustrating the mass transport in an electrodialysis cell pair due to convection and diffusion in the direction of the x coordinate and due to migration and diffusion in z-direction.

Two feed solutions of a concentration C_s^{fc} and C_s^{fd} enter the two compartments of a cell pair at the bottom and leave one compartment as the diluate with the concentration C_s^d and the other as the concentrate solution with the concentration C_s^c. Thus, due to an electrical current flow perpendicular to the membrane surface, i.e. in direction of the coordinate z, the salt concentrations in the compartments are changing. While the solution flows through the compartments in direction of the coordinate x a concentration gradient develops also in the x direction. The flux in x-direction is determined by convection due to a hydrostatic pressure gradient, and any diffusive flux in x-direction can be neglected. The flux in z-direction is determined by migration and diffusion due to an electrical potential gradient and concentration differences between the diluate and concentrate solutions. Concentration gradients in the boundary layers at the membrane surface also provide diffusive fluxes towards the membrane in the diluate and towards the bulk solution in the concentrate. Under steady state conditions all fluxes are constant and at a certain geometrical point in x-direction the flux in z-direction, i.e. perpendicular to the membrane surface, can be described by mass balance considering all fluxes in the diluate solution, the concentrate solution, and the cation- and anion-exchange membranes.

The mass transport in ion-exchange membranes and in electrolyte solutions is the result of gradients in the electrical potential and the electrolyte concentration as pointed out earlier. The electrical potential and concentration gradients in z-direction in an electrodialysis cell pair at a given point in x-direction are shown schematically in Figure 4.3.

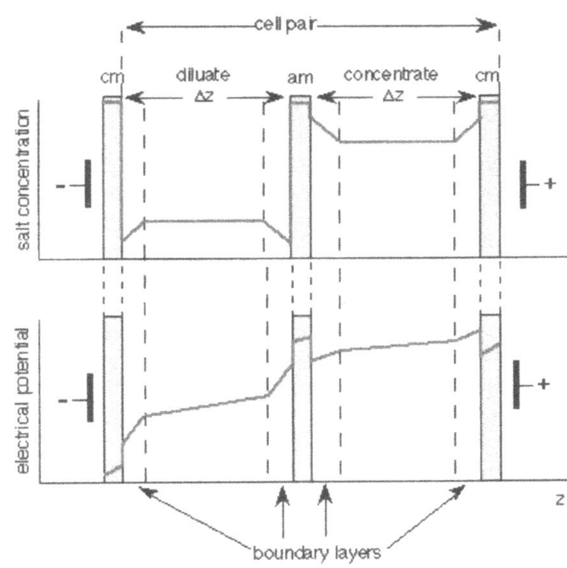

Fig. 4.3 Schematic diagram illustrating the concentration and electrical potential gradients in an electrodialysis cell pair consisting of two geometrically identical compartments.

The diagram shows the cross-section of a cell pair at a certain distance x from the entrance of the cells. At this point it is assumed that the solutions in the cells are well mixed and the concentrations are identical over the entire cross section of a cell except for the two boundary layers at the membrane surfaces. In these boundary layers the flow is laminar. Due to the fact that the flux of the counterion in the membrane is higher than in the solution the concentration in the boundary layer will decrease towards the membranes in the dilute solution and increase towards the membranes in the concentrate solution. In highly permselective ion-exchange membranes the concentration of the mobile ions in the membrane is to a first approximation identical to the fixed ions of the membrane which is equal over the membrane cross-section. Therefore, in the membrane and in the well mixed bulk solutions the ions are transported by migration under the driving force of an electrical potential gradient, and in the laminar boundary layers the ion fluxes are the result of migration and diffusion caused by concentration gradients. The fluxes in the solutions and the membranes are consecutive processes and therefore equal. The driving forces for these fluxes, however, are different, since the transport resistance, i.e. the electrical resistance is different in the membrane and the solutions.

The driving forces for the transport of ions in the solutions and in the membrane are gradients in the ion concentration and the electrical potential. The electrical potential gradients across a cell pair are also shown in Figure 4.3. The total potential difference across the cell pair consists of the Donnan potentials at the membrane surfaces and the potential drops due to the electrical resistance of the membranes and the solutions. The voltage drops due to the Donnan potentials represent an electromotive force and do not act as driving force for the mass transport, nevertheless they must be overcome by the applied electrical potential [K.S. Spiegler, 1958].

2.1.1 Material Balance Between the Diluate and Concentrate Flow Streams

For designing an electrodialysis desalination plant of a given capacity, feed solution composition, membrane properties, stack construction, and product solution properties the required membrane area, the current density, the applied voltage to achieve this current density, the necessary residence time of the feed solution in the stack, i.e. the feed flow velocity, and the process path length must be known. All these data can be calculated by a material balance of the entire mass transport in the stack.

The degree of desalination in an electrodialysis stack is determined by the total amount of ions removed from a feed solution and transferred to the concentrate solutions. It is given by differences between the feed and the concentrate solutions at the entrance in the diluate and concentrate cells and the exit of the cells. It is a function of the residence time of the different solutions in the cells, i.e. the flow velocities of the solutions and the applied current density.

The different concentrations are related by a material balance which takes the volume flow rates in the diluate and concentrate cells and the electric current passing through the stack into account as shown in the following equation:

$$\left(C_s^{fd} - C_s^d\right)Q^d = \left(C_s^c - C_s^{fc}\right)Q^c = \frac{\xi I}{\sum_c z_c v_c F} \tag{4.1}$$

Here C is the salt concentration, I is the total current passing through the cell pair, ξ is the current utilization, z the valence, v the stoichiometric coefficient, F the Faraday constant, Q^d and Q^c are the flow velocities of the diluate and concentrate solutions in the cells parallel to the membrane surface. The superscripts in C_S^{fd} and C_S^{fc} refer to the diluate and concentrate cell inlet concentrations; the superscripts d and c refer to diluate and concentrate, the subscripts s and c refer to salt solution and to cation.

The current utilization ξ is a measure for the amount of the total current which is passing through an electrodialysis stack that can be utilized in a given application and process design for the removal of the ions from a feed stream. The current utilization is always <1 because in any practical electrodialysis process not all of the current passing through the stack is utilized for desalting of the feed solution. Several factors may contribute to incomplete current utilization, such as not perfectly selective membranes, parallel current paths through the stack manifold, water transfer across the membranes due to osmosis and electroosmosis, etc. The current utilization, its causes and consequences will be discussed later.

The mass transport that occurs during electrodialysis in a stack is illustrated in Figure 4.4 which shows a cross-section through a cell pair containing two compartments, i.e. a diluate containing cell and a concentrate containing cell separated by two membranes, i.e. a cation- and an anion-exchange membrane between two electrodes.

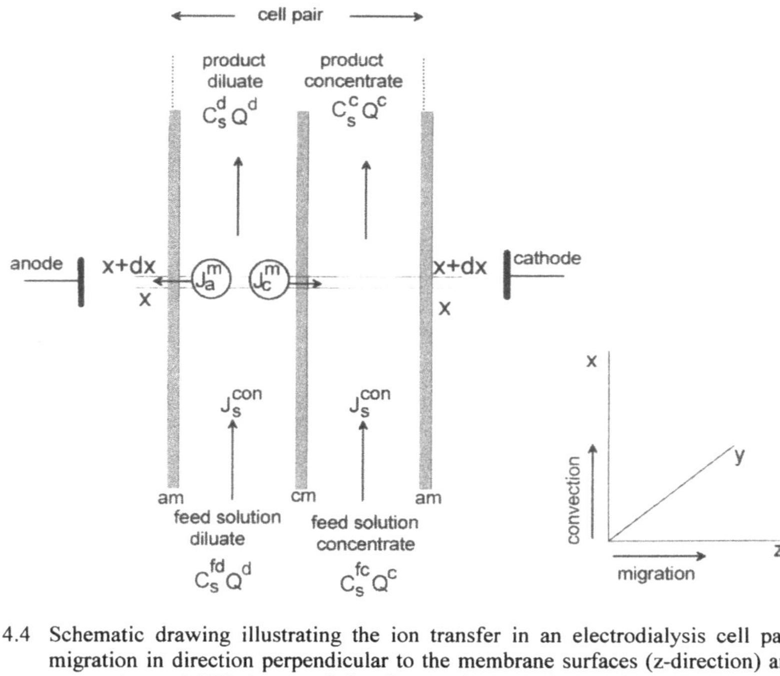

Fig. 4.4 Schematic drawing illustrating the ion transfer in an electrodialysis cell pair due to migration in direction perpendicular to the membrane surfaces (z-direction) and due to convection and diffusion parallel to the membrane surface (x-direction).

The mass transport in the cell pair is the result of fluxes caused by migration of cations and anions from the diluate to the concentrate solution through the membranes in z-direction and by convection of the solutions in x-direction from the entrance to exits of the cells. The migration of ions through the membranes due to the electrical potential driving force is proportional to the electrical current passing through the cell pair. Diffusion of ions in x-direction is extremely small and can be neglected compared to the convective flux in this direction. Diffusion of ions and solvent across the membrane due to concentration differences between the diluate and concentrate solution are in most applications of electrodialysis comparatively small and can be expressed by the current utilization which is always smaller than 1. The driving force for the ion transport, i.e. the electrical potential gradient across the cell pair does not change in x-direction from the cell entrance to the cell exit. The salt concentration, however, changes in both the diluate and the concentrate stream from the cell entrance to the cell exit. Therefore, the resistance of the solutions and thus the current density is changing correspondingly. The resistance in the concentrate is decreasing and that in the diluate is increasing while the solutions pass through the cells. It should be noticed that because of the electroneutrality requirements the absolute values of $|z_c|\nu_c$ and $|z_a|\nu_a$ are equal at any point in the solutions and the cat- and anion concentrations are proportional to the total salt concentration, i.e. $C_c = \nu_c C_s$ and $C_a = \nu_a C_s$. If it is assumed that the membrane is strictly permselective, i.e. T_c^{cm} and $T_a^{am} = 1$ and that $z_c = z_a$, the fluxes of the cations through the cation-exchange membrane and that of anions through the anion-exchange membranes J_c^m and J_a^m are identical and proportional to the total salt flux through both membranes J_s^m, i.e. $J_c^m = J_a^m = \nu_c J_s^m = \nu_a J_s^m$.

At steady state the concentration in a volume element V of the diluate and the concentrate stream is constant with time. It is changing however, in the solution in the cell in x-direction. This concentration change can be calculated by the material balance. The concentration change in the diluate and concentrate solutions along their pathway in the cells from x to x + dx is given by:

$$\frac{\partial\left(V^d C_s^d\right)}{\partial t} = \left(Q^d C_s^d\right)_x - \left(Q^d C_s^d\right)_{x+dx} + dA^m J_s^m = 0 \qquad (4.2)$$

Here C_s is the salt concentration, V is the volume, Q is the volume flow rate, dA^m is a membrane area element which is given by the cell width Y multiplied by the distance dx in direction of the flow stream parallel to the membrane surface, J_s^m is the salt flux through the membrane, t is the time, and x is a directional coordinate; the superscript d refers to diluate.

The term $(Q^d C_s^d)_{x+dx}$ in Equation 4.2 can be expressed as:

$$\left(Q^d C_s^d\right)_{x+dx} = \left(Q^d C_s^d\right)_x + \frac{\partial\left(Q^d C_s^d\right)}{\partial x} dx = \left(Q^d C_s^d\right)_x + \left(Q^d \frac{dC_s^d}{dx} + C_s^d \frac{dQ^d}{dx}\right) dx \qquad (4.3)$$

Combining Equations 4.2 and 4.3 leads to:

$$\frac{\partial\left(V^d C_s^d\right)}{\partial t} = -\left(Q^d \frac{dC_s^d}{dx} + C_s^d \frac{dQ^d}{dx}\right) dx + dA^m J_s^m = 0 \qquad (4.4)$$

At steady state the concentration does not change with time, i.e. $\frac{\partial\left(V^d C_s^d\right)}{\partial t} = 0$ and the volume flow rate in x-direction is approximately constant, i.e. $\frac{dQ^d}{dx} = 0$.

Rearranging Equation 4.4 then leads to:

$$dC_s^d = \frac{dA^m}{Q^d} J_s^m = d^{\Delta} C_s \tag{4.5}$$

Equation 4.5 describes the salt concentration change dC_s^d in the diluate cell element in x-direction under steady state conditions as a function of the volume flow velocity Q^d, and the salt flux J_s^m through the membrane area element dA^m; $d^{\Delta}C_s$ represents the salt concentration change due to the salt flux through the membrane area element dA^m while the solution having a volume flow velocity Q^d is passing this area element.

For strictly permselective membranes and ions with identical valence the salt flux through the membrane is proportional to the current density and given by:

$$J_s^m = \frac{J_c^{cm}}{\nu_c} = \frac{J_a^{am}}{\nu_a} = \frac{\xi\, i}{\sum_c |z_c| \nu_c F} = \frac{\xi\, i}{\sum_a |z_a| \nu_a F} \tag{4.6}$$

Here J_s^m, J_c^{cm} and J_a^{am} are the salt flux through the membranes, the cation flux through the cation-exchange membrane, and the anion flux through the anion-exchange membrane, respectively; i is the current density, ξ is the current utilization, F is the Faraday constant, z is the valence, and ν is the stoichiometric coefficient.

Introducing Equation 4.6 into 4.5 leads to:

$$dC_s^d = \frac{\xi\, i}{\sum_c |z_c| \nu_c F Q^d} dA^m \tag{4.7}$$

Equation 4.7 describes the concentration change in a volume element of the diluate cell as a function of the current density.

The concentration change in the concentrate cell in x-direction can be calculated accordingly and is given by:

$$dC_s^c = -\frac{\xi\, i}{\sum_c |z_c| \nu_c F Q^c} dA^m \tag{4.8}$$

Here C_s^d and C_s^c are the concentrations and Q^d and Q^c volume flow rates in the diluate and concentrate cells, i is the current density, ξ is the current utilization, z is the valence, ν is the stoichiometric coefficient and F the Faraday constant, the subscript c refers to cation, dA^m is an area element of the ion-exchange membrane which is given by the cell width Y and the distance in direction of the flow stream parallel to the membrane surface dx as depicted in Figure 4.4.

In Equations 4.7 and 4.8 it is furthermore assumed that concentration changes in the diluate and concentrate solution due to diffusion of ions and osmotic or electroosmotic water transport are expressed in the current utilization.

To a first approximation, i.e. neglecting concentration changes due to boundary layer effects, the current density through a cell pair can be expressed as a function of solution and membrane properties, the cell dimensions, and the applied voltage. It is given by:

$$i = \frac{U}{\frac{\Delta^d}{\Lambda_s C_s^d} + \frac{\Delta^c}{\Lambda_s C_s^c} + r^{am} + r^{cm}} \tag{4.9}$$

Here U is the applied voltage across the cell pair, Δ^d is the thickness of the diluate and Δ^c that of the concentrate cell, Λ_s is the molar conductivity of the solution, r^{am} and r^{cm} are the resistances of the anion- and cation-exchange membrane, and C_s^d and C_s^c are the bulk concentrations of the diluate and the concentrate solution, respectively.

In most electrodialysis stacks the diluate and concentrate cells have the same geometry as indicated in Figure 4.2 and the flow velocities of diluate and concentrate are kept equal, i.e. $\Delta^d = \Delta^c = \Delta$ and $Q^d = Q^c = Q$. Furthermore, the voltage drop across the cell pair is generally constant over the entire cell length. Under these assumptions the change in the salt concentration in the diluate and concentrate solution in the area element dx is given by combining Equations 4.5, 4.7, 4.8 and 4.9 to:

$$dC_s^d = -dC_s^c = d^{\Delta}C_s = \frac{U}{\frac{\Delta(C_s^c + C_s^d)}{\Lambda_s C_s^c C_s^d} + r^{am} + r^{cm}} \frac{\xi Y dx}{\sum_c |z_c| \nu_c F Q} \tag{4.10}$$

Here $d^{\Delta}C_s$ is the concentration change due to the salt transferred from the diluate to the concentrate solution by migration through the cell pair area element dA^m which is given by the cell width Y multiplied by dx.

The salt transfer from the diluate to the concentrate from the cell entrance to the cell exit is given by the mass balance applied to the salt:

$$C_s^c - C_s^{fc} = C_s^{fd} - C_s^d = {}^{\Delta}C_s \tag{4.11}$$

Here C_s^c and C_s^d are the salt concentrations of the concentrate and the diluate at the exit of the cells, and C_s^{fc} and C_s^{fd} are the concentrations of the concentrate and the diluate at the entrance of the cells.

The change in concentration of the diluate and concentrate solutions from the entrance to the exit of the cell is obtained by integration over the cell length.

Rearranging Equation 4.10 and introducing Equation 4.11 leads to:

$$\frac{d^{\Delta}C_s}{(C_s^{fd} - {}^{\Delta}C_s)(C_s^{fc} + {}^{\Delta}C_s)} + d^{\Delta}C_s \frac{\Lambda_s (r^{am} + r^{cm})}{\Delta(C_s^{fc} + {}^{\Delta}C_s^{fd})} = \frac{\Lambda_s \xi U Y dx}{\sum_c |z_c| \nu_c F Q (C_s^{fc} + C_s^{fd})} \tag{4.12}$$

Integration with the boundary conditions that at the cell entrance is x = 0 and ${}^{\Delta}C_s = 0$ and at the cell outlet is x = X and ${}^{\Delta}C_s = {}^{\Delta}C_s$ leads to:

$$\int_0^{{}^{\Delta}C_s} \frac{d^{\Delta}C_s}{\left(C_s^{fd} - {}^{\Delta}C_s\right)\left(C_s^{fc} + {}^{\Delta}C_s\right)} + \frac{\Lambda_s\left(r^{am} + r^{cm}\right)}{\Delta\left(C_s^{fc} + C_s^{fd}\right)} \int_0^{{}^{\Delta}C_s} d^{\Delta}C_s = \int_0^{C_s} \frac{\Lambda_s \xi\, UY dx}{\sum_c |z_c| \nu_c\, FQ\left(C_s^{fc} + C_s^{fd}\right)} \tag{4.13}$$

Integration of Equation 4.13 leads to:

$$\frac{1}{\left(C_s^{fc} + C_s^{fd}\right)} \ln \frac{C_s^{fd} - {}^{\Delta}C_s}{C_s^{fc} + {}^{\Delta}C_s} + \frac{1}{\left(C_s^{fc} + C_s^{fd}\right)} \ln \frac{C_s^{fd}}{C_s^{fc}} + \frac{\Lambda_s\left(r^{am} + r^{cm}\right)}{\Delta\left(C_s^{fc} + C_s^{fd}\right)} {}^{\Delta}C_s = \frac{\Lambda_s\, \xi\, U\, Y\, X}{\sum_c |z_c| \nu_c\, FQ\left(C_s^{fc} + C_s^{fd}\right)} \tag{4.14}$$

Rearrangement of Equation 4.14 leads to a very simple relation for the product and brine concentration in electrodialysis as a function of various stack designs and operating parameters:

$$\ln \frac{C_s^c C_s^{fd}}{C_s^d C_s^{fc}} + \frac{\Lambda_s\left(r^{am} + r^{cm}\right)\left(C_s^{fd} - C_s^d\right)}{\Delta} = \frac{\Lambda_s \xi\, UYX}{\Delta \sum_c |z_c| \nu_c\, FQ} \tag{4.15}$$

Here C_s^{fd} and C_s^{fc} are the concentrations of the diluate and the concentrate at the cell inlet, C_s^d and C_s^c are the concentrations of the diluate and the concentrate at the cell outlet, Λ_s is the equivalent conductivity of the salt solution, r^{am} and r^{cm} are the area resistances of the anion- and cation-exchange membrane, Δ is the cell thickness, i.e. the distance between the cation- and anion-exchange membrane, Q is the flow velocity of the diluate and the concentrate stream, respectively, U is the voltage drop across a cell pair, ξ is the current utilization, F is the Faraday constant, Y and X are the width and the length of a cell, i.e. $Y \times X = A$, the area of the cation- or anion-exchange membrane in the cell.

Equation 4.15 shows that for given feed solution, membrane properties, flow velocity, cell geometry, and voltage drop the concentration of the brine and the diluate is an exponential function of the cell length. It can be considered as a basis for designing and operating an electrodialysis stack. Significant deviations from the simple Equation 4.15 is obtained when excessive water transfer as the result of electroosmosis or salt diffusion occurs between the diluate and concentrate streams or when the flow velocities and the cell geometry are different. In this case Equation 4.15 must be modified accordingly.

For the calculation of the mass transport in electrodialysis it is convenient to express the concentration of the electrolyte not in moles per unit volume but in equivalent per unit volume. This is done by multiplying the molar concentration by the absolute values of the valence and by the stoichiometric coefficient either of the anion or the cation.

For an electrolyte solution containing more than one salt the equivalent concentration is given by:

$$C_{eq} = \sum_s \frac{|z_c \nu_c| + |z_a \nu_a|}{2} C_s = \sum_c z_c \nu_c C_s = \sum_a |z_a| \nu_a C_s \tag{4.16}$$

Here C_{eq} is the equivalent concentration, C_s is the salt concentration, z the valence of the ions, and ν is the stoichiometric coefficient The subscripts a and c refer to anion and cation.

2.2 Membrane Counterion Selectivity

Feed solutions which are treated by electrodialysis often contain more than one electrolyte, i.e. they contain more than one cation or anion species. The transport rate of coions in an ion-exchange membrane is always very low because of the Donnan exclusion, and the transference number is close to zero. The transport number of counterions in an ion-exchange membrane is always quite high, but it may be rather different for different ions. The desalination effect in electrodialysis of solutions containing more than one electrolyte can be and generally is different for the different salts. The transport rates of different ions through a membrane are proportional to their permeability which is the product of their concentration and mobility in the membrane matrix. In a solution the mobility of the ions is determined by the hydrated ion radius. The mobility of different ions in an aqueous solution do not differ very much from each other. An exception are the H^+- and OH^--ions. Their mobility is ca. 6 times higher than that of other ions because of the different transport mechanism of protons and hydroxide ions. Since in water protons form clusters of hydronium ions as indicated in Figure 2.16 a) they can be transferred from one hydronium ion to the next by a so-called tunnel mechanism as indicated in Figure 2.16 b). This transport mechanism explains the extraordinary high mobility of protons in water. The mobility of different ions in an aqueous solution are listed in Table 2.3. This table shows that small ions such as Li^+ or Na^+ have lower mobility than larger ions such as Ca^{2+} or SO_4^{2-} because of the smaller hydration shell and higher charges of the divalent ions.

The transport of different ions in ion-exchange membranes has been discussed in more detail earlier. In an ion-exchange membrane the concentration of co- and counterions is determined by the Donnan exclusion as expressed by Equation 2.102 which indicates that co-ions are more or less completely excluded from the membrane and their permeability is correspondingly low. Therefore, ion-exchange membranes are highly selective for counterions, i.e. ions that carry the opposite charge than the fixed ions of the membrane. But ion-exchange membranes often also show significant difference in their permeability for different counterions. The permeability of counterions in a membrane is determined by the ion-exchange selectivity and the mobility selectivity. The mobility of different ions is determined mainly by steric effects, i.e. the size of the ions and the cross-linking density of the membrane. The ion-exchange selectivity is determined by electrostatic interactions which is referred to as "electroselectivity" in contrast to the size selectivity [F. Helfferich, 1962]. As a general rule the counterions with the higher valence have a higher ion-exchange selectivity than ions with lower valence, and ions with lower valence and the smaller radius have a higher mobility in ion-exchange membranes.

An exception from this rule is the transport of protons which have, because of the formation of hydronium water clusters, not only exceptionally high permeability in water and in cation-exchange membranes, they are also hardly rejected by anion-exchange membranes for the same reason. In a highly swollen anion-exchange membrane the protons are integrated in water clusters and therefore to a much lesser extent effected by Donnan exclusion than salt cations that are more or less completely excluded from the membrane matrix due to their electric charge. The same mechanism as the one described for the transport of protons also holds true for the transport of hydroxide ions, and thus the permeability of hydroxide ions in a cation-exchange membrane is much higher than that of other salt anions.

2.3 Concentration Polarization and Limiting Current Density

An electric current passing through an electrodialysis cell pair is carried in the solution by both cat- and anions according to their transference numbers which are in aqueous solutions not very different for cat- and anions. In the ion-exchange membranes, however, the current is carried mainly by the counterions which have a transference number in the membrane very close to 1. The separation in electrodialysis is the result of differences in the transport numbers of ions in the solution and in the ion-exchange membranes. At the surface of an anion-exchange membrane facing the diluate solution the concentration of ions in the solution is reduced because of the lower transport number of the anions in the solution than in the membrane. The number of anions transported in the solution to the surface of the anion-exchange membrane by migration, i.e. carried by the electric current, is lower than the number of anions removed through the membrane. Because of the electroneutrality requirements the number of cations is reduced in the boundary layer by migration in the opposite direction. The net result of the different transference numbers of the anions in the solution and in the anion-exchange membrane is a reduction of the electrolyte concentration in the solution at the surface of the membrane. Thus, a concentration gradient is established in the solution between the membrane surface and the well mixed bulk. This concentration gradient results in a diffusive electrolyte transport. A steady state situation is obtained when the additional ions that are needed to balance those removed from the interface due to the faster transport rate in the membrane are supplied by the diffusive transport. The other side of the anion-exchange membrane is facing the concentrate solution. Here, an accumulation of negatively charged ions occurs, because more ions are transferred through the membrane that can be carried away by the electric current due to the lower transport numbers in the solution than in the membrane. Thus, the concentration of ions in the solution at the surface of the membrane increases and a concentration gradient of the salt is established in the solution which results in an additional diffusive salt flux from the membrane into the solution. The concentration gradients and thus the diffusive flux are limited to a thin boundary layer at the membrane surface due to turbulent mixing of the bulk solution. In almost all electrodialysis stacks the diluate and concentrate solutions flow through a channel formed by parallel planar ion-exchange membranes, and velocity gradients extend from the surface of the membrane into the solutions towards the center of the channel. Due to the solution flow velocities and the spacer usually installed between the membranes, the bulk of the solution is well mixed and both the velocity and the concentration gradient are limited to a relatively thin layer at the membrane surface as indicated in Figure 4.5 which shows the profiles of the salt concentration in the solutions near the surface of the cation exchange membrane under steady state conditions and the fluxes of salt, cation and anion. In the membranes cations are transported mainly by migration indicated by J_c^{mig} in Figure 4.5. Anions are more or less completely excluded from the cation-exchange membrane and their flux through the membrane by migration and diffusion in opposite direction is compared to that of the cations generally very small. It is expressed in the current utilization as will be discussed later. In the boundary layers at the membrane surfaces both cation and anions are also transported by migration, however, in opposite directions and thus, generating a concentration gradient leading to a diffusion of salt J_s^{diff} as indicated by the arrows in Figure 4.5. In the diluate side of the membrane the diffusive flux is towards the membrane and at the concentrate side it is towards the bulk solution following the concentration gradients in the boundary layers. In the well mixed bulk solution the transport of ions is by migration only since there are no concentration gradients in the direction perpendicular to the membrane surface. The different fluxes lead to a depletions of salt at the membrane surface in the diluate solution and an increase of the salt concentration at the membrane surface in the concentrate solution. At the anion-exchange membrane boundary layers are developed accordingly [T. Sata et al., 1969]. In Figure 4.5 the symbols J and C denote the flux and the concentration, the superscripts mig and diff refer to migration and diffusion, the superscripts d and c refer to diluate and concentrate solution, and the superscripts b and m refer to bulk phase and membrane surface, respectively, the subscripts a and c refer to anion and cation.

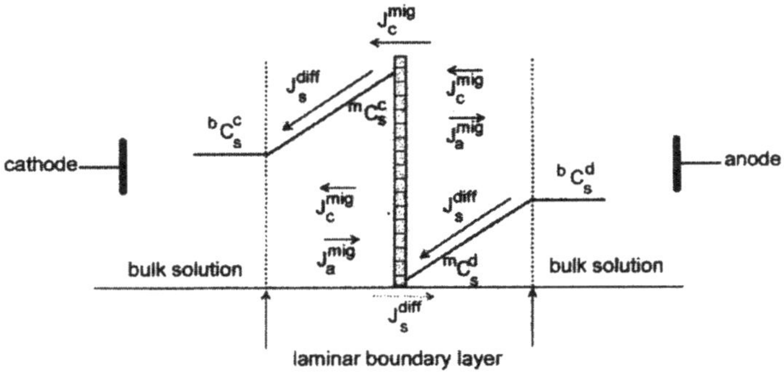

Fig. 4.5 Schematic drawing illustrating concentration profiles of a salt in the boundary layer on both sides of a cation-exchange membrane and the fluxes of cations and anions in the boundary layer, and in the membrane.

The depletion and accumulation of ions at the membrane surfaces in electrodialysis is referred to as concentration polarization. The concentration polarization occurring in electrodialysis, i.e. the concentration profiles at the membrane surface can be calculated by a mass balance taking into account all fluxes in the boundary layer and the hydrodynamic conditions in the flow channel between the membranes. A rigorous treatment of concentration polarization effects is quite complex [R.B. Bird et al., 1965]. To a first approximation, however, concentration polarization phenomena in electrodialysis can be treated by applying the so-called Nernst film model which assumes that the solution in a cell between two membranes can be divided into a bulk phase between two laminar boundary layers at the membrane surfaces [M. Taky et al., 1992]. The bulk solution between the laminar boundary layers is well mixed and has a uniform concentration, whereas the concentration in the boundary layers changes over the thickness. It is assumed that under steady state conditions in the bulk phase the electrical potential gradient is the only driving force for the flux of ions in the direction towards the membranes while in the boundary layer the concentration and electrical potential gradients perpendicular to the membrane surfaces are causing the mass transport. Furthermore, it is assumed in the film model that the boundary layer thickness and the concentration gradients in the solution are constant along the flow channel from the cell entrance to the exit, and there are no viscosity differences between any points in the solution. In a practical electrodialysis stack there will be entrance and exit effects and viscosity as well as concentration differences between the solutions in the entrance and exit region of the cell, and the idealized model hardly exists. Nevertheless, the Nernst model provides a very simple approach to the mathematical treatment of the concentration polarization, which results in an expression that is easy to use in the designing of electrodialysis stacks and that predicts the performance for most practical applications of electrodialysis in desalination adequately [J.H. Choi et al., 2002].

It can be assumed to a first approximation that the transport of cations through a highly permselective cation-exchange membrane is the result of migration only caused by an electrical potential gradient as driving force since there is no significant concentration gradient within the membrane because of the exclusion of the anions.

For this idealized condition the flux of the counterion through an ion-exchange membrane can be calculated from a mass balance which relates the flux through the membrane to the fluxes in the boundary layer. To a first approximation the counterion flux through a highly permselective membrane can be expressed by:

$$J_i^m = J_i^{diff} + J_i^{mig} \tag{4.17}$$

Here J_i^m is the total flux of a counterion through an ion-exchange membrane, J_i^{diff} and J_i^{mig} are the diffusive and migration flux in the boundary layers.

The total flux of the counterion through the membrane is furthermore given by:

$$J_i^m = T_i^m \frac{i}{|z_i|\,F} \tag{4.18}$$

Here T_i^m is the transport number of the counterion in the membrane.

The transport of counterions through the boundary layer at the diluate side of the membrane due to migration and diffusion is given by:

$$J_i^{diff} + J_i^{mig} = -D_i \frac{dC_i^d}{dz} + T_i^s \frac{i}{|z_i|F} \tag{4. 19}$$

Here J is the flux, D is the diffusion coefficient, T is the transport number, F is the Faraday constant, z is the valence, and dz is a directional coordinate, the subscript i refers to cat- or anions; the superscripts mig and diff refer to ion migration and ion diffusion, respectively, and the superscript s refers to the solution.

Introducing Equations 4.18 and 4.19 into Equation 4.17 and rearranging gives the current through an ion-exchange membrane as a function of the ion concentration gradient in the solution, i.e. in the boundary layer, the diffusion coefficient in the solution, and the transference numbers of the ion in the membrane and in the solution:

$$i = \frac{F D_i}{z_i \left(T_i^m - T_i\right)} \frac{dC_i^d}{dz} \tag{4.20}$$

Due to the electroneutrality condition which requires that on a macroscopic scale there is no excess of positive or negative charges and the fluxes of cations and anions are interdependent. In the solution is:

$$z_a \times C_a = z_c \times C_c \tag{4.21}$$

The charges carried by the cations in or out of a solution is identical to the charges carried by the anions.

Furthermore, the concentrations of the cations and anions is proportional to that of the electrolyte, i.e.:

$$C_a = \nu_a \times C_s \text{ and } C_c = \nu_c \times C_s \tag{4.22}$$

For a monovalent salt, or if the concentration is expressed in equivalent, the absolute values of the salt and the ion fluxes are identical, i.e.:

$$|J_s| = |J_a| = |J_c| \tag{4.23}$$

If ions are transported in a system by diffusion only due to a concentration gradient both the cations and the anions are transported in the same direction. If ions are transported by migration due to the driving force of an electrical potential gradient cations and anions move in opposite direct and an electrical current is obtained.

Due to the electroneutrality requirements and the coupling of the ion and salt fluxes the diffusion coefficient of the individual ions must be replaced by an average diffusion coefficient according to Equation 2.165.

Thus, in the solution is:

$$\overline{D}_c = \overline{D}_a = D_s = \frac{D_a D_c (|z_a| + z_c)}{D_a |z_a| + D_c z_c} \tag{4.24}$$

Here D and $\overline{D}$ are the diffusion coefficient and the average diffusion coefficient of the cat- or anion in the solution, D_s is the diffusion coefficient of the salt in the solution.

Furthermore, due to the electroneutrality expressed in Equations 4.21 and 4.22 are the ion concentration gradients in the boundary layer in the diluate cell:

$$\frac{z_c dC_c}{dz} = \frac{z_a C_a}{dz} = \frac{z_c \nu_c C_s}{dz} = \frac{z_a \nu_a C_s}{dz} \tag{4.25}$$

Replacing the diffusion coefficient and the concentration of the individual ions in Equation 4.20 by the salt diffusion coefficient and the salt concentration in the solution according to Equations 4.24 and 4.25 and integrating over the thickness of the laminar boundary layer gives the current density through an electrodialysis cell as a function of the salt concentrations in the bulk solution.

$$i = \frac{FD_s}{z_c \nu_c (T_c^m - T_c)} \left(\frac{\Delta C_s^d}{\Delta z} \right) = \frac{FD_s}{z_c \nu_c (T_c^m - T_c)} \left(\frac{{}^b C_s^d - {}^m C_s^d}{\Delta z} \right) = \frac{FD_s}{z_a \nu_a (T_a^m - T_a)} \left(\frac{{}^b C_s^d - {}^m C_s^d}{\Delta z} \right) \tag{4.26}$$

Here T^m and T are the transport numbers in the membrane and in the solution, D_S is the diffusion coefficient of the salt in the solution, ${}^mC_s^d$ and ${}^bC_s^d$ are the salt concentrations in the solution at the membrane surface and in the well mixed bulk solution, Δz is the thickness of the boundary layer, z is the valence, ν is the stoichiometric coefficient, F the Farady constant, and the subscripts c and a refer to cation and anion.

Equation 4.26 describes the current through an electrodialysis cell as a function of the transport numbers of the ions in the membrane and in the solution, the diffusion coefficient of the salt in the solution and the concentration difference between the solution at the membrane surface and the bulk phase and the thickness of the laminar boundary layer.

Since at constant hydrodynamic conditions for a given solution and ion-exchange membranes D_S, T_c, T_a, T_c^m, T_a^m, Δz, and ${}^bC_s^d$ are constant a change in the current density will result in a change of the concentration in the boundary layer at the membrane surface.

The transport number of the cation in a cation-exchange membrane is generally not very different from the transport number of the anion in the anion-exchange membrane and much larger than its transport number in the solution, i.e. $T_c^m > T_c$. Therefore, the flux of cations through the boundary layer solution to or away from the membrane as a result of the electrical current results in a depletion of salt on the side of the membrane facing the diluate solution, and a concentration of salt on the side of the membrane adjacent to the concentrate solution.

2.4 The Limiting Current Density, Causes and Consequences

The consequences of concentration polarization in electrodialysis are twofold. In the diluate containing cell the salt concentration at the membrane surface is decreased and in the concentrate containing cell increased. Both effects impair the technical feasibility and the economics of the process. When, due to concentration polarization, the salt concentration in the concentrate cell exceeds the solubility limits of the solution constituents, precipitation of salts may occur resulting in an additional electrical resistance and eventual damage of the membrane. When, due to concentration polarization, in the diluate containing cell the salt concentration at the membrane surface is reduced to zero there are no more salt ions available to carry the electric current and the electrical resistance. Thus, the voltage drop across the boundary layer increases drastically resulting in a higher energy consumption and in an electric field enhanced water dissociation [K.S. Spiegler, 1971]. The consequence of the water dissociation is a loss of current utilization and drastic pH-value shifts, with an increase of the pH-value at the surface of the anion-exchange membrane in the concentrate containing cell and a decrease of the pH-value at the surface of the cation-exchange membrane. The increase in the pH-value can lead to a precipitation of multivalent ions on the membrane surface and the decrease of the pH-value can damage the membranes. Thus, in a practical application of electrodialysis the concentration polarization effects should be minimized and especially water dissociation should be avoided. For a given cell design and feed solution and required diluate and concentrate solution concentration polarization can only be reduced effectively by reducing the current density or by decreasing the thickness of the laminar boundary layers at the membrane surfaces as indicated earlier. The thickness of the laminar boundary layer is determined by the hydrodynamic flow conditions which are a function of the cell design and feed flow velocity. When the hydrodynamic flow conditions are kept constant the current density is limited. It will reach a maximum value independent of the electrical potential gradient. This maximum current density is reached when the salt concentration at the membrane surface in the diluate cell becomes 0. This current density is referred to as the limiting current density. Thus i = i_{lim} for ${}^{m}C_{s} \rightarrow 0$:

$$i_{lim} = \frac{FD_s}{z_c\left(T_c^m - T_c\right)}\left(\frac{{}^{b}C_s^d}{\Delta z}\right) \tag{4.27}$$

According to Equation 4.26 which is based on the classical picture of concentration polarization the limiting current cannot be exceeded unless other ions than the salt ions become available for the transport of current in the salt depleted boundary layer. This is the case when water dissociation leads to the production of H^+- and OH^--ions which then will carry the electric charges with the result of a so-called over-limiting current. In the paste it had been assumed that the over-limiting current is mainly due to the migration of the H^+- and OH^--ions. In several experimental studies, however, it has been demonstrated that the over-limiting current is only partly caused by the transport of protons and hydroxyl ions. The major part of the current is still carried by salt ions [T. Aritomo et al., 1996], [J.J. Krol et al., 1999]. The reason for salt transport in the over-limiting current regime is not quite clear and still a subject of intensive discussions in the literature [I. Rubinstein et al., 1979], [R. Simons, 1979], [J.A., Manzanares et all., 1991].

The limiting current can be determined by Equation 4.27 by assuming that the limiting current density is reached when the salt concentration at the membrane surface in the diluate cell goes towards zero. Generally the thickness of the laminar boundary layer in an electrodialysis cell is difficult to measure experimentally. Therefore, boundary layer thickness and diffusion coefficient in Equation 4.27 are generally expressed by a mass transfer coefficient k_s^s which is given by:

$$k_s^s = \frac{D_s}{\Delta z} \tag{4.28}$$

Thus is the limiting current density:

$$i_{lim} = \frac{F\,k_s^s}{z_c\left(T_c^m - T_c\right)}\,{}^{b}C_s^d \tag{4.29}$$

The mass transfer coefficient in the boundary layer is a function of the hydrodynamics of the feed solution, i.e. the flow velocities, the cell geometry, the spacer design, and the salt diffusion coefficient.

2.4.1 Experimental Determination of the Limiting Current Density

The mass transfer coefficient is a function of the Schmidt and Reynolds number [R.B. Bird et al.,1965] and thus a function of the flow velocity in the cell. Because of the spacers the hydrodynamic situation in an electrodialysis cell is generally quite complex and it is difficult to calculate the mass transfer in the boundary layer on a strictly theoretical basis.

Therefore, in praxis generally an empirically derived expression is used to describes the limiting current density as a function to the feed flow velocity in the electrodialysis stack [H.J. Lee, et al., 2002]. The limiting current density is expressed by:

$$i_{lim} = a\,u^{b}\;C_{s}^{d} \tag{4.30}$$

Here C_s^d is the concentration of the solution in the diluate cell, u is the linear flow velocity of the solution through the cells parallel to the membrane surface, and a and b are characteristic constants for a given stack design and must be determined experimentally. This is done in practice by measuring the limiting current density in a given stack configuration and constant feed solution salt concentrations as a function of the feed flow velocity.

Since the laminar boundary layers in a cell are very thin the concentration in the diluate cell is to a first approximation equal to the bulk concentration. Thus is $C_s^d \cong {}^{b}C_s^d$.

The limiting current density can be determined by measuring the current as a function of the applied voltage across a membrane. The procedure is illustrated in Figure 4.6 which shows the current density measured through a cation-exchange membrane (CMX of Tokuyama Soda) in a 0.05 molar NaCl-solution. The slope of the curve is inverse proportional to the electrical resistance of the solutions in the boundary layers at the membrane surfaces. The current versus voltage curve shows a more ore less linear increase of the current with an increase of the applied voltage up to a voltage drop of ca. 0.23 V. At this point the slope of the curve decreases drastically, i.e. the current increases only marginally with the increasing voltage drop. The reason for the drastic increase in the boundary layer resistance is the complete depletion of the salt at the membrane surface facing the diluate solution. The current at the point of the drastic resistance change in the boundary layer is referred to as the limiting current density.

By measuring the limiting current density in a given stack design and diluate concentration the constants a and b can be determined from the slope of the curve which is obtained by plotting the limiting current density versus the solution flow velocity.

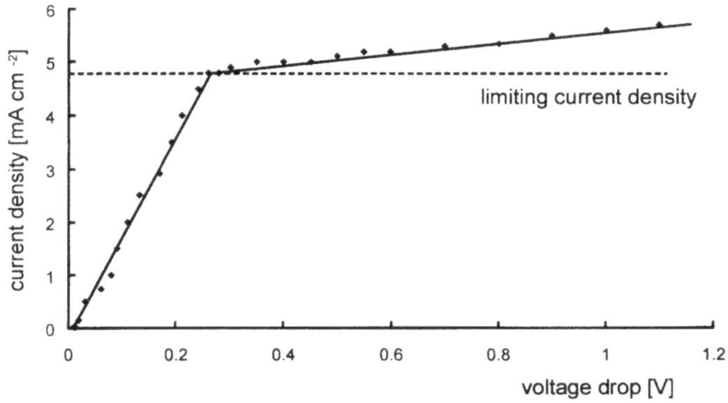

Fig. 4.6 Experimentally determined current versus voltage curve measured with a cation-exchange membrane (CMX of Tokuyama Soda) in a 0.05 molar NaCl-solution.

The measurement of the current versus voltage curve shown in Figure 4.6 is carried out with a single membrane in a laboratory type test cell. Measurements in a multi cell stack often do not show such a clear indication of the change in the resistance when the limiting current density is exceeded. Therefore, Cowan and Brown devised another analytical procedure for determining the limiting current density by plotting the overall resistance versus the reciprocal current density as shown schematically in Figure 4.7 [D.A. Cowan et al., 1959].

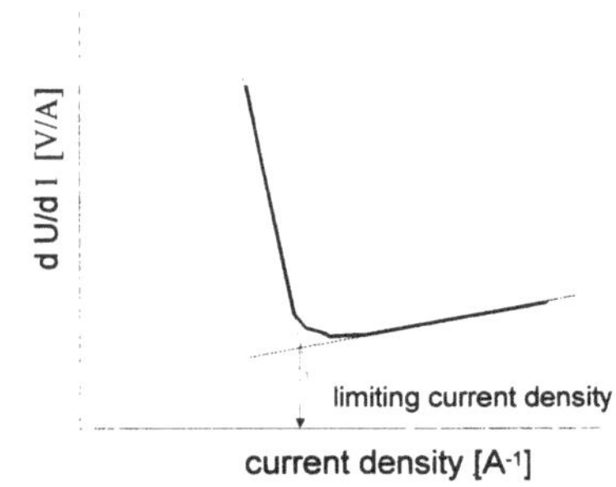

Fig. 4.7 Schematic drawing illustrating the determination of the limiting current density by plotting the electrical resistance of an electrodialysis stack versus the reciprocal current density.

The curve shows a sharp change in the resistance when the limiting current density is reached.

Additionally, a change in the pH-value in the diluate solution can be used to determine the limiting current density. When the limiting current density is exceeded a sharp drop in the pH-value of the diluate solution is observed which is due to the fact that water dissociation occurs mainly at the anion-exchange membrane probably because of the catalytic effect of the tertiary amine groups at the surface of the membrane [R. Simons, 1985]. The protons that migrate from the surface of the anion-exchange membrane are neutralized by hydroxide ions that are generated at the surface of the cation-exchange membrane. But since more water is dissociated at the anion-exchange membrane than at the cation-exchange membrane the overall result is an decrease in the pH-value.

2.4.2 The Overlimiting Current Density

The phenomenon of the overlimiting current density is illustrated in Figure 4.8. This figure shows that in an experimentally determined current density versus voltage curve three distinct regions can be determined. In the first region (indicated by I) the resistance of the cell is fairly constant, i.e. the current density is increasing more or less linearly with the applied voltage according to Ohm's law. When a certain current density is reached the cell resistance increases drastically. The current density where this drastic resistance increase occurs is referred to as the limiting current density. When the limiting current density is reached an increase in the applied voltage does not lead to a significant increase in the current density until the so-called overlimiting current density is reached, and the cell resistance decreases again if the voltage is further increased. When the overlimiting current density is reached the current density increases again linearly with the applied voltage. The limiting current density is determined by the intersection of the first and the second slope of the current versus voltage curve. The begin of the overlimiting current density region is determined by the intersection of the second and the third slope of the current versus voltage curve. The region which is dominated by a high cell resistance is indicated by II and that of the overlimiting current density by III in Figure 4.8. It is interesting to see that the cell resistance in the region below the limiting current density (region I) is about the same as in the region of the overlimiting current density (region III) and significantly lower as in the region between limiting and overlimiting current density (region II).

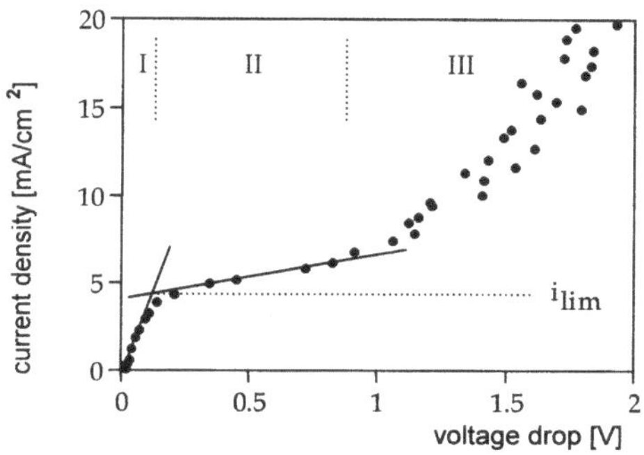

Fig. 4.8 Experimentally determined current versus voltage curve measured in a laboratory electrodialysis stack with a 0.05 molar NaCl-solution [J.J. Krol, 1997].

The reason for this so-called overlimiting current density is still not yet completely understood. A certain amount of the current in the overlimiting current density region is transported by the protons and hydroxyl ions generated by electrodialytic water dissociation at the anion-exchange membrane. However, as has been demonstrated [J.J. Krol et al., 1999] this contribution is very small and generally less than one percent of the overall current. This is illustrated in Figure 4.9

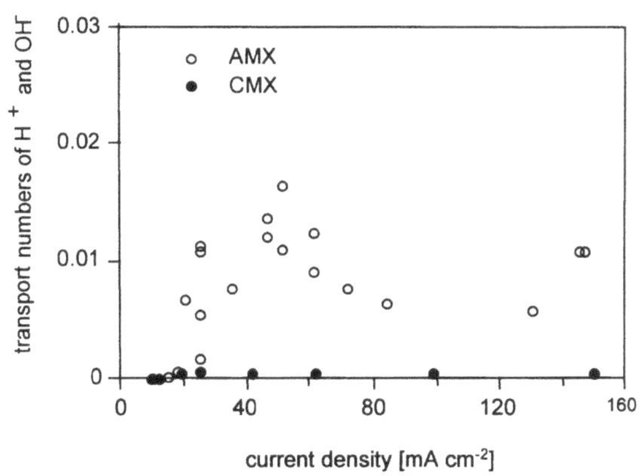

Fig. 4.9 The transport number of protons and hydroxide ions as function of the current density measured in 0.1 n NaCl solution [J.J. Krol 1997].

The Figure 4.9 shows the transport numbers of protons and hydroxide ions measured in a 0.1 n NaCl solution at different current densities which were well above the limiting current density which was ca. 10 mA cm^{-2} for the given system. The graph indicates that in the overlimiting current density region the contribution of the protons and hydroxide ions to the transport of current is negligibly low. Therefore, in addition to the diffusive transport of the salt in the laminar boundary layer a mechanism must exist that transports additional salt towards the membrane surface in the region of the overlimiting current. Based on theoretical studies published in the literature [I. Rubinstein et al., 1984], [I. Rubinstein et al., 1991], a mechanism for the additional salt transport is suggested that is based on electro-convection [J.J. Krol et al., 1997].

2.5 Membrane Fouling and Poisoning

Suspended and colloidal matters, polyelectrolytes, organic anions, and multivalent salts near the saturation level can cause severe problems in electrodialysis due to precipitation on the membrane surfaces or by partial penetration into the membranes. Precipitation of suspended matter, silicates, and salts with low solubility such as calcium carbonates or iron hydroxides may occur within the actual flow channels resulting in high hydrodynamic pressure losses and non-uniform flow distribution in the stack. Precipitation on the surfaces of the membranes also causes an increase of the electrical resistance of the stack and may lead to physical damage of the membranes. Especially organic anions such as humates can precipitate on the anion-exchange membranes as humic acid and cause a sharp increase in the electric resistance. Mechanical cleaning and treatment with dilute bases and acids can generally restore the original properties of the membranes. More severe is the poisoning of membranes by organic anions that are small enough to penetrate the membranes but whose electromobility is so low that they virtually remain inside the membrane causing a drastic increase in the membrane resistance. Certain detergents are also the cause of this type of poisoning which is very difficult to deal with and can best be avoided by a proper pre-treatment of the feed solution.

The various pre-treatment procedures such as precipitation, flocculation or ion-exchange and rinsing cycles can be substantially reduced by a simple but very effective operating mode which is referred to as electrodialysis reversal. In the electrodialysis reversal operating mode which has been developed by Ionics Incorporated [L.R. Siwak, 1992], [W.A., McRae, 1983] the polarity of the electric field applied to the electrodialysis stack as the driving force for the transport of ions is reversed in certain time intervals. Simultaneously the flow streams are reversed, i.e. the diluate cell becomes the concentrate cell and vice versa with the result that matter that has been precipitated at the membrane surface will be redisolved and removed with the flow stream passing through the cell. The principle of the electrodialysis reversal operating mode is illustrated in Figure 4.10. which shows a typical electrodialysis cell formed by a cation- and anion-exchange membrane between two electrodes. If an electric field is applied to a feed solution containing for instance negatively charged particles or large organic anions these components will migrate to the anion-exchange membrane and will be deposited on its surface. If the polarity is reversed the negatively charged components will now migrate away from the anion-exchange membrane back into the bulk solution and the membrane properties are restored.

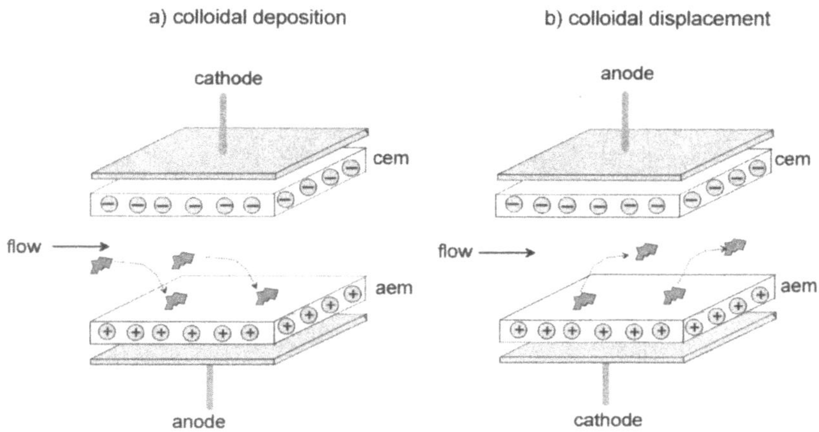

Fig. 4.10 Schematic drawing illustrating the removal of deposited negatively charged colloidal components from the surface of an anion-exchange membrane by reversing the electric field.

This procedure has been very effective not only for the removal of precipitated colloidal materials but also for removing precipitated inorganic salts. In the practical application of electrodialysis in water desalination the flow streams are reversed with the reversed polarity. Thus, the diluate stream becomes the concentrate stream and the concentrate a diluate stream. Generally, the flow streams and the polarity are changed every 30 to 60 minutes. The actual process of changing the flow streams and the polarity takes only a few seconds. During this time the diluate composition does not meet the product specification and is disposed with the brine. Due to the polarity change the productivity of an electrodialysis plant is reduced by ca 1 to 3%.

2.6 Energy Requirements in Electrodialysis

The energy required in an electrodialysis process is an additive of two terms: (1) is the electrical energy to transfer the ionic components from one solution through membranes into another solution and (2) is the energy required to pump the solutions through the electrodialysis unit. Depending on various process parameters, particularly the feed solution concentration, either one of the two terms may be dominating, thus determining the overall energy costs. The energy consumption due to electrode reactions can generally be neglected since more than 200 cell pairs are placed between the two electrodes in a modern electrodialysis stack [J.R. Wilson, 1960].

2.6.1 Minimum Energy Required for the Separation of a Molecular Mixture

In electrodialysis as in any other separation processe a minimum energy is required for the separation of various components from a mixture. This minimum energy is the Gibbs free energy of mixing for instance the energy difference between two solutions of different composition when separated and when mixed. The Gibbs free energy can be related to the chemical potential and thus the activities of the individual component. For the removal of salt from a saline solution the energy of mixing and thus the energy required for the separation can be expressed by the change of the chemical potential of the water. The energy required for removing one mole of water from an infinite large volume of salt solution, the concentration of which is not changed by the removal of the water, is given by:

$$(\Delta G)_{p,T,n_j} = (\Delta \mu_w)_{p,T,n_j} = RT\ \ln \frac{a_w^o}{a_w} \tag{4.31}$$

Here ΔG is the Gibbs free energy change required to remove one mole of water from a solution, n_j is the number of salt molecules in the solution, $\Delta\mu_w$ is the difference in the chemical potential of pure water and water in the salt solution. R is the gas constant, and T the absolute temperature, and a_w^o and a_w are the water activities in pure water and the solution, respectively.

The activity of pure water is unity by definition. The activity of water in a dilute solution is equal to the concentration of water in the solution multiplied by an activity coefficient. Thus is:

$$a_w \cong f_w X_w = f_w \frac{n_w}{n_w + v_d n_s} \tag{4.32}$$

Here a is the activity and X the molar fraction, f_w is a activity coefficient, n is the number of moles, the subscripts w and s refer to water and salt, and v_d is the stoichiometric dissociation coefficient of the salt determining the number of ions in which a salt molecule will dissociate in solution. For NaCl, e.g. v_d is 2. The stoichiometric dissociation coefficient is related to the stoichiometric coefficient of the ions by $v_d = v_c + v_a$, f has the dimension of moles per volume.

A combination of Equations 4.31 and 4.32 leads to:

$$(\Delta G)_{p,T,n_j} = (\Delta \mu_w)_{p,T,n_j} \cong -RT \ln\left(1 + v_d \frac{n_s}{n_w}\right) \tag{4.33}$$

For a dilute solution $\frac{n_s}{n_w}$ is very small compared to n_w and $\ln\left(1 + v_d \frac{n_s}{n_w}\right) \cong v_d \frac{n_s}{n_w}$. Furthermore, to a first approximation the activity of a salt in a dilute solution is equal to the concentrations.

Taking this into account the Gibbs free energy for the removal of one mole of water from a salt solution from an infinite large reservoir at constant pressure and temperature is given by:

$$(\Delta G)_{p,T,n_j} = (\Delta \mu_w)_{p,T,n_j} \cong v_d RT \frac{n_s}{n_w} = v_d RT\ \ln C_s \tag{4.34}$$

Here C_s is the salt concentration.

Equation 4.34 refers to the energy needed to transfer one mole of water from an infinite large volume of solution to a reservoir of pure water. In electrodialysis this is, however, generally not the case. Usually a certain amount of the feed solution with a salt concentration C_s^f is converted into a more dilute and a more concentrated solution C_s^d and C_s^c, respectively. The minimum energy required for this process depends on the concentration of the feed, concentrate and dilute solutions [K.S. Spiegler, 1956]. The change in Gibbs free energy in this process equals the sum of the Gibbs free energy changes of salt and water:

$$\Delta G = \left(n_s^c \mu_s^c + n_s^d \mu_s^d - n_s^f \mu_s^f\right) + \left(n_w^c \mu_w^c + n_w^d \mu_w^d - n_w^f \mu_w^f\right) \qquad (4.35)$$

Here ΔG is the total reversible free energy for the desalination by electrodialysis when a concentrate and a diluate is generated from a feed solution, n is the number of moles, and μ the chemical potential. The subscripts s and w refer to the salt and the water; and the superscripts f, c, and d refer to feed solution, concentrate and diluate, respectively.

For dilute solutions the activity coefficients of all the components in the solution can to a first approximation be assumed to be 1. Then the chemical potentials of the salt and the water are:

$$\mu_s = \mu_s^o + \nu_d \, RT \, C_s \qquad \text{and} \qquad \mu_w = \mu_w^o + \nu_d \, RT \frac{n_s}{n_w} \qquad (4.36)$$

Here μ_s^o and μ_w^o are the standard chemical potentials of salt and water in the solution which are 1 by definition.

The mass balances for the salt and the water leads to :

$$n_s^f = n_s^d + n_s^c \quad \text{and} \quad n_w^f = n_w^d + n_w^c \,, \qquad (4.37)$$

indicating that the moles of salt or water in the feed solution is identical to the sum of the moles in the concentrate and diluate.

Introducing Equations 4.36 and 4.37 into 4.35 leads to:

$$\Delta G = \nu_d RT\left[\left(n_s^c \ln C_s^c + n_s^d \ln C_s^d - \left(n_s^c + n_s^d\right)\ln C_s^f\right) + \left(n_w^c \frac{n_s^d}{n_w^d} + n_w^d \frac{n_s^c}{n_w^c} - \left(n_w^c + n_w^d\right)\frac{n_s^c + n_s^d}{n_w^c + n_w^d}\right)\right] \qquad (4.38)$$

The second term in Equation 4.38 refers to the change of the solution in terms of the change in the water concentration, which is very small for a dilute solution and can be neglected compared to the first term which describes the change in the salt concentration.

For dilute solutions the free energy needed for the separation is given to a first approximation by:

$$\Delta G = \nu_d RT\left(n_s^c \ln \frac{C_s^c}{C_s^f} + n_s^d \ln \frac{C_s^d}{C_s^f}\right) \qquad (4.39)$$

Equation 4.39 gives the change in free energy for the desalination process under the assumption that the water transport due to osmosis or electroosmosis can be neglected.

For practical purposes it is convenient to know the energy per unit of diluate produced from a given feed solution. For this reason Equation 4.39 is modified by introducing the volume of the produced diluate.

For a dilute solution it can be assumed that the total volume and the salt concentration can be expressed by:

$$V \approx \overline{V}_w n_w \qquad \text{and} \qquad C_s = \frac{n_s}{n_w \overline{V}_w} \tag{4.40}$$

Here V is the total volume of the solution and $\overline{V}_w$ is the molar volume of the water.

If water transfer through the membrane due to osmosis and electroosmosis is neglected, the total amount of salt transferred from a feed solution to the concentrate is:

$$\Delta n_s = \overline{V}_w n_w^c \left(C_s^c - C_s^f\right) = \overline{V}_w n_w^d \left(C_s^f - C_s^d\right) \tag{4.41}$$

Here Δn_s is the amount of salt expressed in number of moles transferred between a feed solution and a concentrate.

Introducing Equations 4.40 and 4.41 into Equation 4.39 finally leads to:

$$\frac{\Delta G}{V_d} = \nu_d RT\left(C_s^f - C_s^d\right)\left(\frac{\ln\frac{C_s^f}{C_s^c}}{\frac{C_s^f}{C_s^c}-1} - \frac{\ln\frac{C_s^f}{C_s^d}}{\frac{C_s^f}{C_s^d}-1}\right) \tag{4.42}$$

Here ΔG is the Gibbs free energy required for the production of the diluate, C is the salt concentration, the superscripts f, d, and c refer to feed, diluate, and the concentrate, respectively; V_d is the volume of the diluate and ν_d the stoichiometric dissociation coefficient of the salt.

Equation 4.42 gives the energy required per volume diluate to convert a given feed solution into a diluate and a concentrate.

When $\frac{C_s^f}{C_s^c}$ approaches unity and $\frac{C_s^f}{C_s^d}$ infinity, Equation 4.42 becomes identical with Equation 4.34 because these conditions imply that the salt transfer takes place between the initial solution and pure water and that the volumes of the two reservoirs are so large that no appreciable concentration change occurs during the transfer.

It should be kept in mind that the free energy of mixing represents the minimum energy required in a completely reversible process. In any practical application more energy is needed, and in general the "faster" the process the larger is the amount of irreversibly consumed energy.

As a numerical example for the application of Equation 4.42 the desalination of a NaCl feed solution having a concentration $C_s^f = 50$ mol m^{-3} to a product with a concentration $C_s^d = 5$ mol m^{-3} is considered. This corresponds roughly to the conversion of brackish water to drinking water. From Equation 4.42 the energy requirement at 25° C is calculated to ca.

0.123 kWh per m^3 of drinking water produced. A value of 0.07 kWh per m^3 is obtained by Equation 4.34 for the minimum work necessary to transfer the same amount of water from a large sea water reservoir into a large pure water reservoir without changing the concentrations in both reservoirs.

2.6.2 Practical Energy Requirements in Electrodialysis

The minimum energy requirement for desalting discussed previously refers to an idealized completely reversible process. In an actual electrodialysis unit such conditions never prevail. Additional energy is dissipated in form of heat as a result of the friction encounter by the ions when migrating through the solutions and membranes. In addition to the actual desalination process, i.e. the transfer of ions from a feed to a concentrate solution, energy is also required to pump the solutions through the membrane containing unit, i.e. the electrodialysis stack, and energy is also consumed by the electrode reactions [E.L. Huffmann et al., 1972].

Energy requirements for the transfer of ions from the feed solution to a concentrate

The total energy required in electrodialysis for the actual desalination process is given by the current passing through the electrodialysis stack multiplied with the total voltage drop encountered between the electrodes:

$$E_{des} = I_{st} U_{st}\, t \tag{4.43}$$

Here E_{des} is the energy consumed in a stack for the transfer of ions from a feed to a concentrate solution, I_{st} is the current passing through the stack, U_{st} is the voltage applied across the stack, i.e. between the electrodes, and t is the time of operation.

The voltage drop in an electrodialysis stack arrangement is the result of the resistance and the potentials which are established between solutions of different salt concentrations. The resistance is caused by the friction of the ions with the membrane matrix and the water molecules while being transferred from one solution to another, resulting in an irreversible energy dissipation in the form of heat generation. Moreover, additional energy is also consumed by the electrode processes in the terminal compartments since these do not contribute to the yield of either diluate or concentrate.

In general, the energy loss due to the resistances is much larger than energy losses due to concentration potentials or electrode reactions. It is, therefore, important to use membranes with low electrical resistance and to space the membranes very closely to reduce the energy losses due to the resistance of the cell per unit of salt transferred. An exact calculation of the power requirement for a given electrodialysis process unit takes into account the electrical resistance of the diluate and concentrate solution and of the membranes as well as the concentration potentials. Furthermore, the concentration polarization effect in the boundary layers of the membrane surface due to a depletion of ions leads to additional voltage drops. The magnitude of the concentration polarization in the boundary layer is again a function of the feed solution flow velocity, and the cell and spacer geometry. To minimize concentration polarization effects certain feed flow velocities and thus, a certain energy to overcome feed solution pressure losses in the electrodialysis stack are required.

In electrodialysis generally several hundred cell pairs are installed in parallel between two electrodes. To determine the total energy consumption for the desalination process it is convenient to calculate the energy needed to transfer the ions in one cell pair from a feed to a concentrate solution and then to multiply by the number of cell pairs and add the energy dissipated in the electrode reactions. To calculate the energy requirement for the salt transfer in a cell pair, the total voltage drop across the cell pair has to be multiplied by the electric current used to transfer the salt ions from the feed stream to the concentrate. The total voltage drop consists of three parts:

1. the concentration potential across the membranes between the diluate and concentrate solutions, and between boundary layer and bulk solution concentration;
2. the potential drop due to the resistance of the solutions; and
3. the potential drop due to the resistance of the membranes.

The voltage drop obtained across a cell pair is illustrated in Figure 4.11.

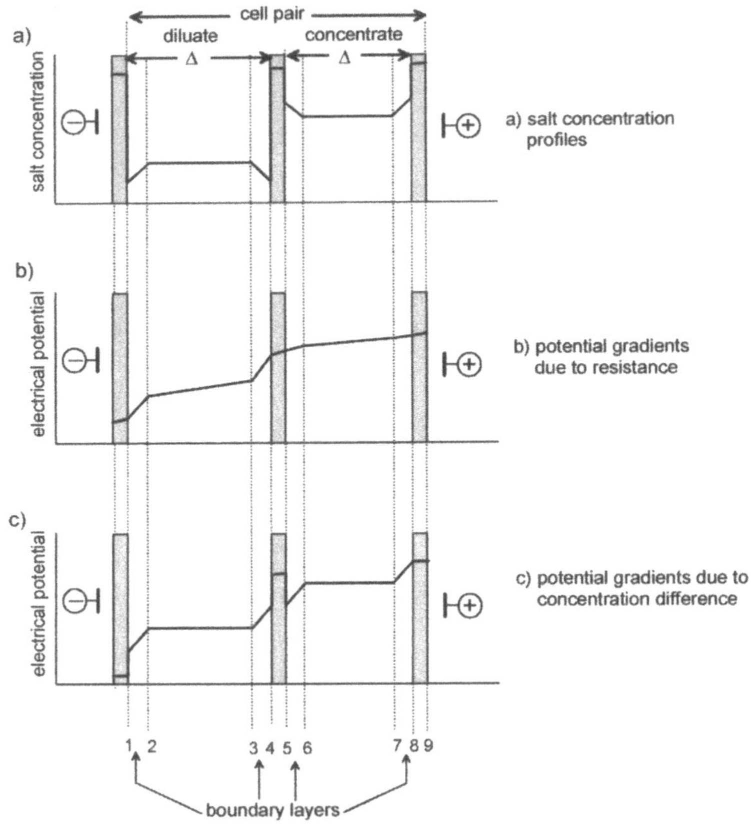

Fig. 4.10 Schematic drawing illustrating a) the salt concentration profiles across a cell pair established at a certain point along the flow channel during the electrodialysis desalination process, b) the voltage drop across a cell pair due to the resistance of the solutions and the membranes, and c) the voltage differences due to ion concentration differences, i.e. concentration potentials. (Points 1 through 9 indicate points of concentration changes).

The cell pair contains the concentrated and the depleted bulk solutions, the two membranes, and the four boundary layers in which the concentrations of the salts in the solutions near the membranes may vary considerably from that in the bulk. The concentration profile across a cell pair at a certain point depends on various process and cell design parameters such as the cell geometry, concentrations of the feed, the concentrate and the diluate solutions at the cell in- and outlet, the mode of operation which can be co- or counter-current. When both solutions flow in the same direction, the mode of operation is referred to as co-current. When they flow in opposite directions, it is referred to as counter-current. For simplicity reasons Figure 4.10 shows a cell pair with co-current flow in geometrically identical cells. The figure shows a) the concentration profiles at a certain point along the flow channel in direction perpendicular to the membrane surfaces that are established in the steady state in a cell pair when a current flows through an electrodialysis stack, b) the voltage drop due to the resistance of the solutions and the membranes of the cell pair, and c) the voltage differences due to ion concentration differences between the different solutions, i.e. the concentration potentials.

Thus, the total voltage drop is the sum of the concentration potentials resembling an electromotive force and the voltage drop due to the resistance. If the cross-section through a cell pair is illustrated by an electric circuit a series of electromotive forces resembling differences in the salt concentrations of the bulk, the boundary, and the membrane phase is obtained in addition to resistances of the bulk and boundary solutions and in the membranes. For simplicity reasons it shall be assumed that the feed is a monovalent aqueous salt solution, i.e. the absolute value of the valence z_i and the stoichiometric coefficient ν_i are one.

The concentration potential between solutions of different concentrations is obtained from the electrochemical equilibrium as discussed earlier. For constant temperature and pressure and a monovalent salt is:

$$\frac{\Delta\varphi}{\Delta z} = \frac{RT}{F}\left(T_a - T_c\right)\int_{z}^{z+\Delta z}\frac{d\ln a_i}{dz} \tag{4.44}$$

The voltage drop due to the resistance in the solution is given by:

$$\Delta U = i\int_{z}^{z+\Delta z}\frac{dz}{\sum_i C_i\Lambda_i} \tag{4.45}$$

and in the membrane by:

$$\Delta U = i\, r^{im} \tag{4.46}$$

Here $\Delta\varphi$ is the concentration potential, ΔU is the overall voltage drop caused by the resistances of the membranes and the solutions and concentration differences between the solutions, T is the transport number, i is the current density, Λ_i the equivalent conductivity, r the area resistance, and z is the directional coordinate perpendicular to the membrane surface; the subscripts a, c and i refer to anion, cation and ion in general; the superscript im refers to ion-exchange membrane.

For a monovalent salt the total voltage drop can explicitly be described by:

$$\Delta U = \frac{RT}{F}\left[\left(T_a^d - T_c^d\right)\int_1^2 \frac{d\ln a_s}{dz} + \left(T_a^d - T_c^d\right)\int_3^4 \frac{d\ln a_s}{dz} + \left(T_a^{am} - T_c^{am}\right)\int_4^5 \frac{d\ln a_s}{dz} + \left(T_a^c - T_c^c\right)\int_5^6 \frac{d\ln a_s}{dz}\right.$$
$$\left. + \left(T_a^c - T_c^c\right)\int_7^8 \frac{d\ln a_s}{dz} + \left(T_a^{cm} - T_c^{cm}\right)\int_8^9 \frac{d\ln a_s}{dz}\right] + i\left[\int_1^2 \frac{dz}{\overline{C}_s^d \Lambda_s} + \frac{\Delta - 2d}{C_s^d \Lambda_s} + \int_3^4 \frac{dz}{\overline{C}_s^d \Lambda_s} + r^{am} + \int_5^6 \frac{dz}{\overline{C}_s^c \Lambda_s}\right. \quad (4.47)$$
$$\left. + \frac{\Delta - 2d}{C_s^c \Lambda_s} + \int_7^8 \frac{dz}{\overline{C}_s^c \Lambda_s} + r^{cm}\right]$$

Here T is the transport number, a is the activity, C is the concentration in the bulk solutions, and $\overline{C}$ is an average concentration in the laminar boundary at the membrane surfaces, Λ is the equivalent conductivity, Δ is the cell thickness, and d the boundary layer thickness, i is the current density, and r is the area resistance, the superscripts d, c, am, and cm refer to the diluate, the concentrate, the anion-exchange membrane, and the cation-exchange membrane, respectively. The subscripts a, c, and s refer to anion, cation, and salt, respectively.

The numbers 1 to 9 are corresponding to the numbers in Figure 4.10 and refer to the integration boundaries of the directional coordinate z. Although Equation 4.47 contains the approximation of some terms, it is a reasonably accurate prediction of the total potential drop across a cell pair [L.H. Shaffer et al., 1966]. However, even for a monovalent salt solution it is already quite complex.

Most of the electrodialysis stacks used in practical applications consist of geometrically identical cells which are operated in co-current flow as indicated in Figure 4.4. Therefore, the following approximations can be made:

- The equivalent conductivity is independent of the concentration over the range of interest.
- The concentration potentials are negligibly low compared to the potential drops caused by the resistance of the solutions and the membranes.
- The concentrate and diluate cells have identical geometry.
- The flow of the solution is co-current and of equal velocity.
- Changes in the resistance of the solutions due to boundary layer effects are also neglected or averaged over the cross-section of the cell.
- The salt activity can be replaced by the concentration.

With these approximations Equation 4.47 simplifies to:

$$\Delta U = i\left[\frac{\Delta}{\Lambda_s}\left(\frac{1}{C_s^d} + \frac{1}{C_s^c}\right) + r^{am} + r^{cm}\right] \quad (4.48)$$

Here Δ is the cell thickness, Λ is the equivalent conductivity of the salt solution, r is the area resistance, the subscript s refers to the salt solution and the superscripts d, c, am and cm refer to the diluate, the concentrate, the anion- and the cation-exchange membranes, respectively.

The total electric current required for the transfer a certain amount of ions in the actual desalting process in a cell pair is given by the current, the time and an efficiency term referred to as current utilization:

$$I = \frac{Q^{d}_{cell} |z| \nu F \left(C^{f}_{s} - C^{d}_{s}\right)}{\xi} \tag{4.49}$$

Here I is the electric current passing through a cell pair, Q is the volume of the diluate, C is the concentration, z is the valence, ν is the stoichiometric salt dissociation constant, F is the Faraday constant, and ξ the current utilization. The subscripts s and cell refer to the salt and the diluate cell, and the superscripts d and f refer to diluate and feed solution.

The current utilization is an expression for the fraction of the total current passing through the cell pair that is actually used for the transfer of ions. In most practical applications the current utilization in a cell pair is close to unity. The factors effecting the current utilization will be discussed in more detail later.

The electrical resistance of a cell pair can be derived from Equation 4.48 to:

$$R = \frac{\Delta U}{I} = \frac{1}{A}\left[\frac{\Delta}{\Lambda_s}\left(\frac{1}{C^{d}_{s}} + \frac{1}{C^{c}_{s}}\right) + r^{am} + r^{cm}\right] \tag{4.50}$$

Here R is the electrical resistance of a cell pair, Δ is the thickness of a cell, Λ the equivalent conductivity, r is the area resistance of the ion-exchange membranes, and C_S the salt concentration. The superscripts d, c, am, and cm refer to diluate, concentrate, anion-exchange membrane, and cation-exchange membrane, respectively, A is the cell pair area given by the width Y of the cell and a distance x+dx in direction of the solution flow parallel to the membrane as indicated in Figure 4.4.

The electrical resistance of the solution is inverse proportional to the salt concentration of the solutions. This concentration is changing, while the solution flows through the stack in direction of the coordinate x from the inlet to the outlet of the cell pair. The concentration in the diluate cell is decreasing from the feed concentration C^{fd}_{s} to product concentration C^{p}_{s}, and that in the concentrate cell is increasing from C^{fc}_{s} to the concentrate concentration C^{C}_{S}. The concentration change is a function of the cell geometry and the flow velocities. For a cell pair with identical cell geometry and equal co-current solution flow velocities and a constant electrical potential driving force across the cell pair, the concentration changes in the solutions and thus the resistance of the solutions can be calculated as a function of the cell length. The average resistance over the entire length of the flow channel is determined by the integral average of the solution concentrations. Thus is:

$$\overline{R} = \frac{1}{A}\left[\frac{\Delta \ln \dfrac{C^{fd}_{s}\,C^{c}_{s}}{C^{fc}_{s}\,C^{d}_{s}}}{\Lambda_s\left(C^{fd}_{s} - C^{d}_{s}\right)} + r^{am} + r^{cm}\right] \tag{4.51}$$

Here C^{fd}_{s} and C^{d}_{s} are the salt concentrations of the diluate cell at the inlet and outlet, C^{fc}_{s} and C^{c}_{s} are the salt concentrations of the concentrate cell at the inlet and outlet, $\overline{R}$ is the average resistance of a cell pair.

A combination of Equations 4.43, 4.49 and 4.51 gives the energy required for the desalination of a certain amount of feed to a desired product and brine concentration:

$$E_{des} = \overline{R} I^2 t = \frac{t}{A}\left[\frac{\Delta \ln \frac{C_s^{fd}}{C_s^{fc}}\frac{C_s^c}{C_s^d}}{\Lambda_s\left(C_s^{fd} - C_s^d\right)} + r^{am} + r^{cm}\right]\left[\frac{Q_{cell}^d |z| \nu F\left(C_s^{fd} - C_s^d\right)}{\xi}\right]^2 \qquad (4.52)$$

Here E_{des} is the desalination cost for the desalination in electrodialysis, I is the total current, t is the time of operation; C_s^{fd} and C_s^{fc} are the equivalent concentrations of the diluate and the concentrate at the cell inlet, C_s^d and C_s^c are the concentrations of the diluate and the concentrate at the cell outlet, Λ_s is the equivalent conductivity of the salt solution, r^{am} and r^{cm} are the area resistances of the anion- and cation-exchange membrane, Δ is the cell thickness, ξ is the current utilization, and Q_{cell}^d is the diluate flow rate in a cell, $\overline{R}$ is the average resistance of a cell pair, and A is the cell pair area.

In the calculation of the desalination energy in Equation 4.52 it was assumed that the total membrane are necessary for a certain desalination capacity is installed in one cell pair. In reality, however, the total membrane area is installed in a number of parallel cell pairs in a stack. To calculate the desalination energy in an electrodialysis stack the resistance and the volume of the product of a cell pair must be multiplied with the number of cell pairs in the stack. The total energy consumption for the desalination in a stack is:

$$E_{des} = N_{cell} \overline{R}_{cell} I^2 t = \frac{N_{cell}\, t}{A}\left[\frac{\Delta \ln \frac{C_s^{fd}}{C_s^{fc}}\frac{C_s^c}{C_s^d}}{\Lambda_s\left(C_s^{fd} - C_s^d\right)} + r^{am} + r^{cm}\right]\left[\frac{Q_{cell}^d |z| \nu F\left(C_s^{fd} - C_s^d\right)}{\xi}\right]^2 \qquad (4.53)$$

Here N_{cell} and $\overline{R}_{cell}$ are the number of cells in a stack and the average resistance of a cell pair.

Equation 4.53 indicates that the energy dissipation due to the resistance of the solutions and membranes is increasing with the current density, since the electrical energy for a given resistance is proportional to the square of the current, whereas the salt transfer is directly proportional to the current. Hence, the power necessary for the production of a given amount of product increases with the current density. The higher the current density the more power is needed to maintain a given production rate. However, the higher the current density the lower is the required membrane area for a given capacity installation. Thus, there is an optimum in the current density and in the installed membrane area which must be determined for each stack design and feed, diluate and concentrate solution. The upper limit for the current density of any given installation is determined by the limiting current density which should not be exceeded.

Equation 4.53 is also the basis for determining the process path-length for converting a certain amount of feed solution into a product of the desired concentration as will be shown later.

Pumping energy requirements

The operation of an electrodialysis unit requires one or more pumps to circulate the diluate, the concentrate, and the electrode rinse solution through the stack. The energy required for pumping these solutions is determined by the volumes of the solutions to be pumped and the pressure drop. It can be expressed by:

$$E_{p,spe} = \frac{E_P}{Q^d t} = k_{eff} \frac{\left(Q^d \Delta p^d + Q^c \Delta p^c + Q^e \Delta p^e\right)}{Q^d} \tag{4.54}$$

Here $E_{p,spec}$ is the total energy for pumping the diluate, the concentrate, and the electrode rinse solution through the stack per unit diluate water, k_{eff} is an efficiency term for the pumps, Q^d, Q^c, and Q^e are the volume flow rates of the diluate, the concentrate, and the electrode rinse solution through the stack.

The energy consumption due to the pressure loss in the electrode rinse solution can be neglected in most practical applications because the volume of the electrode rinse solution is very small compared to the volumes of the diluate and concentrate.

The pressure losses in the various cells are determined by the solution flow velocities and the cell design. The energy requirements for circulating the solution through the system may become a significant or even dominant portion of the total energy consumption for solutions with rather low salt concentration. The total volume of the electrode rinse solution is much smaller than that of the diluate and the concentrate, and the pumping energy for the electrode rinse solutions can generally be neglected.

Energy consumption by electrode reactions

Throughout the desalting process in an electrodialysis unit the electrical charges are carried by ions in the solutions and the membranes, i.e. there is a relative motion of one equivalent of cations and anions for each Faraday electricity passed through the stack. At the electrodes the mechanism of electrical conductance changes abruptly from ionic to electronic. This change is coupled with an electrochemical reaction at the electrode surface. Using inert metal electrodes such as platinum, the transition is accomplished by the addition or subtraction of electrons to or from the ions present in the solution. Positively charged cations are reduced at the negatively charged cathode by receiving electrons, and negatively charged anions are oxidized at the positively charged anode by discharge of electrons.

A typical reaction of a positive charged metal cation at the cathode is:

$$M^{z+} + ze \rightarrow M^o$$

At the anode the reaction is reversed:

$$M^o \rightarrow M^{n+} + ne$$

The reaction of the negatively charged anion at the anode is:

$$X^{n-} - ne \rightarrow X^o$$

At the cathode the reaction is reversed:

$$X^o \rightarrow X^{n-} + ne$$

Here M^{n+} is the metal cation, M^o the uncharged metal,. X^{n-} is the negatively charged anion, and X^o the neutral molecule, n is the number of charges of the ions or the number of electrons, and e indicates the electron.

The uncharged metal will be deposited at the anode if its standard potential is positive (compared to hydrogen which is per definition 0) and dissolved at the anode. If the standard potential of the metal is negative it will react with water forming hydroxide ions, metal ions, and hydrogen which is released at the cathode as gas. At the anode oxygen is released as gas if the standard potential of the anion is less negative than that of oxygen. For the electrode rinse solution generally salts are selected which have anions such as SO_4^{2-} with a standard potential that is less negative than that of oxygen. Thus, the entire electrode reaction evolves to the decomposition of water and the production of oxygen, hydrogen, protons, and hydroxyl ions according to the following reactions at the cathode:

$$z\,H_2O + M^{n+} + n\,e \rightarrow \frac{1}{2} n\,H_2 + n\,OH^- + M^{n+} \tag{4.55}$$

and at the anode:

$$n\,OH^- - n\,e \rightarrow \frac{1}{4} n\,O_2 + \frac{1}{2} n\,H_2O \tag{4.56}$$

Here z is the number of electrons that are transferred from the solution to the electrode.

The energy required for the transfer of electrons from ions to the electrodes and vice versa is the energy for the decomposition of water given by:

$$\Delta G = RT \ln \frac{[a_{H_2}][a_{O_2}]^{\frac{1}{2}}}{[a_{H_2O}]} \tag{4.57}$$

Furthermore is:

$$\frac{\Delta G}{zF} = \Delta U = \frac{RT}{zF} \ln \frac{[a_{H_2}][a_{O_2}]^{\frac{1}{2}}}{[a_{H_2O}]} \tag{4.58}$$

Here ΔG is the energy required for the decomposition of water into oxygen and hydrogen gas, R is the gas constant, F the Faraday constant, z the electrochemical valence, a the activity, and the subscripts H_2, O_2, and H_2O refer to hydrogen, oxygen, and water. ΔU is the so-called decomposition potential [J. Bockris, 1970]. Under standard conditions and a operating temperature is 25°C, the decomposition potential of water is: $\Delta U^o = 1.23$ Volts.

In addition to the energy required for the decomposition potential there is a polarization potential at the electrode, which has to be overcome, and thus the energy requirements of the electrode reactions exceed that of the decomposition potential of water.

Furthermore, other ions such as Cl^- may lead to other electrode reactions and thus to other energy requirements. A commercial electrodialysis stack, however, contains many, i.e. typically 200 to 400 cell pairs arranged in parallel between two electrodes. Thus, the energy consumed by electrode reactions is generally less than 1% of the energy used for the ion transfer and the pumping of the solutions [K.S. Spiegler, 1956].

2.7 Processes Effecting the Efficiency of Electrodialysis

The actual energy required to operate the electrodialysis process exceeds the energy theoretically required not only to overcome the concentration potentials that exist at the membrane solution interfaces and the power that will be dissipated due to the electrical resistance of the membranes and solutions, the process efficiency will also be decreased by the fact that the membranes are not perfectly selective, and that water is transported from the diluate to the concentrate solution due to osmotic and electroosmotic effects, and that current flows through the manifold system without contributing to the transfer of salt from the diluate to the concentrate. There may be other sources that effect the overall efficiency of the process such as water dissociation at the membrane surface or hydraulic leaks between diluate and concentrate cells. All these effects have to be considered when designing and operating an electrodialysis unit.

2.7.1 Current Utilization

In any practical electrodialysis process not all of the current flowing through the stack can be utilized for desalting of the feed solution. Several factors may contribute to incomplete current utilisation in an electrodialysis stack:

- the membranes are not perfectly selective,
- there may be parallel current paths through the stack manifold,
- water transfer across the membranes due to osmosis and electroosmosis,
- at high current densities and low salt concentration H^+- and OH^--ions may be generated and participate in the current-carrying process.
-

The current efficiency loss due to H^+- and OH^--ion generation which are generally referred to as "water splitting" can be avoided by a proper cell design and controlled concentration polarization, and is generally negligibly low as has been shown lately [J.J. Krol et al., 1999]. It must, however, be considered because it might lead to considerable pH-changes in the electrolyte solutions.

The ratio of the salt transported through a membrane from a feed to a concentrate stream in reality to the theoretical current requirement is referred to as Faraday efficiency. This efficiency depends on the membranes used in the process as well as on current leakage through the manifold which again depends on the system design and operating parameters. In addition there is some transport of water through the membranes from the diluate to the concentrate solution due to osmosis and electroosmosis which effects the process. The total current efficiency is therefore defined as the current required in practice to obtain a certain amount of product water of a given quality divided by the theoretically required current. The total efficiency is given by:

$$\eta_{tot} = \eta_F \eta_w = \eta_{ms} \eta_l \eta_w \qquad (4.59)$$

Here η is an efficiency term, and the superscripts tot, F, ms, l, and w refer to total efficiency, Faraday efficiency, membrane permselectivity, current leakage through the manifold, and water transport through the membranes. The efficiency terms η_{ms}, η_l, and η_w are all defined in such a way that under ideal conditions they approach 1.

The current efficiency terms as defined above give a relation between the change of the salt concentration in the diluate cell as the function of the current passing through this cell. In electrodialysis the current that is passing through one cell pair of a multi-cell stack unit will also pass through all cells installed in parallel between the electrodes. That means that the current flowing between the electrodes will be used many times over, and its utilization is directly proportional to the number of cell pairs in a stack. The total current utilization in an electrodialysis stack is expressed by the net removal of salt from a feed volume. It is referred to as Faraday utilization and given by:

$$\nu_c z_c F\left(C_s^f - C_s^d\right)\frac{dV^d}{dt} = I\,\eta_l\left[N_c\left(T_c^{cm} - T_c\right) + N_a\left(T_a^{am} - T_a\right)\right] \qquad (4.60)$$

The last term in Equation 4.60 is referred to as the Faraday current utilization.

$$\xi_F = \eta_l\left[N_c\left(T_c^{cm} - T_c\right) + N_a\left(T_a^{am} - T_a\right)\right] \qquad (4.61)$$

Here ξ_F is the Faraday current utilization, C is the salt concentration, V^d is the volume of solution passing through the diluate cell, I is the current passing through the stack, η_l is an efficiency term referring to current leakage through the manifold, z is the valence, F is the Faraday constant, ν_c is the stoichiometric coefficient referring to the number of cations in the electrolyte. For a monovalent salt both z and ν are one. N is the number of membranes in a stack, T is the transport number, the superscripts f, d, cm, and am refer to feed solution, diluate solution, cation-exchange membrane, and anion-exchange membrane, respectively. The subscripts c and a refer to cation and anion.

Introducing the membrane permselectivity as defined by Equations 2.173 and 2.174 for the transport numbers gives:

$$T_c^{cm} - T_c = \Psi^{cm} T_a \qquad (4.62)$$

and

$$T_a^{am} - T_a = \Psi^{am} T_c \qquad (4.63)$$

Combination with of Equations 4.61 to 4.63 leads to:

$$\xi_F = \eta_l\left[N_c \Psi^{cm} T_a + N_a \Psi^{am} T_c\right] \qquad (4.64)$$

Here Ψ^{cm} and Ψ^{am} are the permselectivities of the cation- and anion-exchange membrane, respectively. The membrane permselectivity describes the degree to which it passes an ion of one charge and discriminates against an ion of the opposite charge. An ideal permselective ion-exchange membrane would transmit counterions only and the permsectivity approaches unity. The permselectivity approaches zero when the transference number within the membrane is identical to that in the electrolyte solution, i.e. for a cation-exchange membrane $\Psi^{cm} = 1$, if $T_c^{cm} = 1$ and $\Psi^{cm} = 0$, if $T_c^{cm} = T_c$. Since $T_a + T_c = 1$ and the number of anion-exchange membrane is equal to the number of cation-exchange membranes the Faraday current utilization can be expressed to a first approximation by:

$$\xi_F = \eta_l N_{cell}\left(\frac{\Psi^{cm} + \Psi^{am}}{2}\right) \tag{4.65}$$

Here N_{cell} is the number of cell pairs in a stack.

In addition to the Faraday utilization an overall current utilization can be defined as being a function of the number of cell pairs, membrane selectivity, water transfer, and manifold current flow as follows:

$$\xi_{tot} = \xi_F \, \eta_w \tag{4.66}$$

Here ξ_{tot} is the overall current utilization in an electrodialysis stack, ξ_F and η_w are efficiency terms referring to the water transport between the diluate and concentrate flow streams.

The water transfer in electrodialysis effects the efficiency of the process by reducing the amount of diluate obtained from a given amount of feed solution. If convective flow between cells due to hydrostatic pressure differences is excluded, water is transferred by osmosis and by electroosmotic effects. In the dilute solution the osmotic water transport may be neglected and the efficiency loss due to water transport may be expressed by the water transport numbers as described in Equation 2.171. To a first approximation the water transport number is given by:

$$\eta_w = 1 - \left(T_w^{am} + T_w^{cm}\right)\overline{V}_w\left(C_s^f - C_s^d\right) \tag{4.67}$$

Combining Equations 4.65 to 4.67 leads to:

$$\xi_{tot} = \eta_l N_{cell}\left(\frac{\Psi^{cm} + \Psi^{am}}{2}\right)\left[1 - \left(T_w^{am} + T_w^{cm}\right)\overline{V}_w\left(C_s^f - C_s^d\right)\right] \tag{4.68}$$

Equation 4.68 gives the total current utilization as a function of the number of cell pairs in the stack N_{cell}, the cation- and anion-exchange membrane permselectivities Ψ^{cm} and Ψ^{am}, the water transport numbers of the cation- and anion-exchange membranes T_w^{cm} and T_w^{am}, the number of ions transferred from the feed to the diluate multiplied with the molar volume of the water $\overline{V}_w$, and the current passing through the manifold assuming that the electrolyte is a monovalent, completely dissociated salt.

For relatively dilute feed solutions, i.e. $C_s^f < 0.1$ mol L^{-1}, the efficiency loss due to water transfer is very low, i.e. $\eta_w \approx 1$. However, for higher feed solution salt concentration the water transfer may effect the efficiency of electrodialysis quite significantly.

The current leakage through the manifold system depends on the difference between the diluate and concentrate salt concentration and on the stack design. In a well designed stack and for concentration differences between the concentrate and the diluate of less than a factor of 10 the efficiency loss due to current leakage can be neglected, i.e. $\eta_l \approx 1$.

However, at very high salt concentrations in the concentrate and low salt concentrations in the diluate not only the current leakage through the manifold can effect the overall current utilization severely but osmotic effects and salt diffusion must also be taken into account.

3 ELECTRODIALYSIS WITH BIPOLAR MEMBRANES

In electrodialysis with bipolar membranes the water dissociation in bipolar membranes induced by an electrical potential gradient is utilized to produce acids and bases from a corresponding salt solution. The principle of the process is illustrated in Figure 4.11

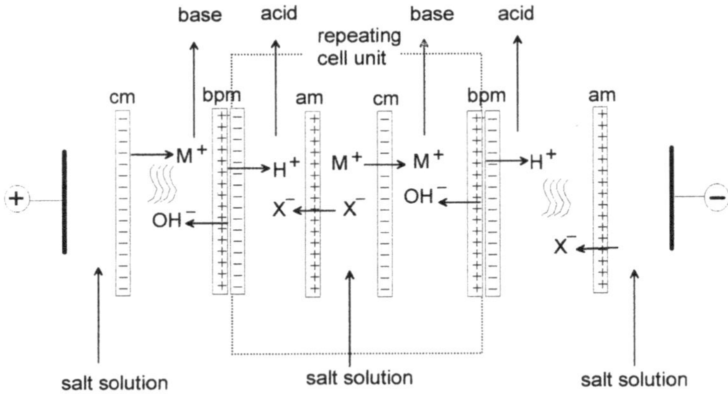

Fig. 4.11 Schematic drawing illustrating the production of acids and bases from the corresponding salts by electrodialysis with bipolar membranes.

The schematic drawing shows bipolar membranes and cation-exchange and anion-exchange membranes arranged in parallel between two electrodes to form individual compartments. If a salt solution is introduced in the middle compartment and an electrical potential difference between the electrodes is established, the cations in the salt solution will migrate towards the cathode. They will permeate the cation-exchange membrane and form a base with the hydroxide ions generated within the bipolar membrane. On the other side of the bipolar membrane the protons, which are generated simultaneously with the hydroxide ions by water dissociation, form an acid with the anions migrating from the salt solution through the anion-exchange membrane towards the anode. The net result of the entire process is the production of an acid and a base from the corresponding salt solution in an electrodialysis cell arrangement consisting of 3 individual compartments, and a bipolar, a cation-exchange and an anion-exchange membrane. As in conventional electrodialysis the 3 compartment cell system can be stacked as a repeating unit between electrodes.

The key element in this process is the bipolar membrane. Its performance determines to a large extent the technical feasibility and the economics of the process[K.N., Mani, 1991]. Therefore, some fundamental aspects of the water dissociation mechanism shall be briefly reviewed.

3.1 The Principle of Water Dissociation in Bipolar Membranes

The process of electrodialytic water dissociation is illustrated in Figure 4.12, which shows the schematic diagram of a bipolar membrane consisting of a cation- and an anion-exchange layer arranged in parallel between two electrodes.

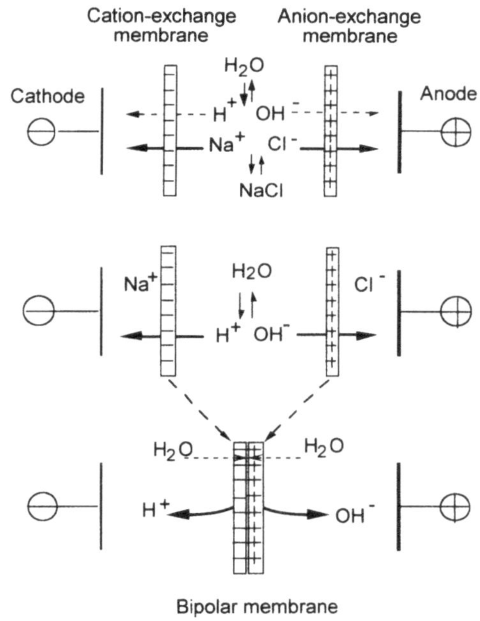

Fig. 4.12 Schematic diagram illustrating the principle of electrodialytic water dissociation in bipolar membranes.

If an electrical potential difference is established between the electrodes, charged species are removed from the phase between the two ion-exchange layers. When all salt ions are removed from the phase between the two ion-exchange layers further transport of electrical charges can be accomplished only by protons and hydroxide ions, which are available due to the dissociation equilibrium of the water available in the phase in a concentration of ca. 1×10^{-7} mol L^{-1}. Protons and hydroxide ions removed from the phase between the cation- and anion-exchange membranes are replenished by the water dissociation equilibrium given by:

$$2H_2O \Leftrightarrow H_3O^+ + OH^-$$

If the distance between the cation and the anion-exchange membrane goes towards 0 a bipolar membrane is obtained.

The Gibbs free energy required for the production of acids and bases in a bipolar membrane at constant temperature and pressure can be calculated from the concentration chain of solutions with different H^+-ion activities as discussed earlier. For the water dissociation in a bipolar membrane the Gibbs free energy can be calculated from the difference in the pH-values of the solution in the phases between the cation-and anion-exchange layer of the bipolar membrane and the solutions outside the bipolar membrane and is given by:

$$\Delta G = F\ \Delta\varphi = 2.3\ RT\ \Delta pH \tag{4.69}$$

Here ΔG is the Gibbs free energy, $\Delta\varphi$ is the electrical potential difference between the solution in the bipolar membrane and the solutions outside the bipolar membrane, R is the gas constant, T is the absolute temperature, F is the Faraday constant, and ΔpH is the difference between the pH-value of the two solutions separated by the bipolar membrane.

For 1n acid and base solutions in the two phases separated by the membrane potential difference at 25°C is $\Delta\varphi = 0.828$ V and the molar Gibb's free energy is $\Delta G = 0.0222$ kWh. The actual potential drop across the bipolar membrane is higher than the calculated value because of irreversible effects due to the electrical resistance of the membranes and the solutions.

The overall economics of the process are determined by the energy and the membrane area requirements. Thus, for achieving minimal process costs the electrical resistance of the membrane should be as low as possible and the current density should be as high as possible. The actual mechanism of water dissociation which to a very large extent determines the overall efficiency of the process is still subject of a rather controversial discussion [S.F Timashev et al., 1981], [Zabolotsky et al., 1988], [Simons, 1985].

The electrical resistance of an ion-exchange membrane is inverse proportional to the number of mobile ions in the membrane, which is determined by the concentration and the degree of dissociation of the fixed charges of the ion-exchange resin. In the bipolar membrane, strong acid and strong base groups such as sulfonic acid or quarternized amines, are used in the cation- and anion-exchange layers of the membrane in a concentration of 1 to 2 milliequivalent per gram dry polymer resulting in a specific resistance σ of 50 to 100 Ω·cm. Assuming a thickness of 100 µm for the cation- and anion-exchange layer respectively, the total area resistance r of the ion-exchange layers of the bipolar membrane is in the order of 1 - 2 $\Omega \cdot cm^2$.

The electrical resistance of the transition layer between the two ion-exchange layers of a bipolar membrane is assumed to consist of de-ionized water and can be calculated by:

$$r_{in} = \frac{2l}{\kappa} = \frac{2\lambda}{F\left(C_{H^+} u_{H^+} + C_{OH^-} u_{OH^-}\right)} \tag{4.70}$$

Here r is the area resistance, 2λ is the thickness of the transition layer, κ is the specific conductivity of the transition layer, C is the concentration, u is the ion mobility, and F is the Faraday constant; the subscript in refers to the layer between the ion-exchange layers of the bipolar membrane, and the subscripts H^+ and OH^- refer to the H^+- and OH^--ions of the solution between the cation- and anion-exchange layer of the bipolar membrane.

The concentrations of H^+- and OH^--ions in the transition layer between the two ion-exchange layers are equal and in the order of 10^{-7} mol L^{-1}. The electrical resistance of the transition layer can be calculated by Equation 4.70. For minimizing the area resistance of the bipolar membrane the transition layer should be as thin as possible. For instance, a transition layer of 1 nm thickness would have an area resistance of ca. 1.8 Ω cm^2, which would then be of about the same order of magnitude as that of the ion-exchange layers of the bipolar membrane. The above consideration, however, is correct only if all ions removed from the transition layer by the electric current across the bipolar membrane are replenished by the water dissociation, i.e. the ion fluxes from the transition layer of the bipolar membrane into the outer phases cannot exceed the rate of their generation. Thus, the maximum flux of H^+- and OH^--ions of the bipolar membrane is given by:

$$J_{H^+} = J_{OH^-} = k_d \, C_{H_2O} \, 2\lambda \tag{4.71}$$

Here is J the maximum ion flux from the bipolar membrane into the outer phases, k_d is the water dissociation rate constant, C is the concentration in the transition layer, and 2λ is the thickness of the transition layer. The subscripts H^+, OH^-, and H_2O refer to H^+-, OH^--ions and water respectively. The water dissociation rate constant k_d at 25° C is $2.5 \cdot 10^{-5}$ s^{-1}. The water concentration in pure water is 55.5 mol L^{-1}. Assuming a thickness of the transition layer of 1 nm the maximum ion fluxes $J_{H^+}$and J_{OH^-}, respectively, calculated by Equation 4.71 would be 14×10^{-9} mol m^{-2} s^{-1}. The electrical current density i is given by:

$$i = F \sum z_i J_i . \tag{4.72}$$

Thus, the ion fluxes J_{H^+} and J_{OH^-} would correspond with a current density of ca. 1.4×10^{-4} A m^{-2}. A current density exceeding this value would lead to a depletion of ions in the transition layer and thus to a drastic increase in its electrical resistance. In practice, however, bipolar membranes can be operated at current densities in excess of 1000 A m^{-2} [J.J. Krol et al., 1998].

This means that the simple model of a bipolar membrane is incorrect. Either the water dissociation rate in the bipolar membrane is by several orders of magnitude faster than in free solution or the transition layer is much thicker, i.e. in the range of several mm. That, however, would mean that the specific resistance of the transition layer is lower by several orders of magnitude than that of pure water. From scanning electron microscope photographs and from general observation and theoretical consideration [A. Mauro, 1962] it can be concluded that the thickness of the transition layer is less than 10 nm. This means that in bipolar membranes the water dissociation is at least 10^6 times faster than in a solution.

3.1.1 Models for the Accelerated Water Dissociation in Bipolar Membranes

To explain the accelerated water dissociation which is observed in the practical application of bipolar membranes various models are suggested in the literature. Two of these models shall briefly be discussed. The first model is based on the so-called second Wien effect which postulates an increase in the water dissociation constant by a strong electric field. The second model is based on a catalytically enhanced protonation and deprotonation of weak basic or weak acidic groups. Both models have to assume a certain thickness and structure of the transition layer between the cation- and anion-exchange layers of the bipolar membrane.

Determination of the transition layer thickness

As far as the structure of the transition region between the two ion-exchange layers of a bipolar membrane is concerned two hypotheses exist. One postulates that the distance between the cation- and anion-exchange layer is zero and that the transition region is located within the two ion-exchange layers. Adjacent to the contacting surfaces of the two ion-exchange layers a region is obtained where the concentration of counterions is decreased and uncompensated fixed charges exist. This region is referred to as the "space charge region" [A. Mauro, 1962] or "depletion layer" [R. Simons, 1979]. In correspondence to solid state physics nomenclature, the transition phase in this case is called the "abrupt junction". The second hypothesis assumes a thin neutral region between the two ion-exchange layers [R. Simons et al., 1978]. Both models can be used to estimate the transition phase thickness.

If it is assumed that during operation all salt ions are removed from the transition phase region and the transition layer is neutral, its thickness can then be calculated from the electric resistance of the bipolar membrane. Experimentally the electrical resistance of a bipolar membrane can be measured and is in the order of 10 Ω cm^2. As a first approximation the electrical resistance of a bipolar membrane r^{bm} can be considered as three resistances in series, i.e. the resistance of the cation-exchange layer r^{cm}, the resistance of the transition phase region r^{tr} , and that of the anion-exchange layer r^{am}. Thus is:

$$r^{tr} = r^{bm} - (r^{cm} + r^{am}) \qquad (4.73)$$

The resistances of the cation- and the anion- exchange membrane can be measured and are 1 to 2 Ω cm^2 each. If the resistance of the bipolar membrane is about 10 Ω cm^2 the resistance of the transition layer calculated. It is in the order of 6 to 8 Ω cm^2. The resistance of the transition layer is also a function of the concentration and the mobility of the protons and hydroxyl ions. The thickness of the transition layer can be calculated by:

$$r^{tr} = \frac{2\lambda}{F\left(C_{H^+}u_{H^+} + C_{OH^-}u_{OH^-}\right)} \qquad (4.74)$$

and

$$2\lambda = r^{tr}F\left(C_{H^+}u_{H^+} + C_{OH^-}u_{OH^-}\right) \qquad (4.75)$$

Here F is the Faraday constant, r^{tr} is transition layer resistance, 2λ is the thickness of the transition layer, u is the ion mobility, and C is the concentration, the subscripts H^+ and OH^- refer to protons and hydroxide ions, respectively.

The thickness of the transition region can thus be estimated by introducing the ion mobility and the concentration of the protons and hydroxide ions in pure water and the assumed resistance for the cation- and anion-exchange layers into Equation 4.75. Taking appropriate values for the ion mobility and the resistance from the literature [P.W. Atkins, 1994], [F. Maletzki et al., 1992] results in a transition layer thickness of ca. 4 nm.

The thickness of the transition layer can also be calculated by a continuum model [A. Mauro, 1962], [W. Shockley, 1949]. The model is based on an analogy between the p-n semi-conductor junction and the fixed charged membranes, i.e. the fixed charges in ion-exchange membranes play the same role as "doping" ions in semiconductors. By applying the Maxwell-Boltzmann and the Poisson equations to ion-exchange membranes, concentration and potential profiles in the transition layer and the thickness of the transition layer can be calculated as a function of the fixed charge density of the ion-exchange layers, the potential difference between the charged layers and the transition layer, the permittivity of the transition layer [H. Strathmann et al., 1997]. When no external potential is applied the thickness of the transition layer is given by:

$$2\lambda = 2\sqrt{\frac{\Delta\varphi^{tr}\varepsilon}{C_{fix}F}} \tag{4.76}$$

Here is ε the permittivity of the transition layer, C_{fix} is the concentration of the fixed charges of the ion-exchange layers and $\Delta\varphi^{tr}$ is the potential difference across the space charge layer which is given by:

$$\Delta\varphi^{tr} = 2\,|\varphi_{Don}| \tag{4.77}$$

Assuming the system contains only monovalent ions and that the concentration of the solution in the transition region C^{tr} is much smaller than the fixed ion concentration $C_{fix,}$ then is:

$$C^{m}_{cou} \cong C_{fix} \tag{4.78}$$

The Donnan potentials between the solution in the transition region and the cation- and anion-exchange layer is:

$$\varphi_{don} = -\frac{RT}{F}\ln\frac{C_i^{tr}}{C_{fix}} = \frac{RT}{F}\ln\frac{C_{fix}}{C_i^{tr}} \tag{4.79}$$

The potential difference across the transition region $\Delta\varphi^{tr}$ can be determined by introducing Equation 4.79 into Equation 4.77:

$$\Delta\varphi^{tr} = 2\,\varphi_{Don} = 2\frac{RT}{F}\ln\frac{C_{fix}}{C_i^{tr}} \tag{4.80}$$

With a fixed charge density C_{fix}= 1.0 mol L^{-1} the calculated potential difference across the transition region at T = 20°C and a proton and hydroxide ion concentration of 10^{-7} mol L^{-1} the potential drop across the transition region is 0.814 V. If a relative permittivity of water of ε_r = 78.5 is assumed the thickness of the transition region calculated according to Equation 4.76 is ca. 4.8 nm. This seems to be a maximum value since the relative permittivity of the transition region is most likely significantly lower.

Equation 4.76, however, describes the thickness of the transition region if no external electrical field is applied. When an external voltage U_{ex} is established across the transition layer of the bipolar membrane the voltage drop across the depletion layer and the thickness of the transition region will be changed accordingly to:

$$\Delta\,\varphi^{tr} = 2\Delta\,\varphi_{Don} + U_{ex} \tag{4.81}$$

and

$$2\lambda = 2\sqrt{\frac{(2\Delta\varphi_{Don} + U_{ex})\varepsilon_r\varepsilon_0}{C_{fix}F}} \tag{4.82}$$

Here 2λ is the thickness of the transition region, $\Delta\varphi_{don}$ is the Donnan potential between the transition region and the ion-exchange layers of the bipolar membrane, U_{ex} is the externally applied voltage drop across the transition region, C_{fix} is the fixed ion concentration of the bipolar membrane, F is the Faraday constant, and ε_r and ε_0 are the relative permittivity and the dielectric constant in a vacuum.

For a forward bias U_{ex} is positive and thus 2λ, i.e. the thickness of the transition region decreases and for a reverse bias 2λ increases. In practical applications bipolar membranes are operated in reverse bias. Therefore, the thickness of the transition region is increasing with the externally applied voltage. However, the external applied voltage drop across the transition region is difficult to measure. Generally, the voltage drop across the entire bipolar membrane U^{bm} is measured. However, because of the relatively high conductivity of the ion-exchange layers of the bipolar membrane it can be assumed that for relatively low current densities the external applied voltage across the bipolar membrane is not very different from the voltage drop across the transition region, i.e. $U_{ex} \cong U^{bm}$. The thickness of the transition layer depends furthermore on the relative permittivity of the transition layer which is also unknown and can only be estimated. Its value is probably somewhere between 10, which is the relative permittivity of a hydrophilic polymer, and 78, which is the relative permittivity of pure water. Taking a value of 45 for the relative permittivity in the transition layer and the voltage drop across the bipolar membrane the transition region thickness can be estimated from the theoretical considerations expressed in Equation 4.82 to be in the range of a few nanometers, which is well in agreement with the estimation that is based on the resistance of the transition region expressed in Equation 4.75.

The accelerated water dissociation due to the second Wien effect

One possible explanation for the experimentally determined accelerated water dissociation in bipolar membranes is based on the second Wien effect which describes the influence of a strong electric field on the water dissociation rate constant while it is assumed that the recombination rate of the protons and hydroxide ions is not effected by the electric field. It is in the range of ca. 1×10^9 mol^{-1}s^{-1} [F. Kortüm, 1957]. The dissociation rate increase due to the electric field effect can be expressed as the ratio of the water dissociation rate under the influence of an electric field to the dissociation rate without the influence of an electric field [L. Onsager, 1934]. For electrical field densities in excess of 1×10^8 Vm^{-1} this ratio is given according to the second Wien effect by:

$$\frac{k_d(E)}{k_d(0)} = \left(\frac{2}{\pi}\right)^{0.5} (8b)^{-0.75} e^{(8b)^{0.5}} \tag{4.83}$$

For a monovalent electrolyte the constant b is given by:

$$b = 0.096\frac{E}{\varepsilon_r T^2} = 0.048\frac{U\left(C_{fix}F\right)^{0.5}}{\varepsilon_r T^2\left[\left(2\Delta\varphi_{Don} + U_{ex}\right)\varepsilon_r\varepsilon_0\right]^{0.5}} \tag{4.84}$$

Here E is the electric field density which is given by the voltage drop across the transition region U divided by the thickness of the transition region 2λ; ε_r the relative permittivity in the transition region, ε_0 is the permittivity in a vacuum, $k_d(E)$ and $k_d(0)$ are the dissociation rate constants under the influence of an electric field and without an electric field, respectively, φ_{Don} is the Donnan potential between the ion-exchange layers of the bipolar membrane and the transition region, U_{ex} is the externally applied potential, and T is the temperature.

The maximum current density that can be achieved is proportional to the flux of protons and hydroxide ions generated by the water dissociation which is given in Equation 4.71. Thus is:

$$i_{max} = F\, k_d(E)\, C_{H_2O}\, 2\lambda \tag{4.85}$$

Here i_{max} is the maximum current density, F is the Faraday constant, $k_d(E)$ is the water dissociation rate constant under influence of an electrical field, C_{H_2O} is the water concentration in the transition region, and 2λ the thickness of the transition region.

To achieve the experimentally observed current densities of more than 2000 Am^{-2} the water dissociation rate constant must be in the order of 100, i.e. $k_d(E) = 100$ when the water concentration in the transition region is in the order of 10 000 to 30 000 mol m^{-3}. With the assumption that the main voltage drop across the membrane is in the transition region between the two ion-exchange layers and a given relative Permittivity, the increase in the water dissociation rate due to an applied external voltage can be calculated by Equations 4.83 and 4.84. As indicated before, a critical parameter in Equation 4.84 is the relative permittivity of the transition region. Assuming an externally applied voltage of 1 V, a relative permittivity in the transition region of 45, and a fixed ion concentration of 10^3 mol m^{-1} the increase in the water dissociation rate would be about a factor of 5×10^3. If it is assumed that the relative permittivity would be that of pure water, i.e. 78 the water dissociation would be increased by a factor of less than 100. For transition region permittivity of 20, i.e. the permittivity of hydrophilic polymer the dissociation rate would be increased by a factor of 5×10^6. This value is about the same order of magnitude observed experimentally in bipolar membranes.

Although the model for the increased water dissociation in bipolar membranes based on the second Wien effect is suited for qualitative description of the experimentally observed high water dissociation rate. It has, however, several weaknesses:

- The model based on the second Wien effect meets the experimental values only when the relative permittivity of the transition region is less than 20.
- The application of the Onsager theory which has been developed for homogeneous solutions and field strength of less than 5×10^7 Vm^{-1} to bipolar membrane seems questionable.
- According to the Onsager theory the increase of the water dissociation should be identical at the cation- and anion-exchange membrane but this is not the case as has been shown experimentally [I. Rubinstein et al., 1984].

The accelerated water dissociation due to a reversible proton transfer reaction

Theoretical considerations and experimental evidence strongly support a hypothesis that the accelerated water dissociation is also caused by a reversible proton transfer reaction between the charged groups and water. This hypothesis is based on studies carried out at monopolar membranes which showed that the water dissociation in electrodialysis is obtained mainly at the anion-exchange membrane [I. Rubinstein et al. 1984]. Based on this general observation it can be assumed that in the presence of ionic groups the water dissociation rate constant may be several orders of magnitude higher than in free solution and that with anion-exchange membranes the water dissociation is caused by a reversible protonation of weakly basic groups such as tertiary amines [R. Simons, 1979, 1984, 1985, 1993].

In case of the bipolar membrane both the cation- and the anion-exchange groups of the membrane polymer adjacent to the transition layer can react with the water molecules at the membrane surface of the ion-exchange layers facing the transition region. Thus, the water dissociation in a bipolar membrane may be the result of the following reactions:

$$2H_2O \underset{k_{-1}}{\overset{k_1}{\Leftrightarrow}} H_3O^+ + OH^- \quad (a) \tag{4.86}$$

$$B + H_2O \underset{k_{-2}}{\overset{k_2}{\Leftrightarrow}} BH^+ + OH^- \quad (b) \quad \text{and} \quad BH^+ + H_2O \underset{k_{-3}}{\overset{k_3}{\Leftrightarrow}} B + H_3O^+ \quad (c) \tag{4.87}$$

$$A^- + H_2O \underset{k_{-4}}{\overset{k_4}{\Leftrightarrow}} AH + OH^- \quad (d) \quad \text{and} \quad AH + H_2O \underset{k_{-5}}{\overset{k_5}{\Leftrightarrow}} A^- + H_3O^+ \quad (e), \tag{4.88}$$

Here B is a neutral base and AH a neutral acid.

It is assumed that the accelerated water dissociation in a bipolar membrane is the result of the chemical reaction expressed in the Equations 4.87 or 4.88 and can be rationalized by the so-called chemical reaction model [S. Mafé et al., 1997], [S. Mafé et al., 2000], [P. Ramirez et al., 1991]. The reaction model describes the water dissociation by an empirical equation which relates the water dissociation constant to the chemical reaction, i. e. the reversible protonation reaction of the water with a functional group, i.e. a weak acidic or weak basic group of the polymer in the transition region. According to the chemical reaction model the water dissociation rate constant is given by:

$$k_d^{cr}(E) = k_d^{cr}(0)\exp\left[\frac{\alpha F}{RT}E\right] \tag{4.89}$$

Here $k_d^{cr}(E)$ and $k_d^{cr}(0)$ are the water dissociation rate constants due to a reversible protonation with and without an externally applied electric field, F is the Faraday constant, E is an externally applied electric field, and α is a characteristic parameter with the dimensions of length.

In the reaction model still an external field is assumed necessary to achieve an increase in the water dissociation rate constant.

The observed temperature dependency of the water dissociation constant can be expressed by the Arrehnius relation [P. Ramirez et al.,1992] and is given by:

$$k_d^{cr}(0) = B\exp\left[-\frac{E_A}{RT}\right] \tag{4.90}$$

Here B is the pre-exponential factor and E_A is the activation energy.

The maximum current density that can be achieved in bipolar membrane water dissociation is given by:

$$i_{max} = F k_d^{cr} C_{cag} 2\lambda = F C_{cag} B\exp\left[-\frac{E_A}{RT}\right]\exp\left[\frac{\alpha F}{RT}E\right] \tag{4.91}$$

Here i_{max} is the maximum current density, k_d^{cr} is the water dissociation rate constant, C_{cag} is the concentration of catalytically active groups, E_A is the activation energy, E is an externally applied electric field, B is the pre-exponential factor, 2λ is the thickness of the transition region, and α is a characteristic parameter with the dimensions of length. The symbols F, R, and T have been defined previously.

With Equation 4.91 water dissociation rates and current densities obtained in experimental studies with bipolar membranes and also their temperature dependency can be explained. Nevertheless, the chemical reaction model has again several weaknesses.

- The model is based on three parameters, i.e. the concentration of the catalytic groups C_{cag}, the pre-exponential factor B, and the constant α, that are difficult to determine by independent measurements.
- There is no mechanistic interpretation of the catalytic reaction mechanisms and especially the constant α must be considered to be just a fitting parameter.
- The water concentration in the transition region is not considered and assumed to be always high enough for the catalytic reversible proton transfer reaction. This, however, is not the case as will be shown later.

The extended proton transfer reaction model

There are other models described in the literature [V.I. Zabolotsky et al., 1984, 1986,1988], [H.D. Hurwitz et al., 1998]. A model suggested by Rapp [H.J. Rapp, 1995], and [H. Strathmann et al., 1997] is also based on a reversible protonation reaction and a complete mass balance considering all reactions and the mass transport in the transition region. This model provides some information concerning the actual mechanism of the reversible protonation and also reduces the fitting parameters required by the chemical reaction model.

The model is developed under the following assumptions:

- The water dissociation occurs in the transition region of the anion-exchange layer of the membrane.
- The water dissociation is accelerated by protonation and de-protonation of weakly basic groups.
- The water dissociation may also be accelerated by the electric field according to the second Wien effect.
- The voltage drop across the ion-exchange layers of the bipolar membrane are neglected so that $U_{ex} = U^{bm}$.
- The generated protons and hydroxyl ions are removed from the transition region by migration and the electric current is calculated from this migration flux.
- The driving force for the migration of protons and hydroxyl ions is U_{ex}.
- Water is transported into the transition region by diffusion.

For each species i, which participates in the reaction and the mass transport in the transition region of the simplified system the following mass balance is valid:

$$2\lambda \frac{dC_i}{dt} = \left(J_i^{diff} + J_i^{mig}\right) + 2\lambda r_i \qquad (4.92)$$

Here are r_i the reaction rates of the different components in the transition region, i.e. the protons, the hydroxyl ions, the water, and the protonated and the neutral base, C_i is the concentration of the different components in the transition region, and J_i^{diff} and J_i^{mig} are the fluxes of the different components out of or into the transition region by diffusion and migration.

The reaction rates r_i for the different components of the transition region are:

$$r_{H_3O^+} = k_1 C_{H_2O} C_{H_2O} - k_{-1} C_{H_3O^+} C_{OH^-} + k_3 C_{H_2O} C_{BH^+} - k_{-3} C_{H_3O^+} C_B \quad (4.93)$$

$$r_{OH^-} = k_1 C_{H_2O} C_{H_2O} - k_{-1} C_{H_3O^+} C_{OH^-} + k_2 C_{H_2O} C_B - k_{-2} C_{OH^-} C_{BH^+} \quad (4.94)$$

$$r_B = k_3 C_{H_2O} C_{BH^+} - k_{-3} C_{H_3O^+} C_B + k_{-2} C_{OH^-} C_{BH^+} - k_2 C_{H_2O} C_B \quad (4.95)$$

$$r_{BH^+} = -r_B = -k_3 C_{H_2O} C_{BH^+} + k_{-3} C_{H_3O^+} C_B - k_{-2} C_{OH^-} C_{BH^+} + k_2 C_{H_2O} C_B \quad (4.96)$$

$$r_{H_2O} = k_{-1} C_{H_3O} C_{OH^-} - k_1 C_{H_2O} C_{H_2O} + k_{-2} C_{OH^-} C_{BH^+} - k_2 C_{H_2O} C_B + k_{-3} C_{H_3O^+} C_B - k_3 C_{H_2O} C_{BH^+} \quad (4.97)$$

To determine the different reaction rates described in Equations 4.93 to 4.97 the different reaction rate constants must be known.

The reaction constants k_1 and k_{-1} can be obtained from the dissociation in equilibrium in pure water.

The reaction rates for the water dissociation r_1 and r_{-1} are:

$$r_1 = k_1 [H_2O][H_2O] = k_1^{'} [H_2O] \quad \text{and} \quad r_{-1} = k_{-1} [H_3O^+][OH^-] \quad (4.98)$$

In equilibrium the reaction rates r_1 and r_{-1} are equal. Thus is:

$$\frac{k_1^{'}}{k_{-1}} = \frac{[H_3O^+][OH^-]}{[H_2O]} = \frac{K_w}{[H_2O]} \quad (4.99)$$

Here K_w is the water equilibrium constant which at 25° C is 10^{-14} mol^2 L^{-2}, $k_1^{'}$ is the rate constant for the water dissociation as defined by Equation 4.98 is 2×10^{-5} s^{-1} [R. Simons et al., 1978], and k^{-1} is the ion recombination rate which is calculated from Equation 4.99 to 1.1×10^{11} Lmol^{-1}s^{-1}.

With Equation 4.98 the value of the reaction rate constant k_1 is calculated to be: 3.63×10^{-7} L mol^{-1} s^{-1}.

If a reversible catalytic protonation reaction according to the reaction b) and c) in Equation 4.87 is assumed the rate constants k_2 and k_3 can be calculated as function of the pK_b-value of the catalytic active ionic group in the transition region as follows:

In case of reaction b) in Equation 4.87 is:

$$\frac{[BH^+][OH^-]}{[B]} = \frac{k_2 [H_2O]}{k_{-2}} = \frac{k_2^{'}}{k_{-2}} = K_b \quad (4.100)$$

Furthermore is:

$$k_2' = k_{-2} \times K_b = k_{-2} \times 10^{-pK} \quad \text{and} \quad pK_b = -\log K_b \tag{4.101}$$

In case of reaction c) in Equation 4.87 is:

$$\frac{[BH^+][H_3O^+]}{[BH^+]} = \frac{k_3[H_2O]}{k_{-3}} = \frac{k_3'}{k_{-3}} = K_a \tag{4.102}$$

and

$$k_3' = k_{-3} \times K_a = k_{-3} \times -pK_a = k_{-3} \times 10^{-pK_a} \tag{4.103}$$

With $pK_a + pK_b = 14$ Equation 4.103 becomes:

$$k_3' = k_{-3}\, 10^{-(14-pK_b)} \tag{4.104}$$

The rate constants k_2 and k_3 for the reversible protonation reaction described in Equation 4.87 are:

$$k_2 = \frac{k_2'}{[H_2O]} = \frac{k_{-2}10^{-pK_b}}{[H_2O]} \tag{4.105}$$

and

$$k_3 = \frac{k_3'}{[H_2O]} = \frac{k_{-3}10^{(14-pK_b)}}{[H_2O]} \tag{4.106}$$

The recombination rate constants k_{-2} and k_{-3} in aqueous solutions are in the range of 10^{11} L mol^{-1} s^{-1}, and it is assumed that they are not influenced by the electric field.

The rate constants k_2 and k_3 are only a function of the pK_b-value of the ionic groups and can be calculated from Equations 4.105 and 4.106.

The results of these calculations are shown in Figure 4.13 which shows the values of k_2 and k_3 as a function of the pK_b -value of the ionic groups in the transition region.

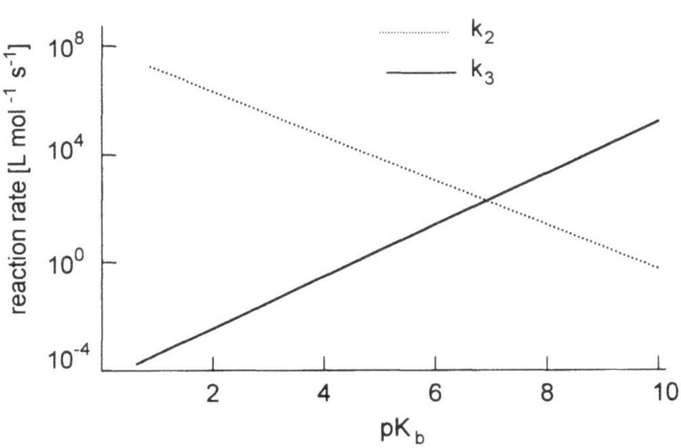

Fig. 4.13 Calculated values for the reaction rates k_2 and k_3 as function of the pK_b-value of the catalytic group in the reaction region.

It can be seen that for strongly basic groups (pK_b <1) k_2 is very high and k_3 is very small. This means that the protonation is very fast but the deprotonation is very slow, so that the groups remain in the protonated form. For strongly acidic groups (pK_a <1) all groups remain in the dissociated, deprotonated form. With increasing pK_b of the groups k_2 is decreasing while k_3 is increasing and at $pK_b = 7$ the deprotonation is as fast as the protonation and $k_2 = k_3 = 100$. This means that the dissociation is ca. 27×10^7 times faster than the dissociation rate in free water and that the proton transfer reactions are capable of explaining the enhanced water dissociation if there is always enough water in the transition region.

The model of the diffusive water transport into the membrane transition region can be described by Fick's law. If it is assumed that:

- water is diffusing from both sides of the bipolar membrane into the transition region,
- the concentration of the water in the bulk solutions on both sides of the bipolar membrane is approximately the same, and
- the diffusion coefficients of water in the cation- and anion-exchange layers of the bipolar membrane are identical.

Under these conditions the diffusive water flux is given by:

$$J^{diff}_{H_2O} = 2\left[-D^m_{H_2O}\frac{dC_{H_2O}}{dz}\right] = 2\left[-D^m_{H_2O}\frac{C^b_{H_2O} - C^{tr}_{H_2O}}{d^{diff}}\right] \qquad (4.107)$$

Here $J^{diff}_{H_2O}$ is the diffusive water flux from both bulk solutions into the transition region of the bipolar membrane, $D^m_{H_2O}$ is the diffusion coefficient of the water in the cation- and anion-exchange layers of the bipolar membrane, $C^{tr}_{H_2O}$ and $C^b_{H_2O}$ are the water concentrations in the transition region and the bulk solution, z is a directional coordinate, and d^{diff} is the thickness of the diffusion path length.

Assuming that the thickness of the bipolar membrane is much larger than that of the transition region, i.e. $\lambda \ll d^{bm}$ and that diffusivity of water in the cation-exchange layer is identical to that in the anion-exchange layer the diffusive path length can be expressed by:

$$d^{diff} = \frac{d^{bm}}{2} - \lambda \cong \frac{d^{bm}}{2} \qquad (4.108)$$

Here d^{bm} is the thickness of the bipolar membrane.

Introducing Equation 4.108 into 4.107 leads to:

$$J^{diff}_{H_2O} = 4\left[D^m_{H_2O}\frac{C^b_{H_2O} - C^{tr}_{H_2O}}{d^{bm}}\right] \qquad (4.109)$$

Here $J^{diff}_{H_2O}$ is the water flux into the transition layer from the bulk phases, $D^m_{H_2O}$ is the diffusion coefficient, z is a direction coordinate, d^{bm} is the thickness of the bipolar membrane and $C^{tr}_{H_2O}$ and $C^b_{H_2O}$ are the water concentrations in the transition region and the bulk solution, respectively.

The migration of protons and hydroxideions from the transition region of the bipolar membrane into the bulk solutions can be calculated:

$$J^{mig}_{H_3O^+} = u^{tr}_{H_3O^+} C^{tr}_{H_3O^+} \frac{U_{ex}}{2\lambda} \tag{4.110}$$

and

$$J^{mig}_{OH^-} = u^{tr}_{OH^-} C^{tr}_{OH^-} \frac{U_{ex}}{2\lambda} \tag{4.111}$$

Here $J^{mig}_{H_3O^+}$ and $J^{mig}_{OH^-}$ are migration fluxes of the protons and hydroxide ions from the transition region into the bulk solution, $u^{tr}_{H_3O^+}$, $u^{tr}_{OH^-}$, $C^{tr}_{H_3O^+}$, and $C^{tr}_{OH^-}$ are the mobility of the protons and the hydroxide ions and their concentrations in the transition region of the bipolar membrane, respectively, and U_{ex} is the externally applied voltage.

By introducing the derived relations for the mass fluxes, i.e. Equations 4.09 to 4.11 and the reaction rates, i.e. Equations 4.93 to 97 into the mass balance Equation 4.92, the change in the concentrations of the different components in the transition region can be calculated by the following set of coupled differential equations:

$$\frac{dC^{tr}_{H_3O^+}}{dt} = k_1 C_{H_2O} C_{H_2O} - k_{-1} C_{H_3O^+} C_{OH^-} + k_3 C_{H_2O} C_{BH^+} - k_{-3} C_{H_3O^+} C_B - \frac{u_{H_3O^+} C_{H_3O^+}}{2\lambda} \frac{U_{ex}}{2\lambda} \tag{4.112}$$

$$\frac{dC^{tr}_{OH^-}}{dt} = k_1 C_{H_2O} C_{H_2O} - k_{-1} C_{H_3O^+} C_{OH^-} + k_2 C_{H_2O} C_B - k_{-2} C_{OH^-} C_{BH^+} - \frac{u_{H_3O^+} C_{H_3O^+}}{2\lambda} \frac{U_{ex}}{2\lambda} \tag{4.113}$$

$$\frac{dC^{tr}_B}{dt} = k_3 C_{H_2O} C_{BH^+} - k_{-3} C_{H_3O^+} C_B + k_{-2} C_{OH^-} C_{BH^+} - k_2 C_{H_2O} C_B \tag{4.114}$$

$$\frac{dC^{tr}_{BH^+}}{dt} = -\frac{dC^{tr}_B}{dt} = -k_3 C_{H_2O} C_{BH^+} + k_{-3} C_{H_3O^+} C_B - k_{-2} C_{OH^-} C_{BH^+} + k_2 C_{H_2O} C_B \tag{4.115}$$

$$\frac{dC^{tr}_{H_2O}}{dt} = k_{-1} C_{H_3O^+} C_{OH^-} - k_1 C_{H_2O} C_{H_2O} + k_{-2} C_{OH^-} C_{BH^+} - k_2 C_{H_2O} C_B + k_{-3} C_{H_3O^+} C_B - k_3 C_{H_2O} C_{BH^+} + \frac{2 D^m_{H_2O} \left(C^b_{H_2O} - C^{tr}_{H_2O}\right)}{d^m \lambda} \tag{4.116}$$

By a numerical solution of the Equations 4.112 to 4.116 the time dependent concentration change of all components can be determined. By introducing data for the various constants from the literature, and realistic values for the applied voltage, and for the initial concentrations of the different components in the transition region it was shown [H.J. Rapp, 1995] that the concentrations of all components in the transition region reach a steady state within a very short time, i.e. within ca. 1×10^{-6} to 1×10^{-5} s. An assessment of the various parameters in the stationary state of the system showed furthermore, that the concentration of the protonated and deprotonated groups BH^+ and B of the catalytic active component changes drastically with the pK_b-value of this component. At a pK_b-value of 7 the concentration of the protonated and deprotonated groups are equally high, and the highest water dissociation rate is obtained.

This becomes also apparent when the current density is shown as a function of the applied external voltage at different pK_b-values. The current density can be calculated either from the proton or hydroxide-ion fluxes since both fluxes are identical because of the electroneutrality requirement. Because of their higher mobility the protons will first move faster than the hydroxideions. This disturbs the electroneutrality and leads to an additional electric field which increases the migration of the hydroxide-ions and slows down the protons. To take this phenomenon into account an average mobility for both ions can be calculated.

If the cation- and anion-exchange layers of the bipolar membrane are strictly permselective the current density is given by:

$$i = F\, u^{tr}_{OH^-} C^{tr}_{OH^-} \frac{U_{ex}}{2\lambda} \qquad \text{or} \qquad i = F\, u^{tr}_{H^+} C^{tr}_{H^+} \frac{U_{ex}}{2\lambda} \tag{4.117}$$

Here $u^{tr}_{OH^-}$ and $u^{tr}_{H^+}$ are the mobility and $C^{tr}_{OH^-}$ and $C^{r}_{H^+}$ the concentration of the H^+- and the OH^--ions in the transition region of the bipolar membrane. According to Equations 4.112 and 4.113 the concentration of the H^+- and the OH^- -ions in the transition region depends on the concentration of the catalytic groups B and BH^+ in the transition region. This concentration, however, is a strong function of the pK_b-value as shown earlier. Therefore, the current density obtained at a given applied voltage is a function of the pK_b-value.

The concentration of the H^+- and the OH^- -ions in the transition region depends also on the concentration of the water in the transition region which, according to Equation 4.116, depends mainly on the diffusion of the water from the bulk solutions into the transition region and is determined by the diffusion coefficient of the water in the ion-exchange layers of the bipolar membrane. Therefore, the current density obtained at a given applied voltage is a function of the water diffusion coefficient in the ion-exchange layers of the bipolar membrane.

All constants in Equations 4.112 to 4.117 can be determined by independent measurements with the exception of the thickness of the transition region which can be estimated according to Equations 4.75 and 4.82. The only parameter which is difficult to estimate is the concentration of the catalytically active component in the transition region. Based on the suggested model the current density obtained in bipolar membranes as a function of the applied voltage has be calculated using the concentration of the catalytically active component as a "fitting" parameter and compared with experimentally determined values and a good agreement between the calculated data and data determined experimentally with various bipolar membranes was obtained [H.J. Rapp, 1995].

The extended reaction model also provides a reasonable explanation for the observed limitation of the current density due to the limited water transport into the transition region. This was also confirmed by experimental studies of the overlimiting current densities with mono polar and bipolar membranes [T. Aritomi et al., 1996].

The main conclusion that can be drawn from the extended reaction model are:

- The steady state of the water dissociation in the transition region of the bipolar membrane is reach within ca. 1×10^{5} s.
- The current density obtained at constant voltage decreases with decreasing pK_b-value.

- The current density obtained at constant voltage decreases with decreasing water diffusion coefficient in the bipolar membrane and reaches a limiting value.

This is illustrated in Figure 4.14 which shows schematically the current density as a function of the applied voltage of a bipolar membrane for various pK_b-values and b) for different diffusion coefficients. The schematic curves indicate the strong dependency of the current density on pK_b-value of the catalytic group in the transition layer and the water diffusion coefficient in the membrane which is predicted by the reaction model.

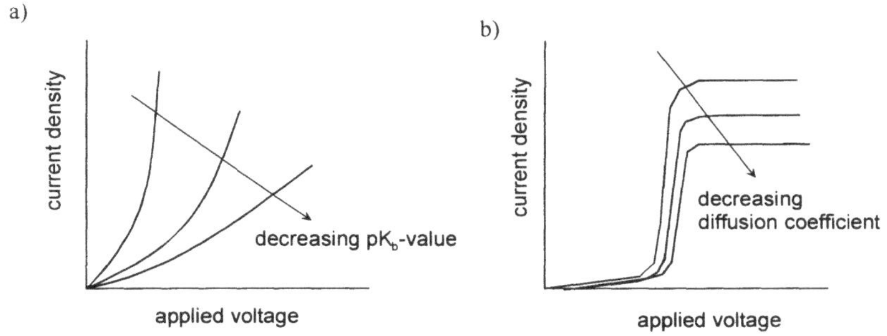

Fig. 4.14 Schematic diagram illustrating the dependency of the current density on the pK_b-values of the catalytically active group in the transition region of the bipolar membrane and the dependency of the water diffusion coefficients in the the ion-exchange layers of the bipolar membrane by showing current density versus applied voltage curves obtained a) by assuming different pK_b-values and b) different water diffusion coefficients.

The model also indicates that the protonation and deprotonation can take place at basic as well as at acidic groups. The pK_b or pK_a value of these groups, however, should be in a range between 5 and 9, respectively. Uncertainty still exists about the value of the relative permittivity ε_r in the reaction zone and the influence of the electric field on the rate constants.

3.1.2 Mass Transport in Electrodialysis with Bipolar Membranes

The mass transport in an electrodialysis stack with bipolar membranes is very similar to the mass transport in conventional electrodialysis. With the exception that the cell unit generally contains three different membranes, i.e. a bipolar membrane, a cation-exchange membrane, and an anion-exchange membrane separating three flow streams, i.e. the salt solution, the produced acid, and the produced base as indicated in Figure 4.11.

The material balance between the different flow streams

The mass balance between the different flow streams in an electrodialysis stack with bipolar membranes can be determined by applying the same basic relations used to characterize the mass transport in a conventional electrodialysis stack. The main difference between the mass transport in a conventional electrodialysis stack and a stack with bipolar membranes is that an acid and a base stream is obtained instead of a concentrate stream. In all three flow streams the ion concentrations change from the entrance to the exit of the cells.

The change in concentration of the salt, the acid, and the base in an electrodialysis stack with bipolar membranes is as in a conventional electrodialysis stack a function of the applied current density, the membrane area in the cell unit of, and the flow velocities in the cells. The determination of the concentration of the different components is obtained by a mass balance which is described in principle in Equations 4.2 to 4.15 for a conventional electrodialysis stack.

Limiting current density in electrodialysis with bipolar membranes

Acid and base production by electrodialysis with bipolar membranes is just as conventional electrodialysis effected by concentration polarization at the membrane surfaces and by a limiting current density. However, this limiting current density is not identical with the limiting current density in conventional electrodialysis as defined in Equation 4.26. In electrodialysis with bipolar membranes the limiting current density is the result of the water depletion in the transition region of the bipolar membrane due to a low water diffusion coefficient in the ion-exchange layers of the bipolar membrane as discussed earlier. The limiting current density in electrodialysis with bipolar membranes determines not only the required membrane area for a given capacity plant, it also effects the energy consumption for the production of acids and bases and is a critical parameter in the practical application of bipolar membranes.

3.1.3 Energy Consumption in Electrodialysis with Bipolar Membranes

The energy required in the production of acids and bases by electrodialysis with bipolar membranes is an additive of two terms:

- The energy required for the water dissociation in the bipolar membrane.
- The electrical energy required to transfer the salt ions from the feed solution and protons and hydroxide ions from the transition region of the bipolar membrane into the acid and base solutions.

The energy required for the dissociation of the water is reversible while the energy required for the transfer of ions is irreversible and is determined by the electrical resistance of the solutions and the membranes. This part of the electrical energy is converted to heat dissipated to the surrounding.

For an accurate calculation of the energy consumption in electrodialysis with bipolar membranes concentration polarization at the membrane surfaces due to boundary layer effects and the concentration potentials between the different solutions must be considered. However, the potential drops due to these effects are small compared to the potential drops caused by the resistances of the solutions and membranes and the water dissociation potential, and can in many practical applications be neglected.

The reversible water dissociation energy

The reversible, i.e. the minimum energy that is required to dissociate water into protons and hydroxide ions at constant temperature and pressure can be calculated by the Nernst-Equation for a concentration chain of solutions with different H^+-ion activities, i.e. pH-values. The reversible Gibbs free energy required for the production of acids and bases in a bipolar membrane is:

$$(\Delta G)_{p,T} = RT \ln \frac{C^{tr}_{H^+} C^{tr}_{OH^-}}{C^{as}_{H^+} C^{bs}_{OH^-}} = F\,\Delta U_c = 2.3\,RT\,\Delta pH \qquad (4.118)$$

Here ΔG is the Gibbs free energy, C is the concentration, F is the Faraday constant, R is the gas constant, T is the absolute temperature, the super scripts tr, as, and bs refer to the transition region, the acid, and the base in the two solutions separated by the bipolar membrane, ΔU_c and ΔpH are the differences in the electrical potential and the pH-value between the two solutions separated by the bipolar membrane, i.e. between the acid and the base.

For 1n acid and base solutions in the two solutions separated by the bipolar membrane ΔU_c is 0.828 V and ΔG is 0.0222 kWh mol^{-1} at 25°C. This is the reversible energy to convert one mole of water into one mole of protons and one mole of hydroxide ions, i.e. 1n acid and base.

Practical energy requirements in electrodialysis with bipolar membranes

The production of an acid and a base requires, in addition to the reversible energy for the water dissociation, energy for the transport of salt ions from the salt solution through the cat- and anion-exchange membranes, and protons and hydroxide ions form the bipolar membrane into the acid and base solutions. The total energy required in electrodialytic water dissociation in a practical process is given be the current passing through the stack multiplied with the total voltage drop encountered between the electrodes:

$$E = I\,\Delta U\,t \qquad (4.119)$$

Here E is the energy consumed in the production of acids and bases in electrodialysis with bipolar membranes, I is the current passing through the stack, ΔU is the voltage applied across the stack, i.e. between the electrodes, and t is the time of operation.

The voltage drop in a stack arrangement is the result of the electrical resistance and the potential differences which are established between solutions of different ion concentrations. This voltage drop can be calculated from the intrinsic electrical properties of the solutions and the membranes used in the process.

To determine the total energy consumption for the process it is convenient to calculate the energy in one cell unit and then to multiply it by the number of cell units and to add eventually the energy dissipated in the electrode reactions. To calculate the energy requirement for the water dissociation in one cell unit, the total voltage drop across the cell unit and the electric current must be determined. The electrical current is proportional to transfer rate of the salt ions from the feed stream to the acid and base compartments. The electrical current is also proportional to the production of the acid and the base. The total voltage drop consists of three parts:

- the potential drop across the bipolar membranes due to the water dissociation equilibrium,
- the potential drop due to the electrical resistance of the solutions and the membranes,
- and the potential drop due to the concentration differences between the different solutions.

To determine the voltage drop across a three cell unit it is assumed for simplicity reasons that the produced acids and bases as well as the salt in the feed solution are monovalent.

The resistance of the ionic solutions is inverse proportional to their concentration. In electrodialysis the concentrations of the solutions change while flowing through the cells of the stack. The applied potential is generally kept constant over the entire cell length. Therefore, the current density is changing from the stack entrance to the stack exit. The current density used for the determination of the energy consumption in electrodialysis is an average value between the current density at the stack entrance and stack exit.

The voltage drop due to the resistance in the solutions is given by:

$$\Delta U_{sol} = i \int_z^{\Delta z} \frac{dz}{\sum_i C_i \Lambda_i} = i \frac{dz}{\dfrac{C_i^{fi} - C_i^{pi}}{\ln\left(\dfrac{C_i^{fi}}{C_i^{pi}}\right)} \Lambda_i} \tag{4.120}$$

and the voltage drop due to the resistance of the membranes is given by:

$$\Delta U_m = i\left(r^{am} + r^{cm} + r^{bm}\right) \tag{4.121}$$

Here ΔU_m is the voltage drop due to the electrical resistance in the membranes, i is the current density, Λ_i and C_i are the equivalent conductivity and the concentration, respectively, of the component i, r is the area resistance, and z is the directional coordinate perpendicular to the membrane surface, the subscript m refers to membranes in general; the superscripts sol, cm, am, and bm refer to the solution, and the cation-exchange, the anion-exchange and the bipolar membrane, respectively, The superscripts fi and pi refer to the feed at the cell entrance and the product at the cell outlet.

The resistance of the bipolar membrane is composed of to parts: One is the resistance of the ion-exchange layers, r^{bml} which can be estimated from the fixed charge density and the thickness of the cation- and anion-exchange layers of the bipolar membrane and the resistance of the transition region r^{tr} which is given by the specific conductivity of the transition region and its thickness:

$$r^{tr} = \frac{2\lambda}{\kappa} = \frac{2\lambda}{\left(C_{H^+}\Lambda_{H^+} + C_{OH^-}\Lambda_{OH^-}\right)} \tag{4.122}$$

Here r^{tr} is the area resistance of the transition region, 2λ is the thickness of the transition region, κ is the specific conductivity of the transition region, and C and Λ are the concentration and equivalent conductivity; the subscripts H^+ and OH^- refer to proton and hydroxide ion.

By assuming that the transition region contains only water and its thickness is 2 to 4 nm the resistance of the transition region can be estimated.

Additional assumption that are made to determine the total voltage drop across the three cell unit of electrodialysis arrangement with bipolar membranes are:

- The boundary layer and concentration potentials between solution of different ion concentrations can be neglected.
- The geometry of all three cells is identical and the flow of the different solutions is co-current and of equal velocity.
- The equivalent conductivity is independent of concentration over the range of interest.

With these approximations the voltage drop across a cell unit can be expressed by introducing the water dissociation energy and the membrane resistance expressed in Equations 4.121 and 4.122 into Equation 4.120 and integration over the cell length which leads to:

$$\Delta U_{cell} = i\left[\Delta\left(\sum_i \frac{\ln\left(\frac{C_i^{fi}}{C_i^{pi}}\right)}{\Lambda_i\left(C_i^{fi}-C_i^{pi}\right)}\right) + r^{tr} + r^{am} + r^{cm} + r^{bml} + \frac{RT}{iF}\ln\frac{C_{H^+}^{tr}\,C_{OH^-}^{tr}}{\frac{C_{H^+}^{fH^+}-C_{H^+}^{pH^+}}{\ln\frac{C_{H^+}^{fH^+}}{C_{H^+}^{pH^+}}}\;\frac{C_{OH^-}^{fOH^-}-C_{OH^-}^{pOH^-}}{\ln\frac{C_{OH^-}^{fOH^-}}{C_{OH^-}^{pOH^-}}}}\right] \quad (4.123)$$

Here ΔU_{cell} is the voltage drop across a cell unit, Δ is the thickness of the cells, C denotes the concentration, Λ is the equivalent conductivity of the solutions, r is the area resistance and the constants R, T, and F have their usual meaning; the subscript i refers to salt, acid, and base; the superscripts tr, am, cm, and bml refer to transition region, the anion- and cation-exchange membrane, and the ion-exchange layers of the bipolar membrane, the superscripts fi and pi refer to the component i in the feed solution at the cell entrance and the product solution at the cell exit.

The total electric energy required for the transfer of ions, i.e. the actual conversion of salt into water in a cell unit is given by the voltage drop multiplied by the current passing through the cell unit and the time:

$$E_{pro} = \Delta U_{tot}\, I\, t = N\, \Delta U\, I\, t \quad (4.124)$$

Here E_{pro} is the energy for the production of an acid and a base from the corresponding salt, ΔU_{tot} is the total voltage across a cell unit, I is the current, N is the number of cell units in a stack, and t is the time.

The current I is related to the current density i by the cell unit area A, i.e. $I = i\, A$. The electric current required for the conversion of a certain amount of feed solution containing a monovalent salt into an acid and a base is directly proportional to the number of ions transferred from the salt solution containing flow stream to the acid and base containing cells. This current is given by:

$$I = \frac{V_p F\, C_{ac}}{\xi\, t} \quad (4.125)$$

Here I is the electric current passing through a cell unit, V_p is the volume of the product, C_{ac} is the concentration of the produced acid, F is the Faraday constant, t is the time, and ξ is the current utilization.

The current utilization is an expression for the fraction of the total current passing through the cell unit that is actually used for the transfer of ions. The factors effecting the current utilization will be discussed in more detail later.

Introducing Equations 4.123 and 4.125 into Equation 4.124 gives a relation for the total energy consumption in an acid and base producing electrodialysis process with bipolar membranes:

$$E_{pro} = N_{st}\, A_{cell} \left(\frac{\Delta}{\sum_i \Lambda_i \overline{C}_i} + r^{tr} + r^{am} + r^{cm} + r^{bml} + \frac{2.3\, RT\, \Delta pH}{i\, F} \right) \left(\frac{Q_p F \left(C_p^{fp} - C_p^{p} \right)}{A_{cell}\, N_{st}\, \xi} \right)^2 t \qquad (4.126)$$

Here is E_{pro} the energy required for the production of an acid or base from a given salt solution, Q_p is the flow rate of the product, F the Faraday constant, R the gas constant, T the temperature, t the time, $\overline{C}$ the concentration, Δ the cell thickness, Λ the equivalent conductivity, A_{cell} the effective area of the cell unit, N_{st} is the number of cell units in a stack, r the area resistance, i is the current density, and ξ the current utilization; ΔpH is the difference in the pH of the two solutions separated by the bipolar membrane, the superscripts tr, am, cm, and bml refer to transition region, anion-exchange membrane, cation-exchange membrane, and the ion-exchange layers of the bipolar membrane, respectively. The subscript i refers to the salt, the acid, and the base. It should be noted that $\overline{C}$ as well as ΔpH are average values which are given by the logarithmic average as expressed in Equation 4.123.

Based on Equation 4.126 the energy consumption for the acid and base production with bipolar membranes can be calculated as a function of the current density, and the feed and product solution properties. All parameters in Equation 4.126 can be determined in independent measurements or obtained from literature with the exception of the current utilization which must be determined for a given stack design and application in laboratory or pilot plant tests.

The current utilization in electrodialysis with bipolar membranes

The actual energy required to operate the electrodialytic water dissociation process exceeds the energy theoretically required for the dissociation of the water and the transport of ions between the different solutions. The process efficiency will also be effected by the fact that the membranes are not perfectly selective [F.G. Wilhelm et al., 2000] and that the current flows through the manifold system without contributing to the transfer of ions in the stack [El Moussanoui et al., 1994]. There may be other sources that effect the overall efficiency of the process such as hydraulic leaks between the cells. All these effects are expressed in the current utilization ξ which is defined as the current required in practice to obtain a certain amount of product divided by the theoretically required current. The total current utilization is given by:

$$\xi_{tot} = \eta_l N \eta_m \qquad (4.127)$$

Here ξ is the current utilization, η is an efficiency term, N is the number of cell units in a stack, and the superscripts tot, l, and m refer to total efficiency, current leakage through the manifold and membrane selectivity efficiency. The efficiency terms η_m and η_l are defined in such a way that under ideal conditions they approach 1.

The current utilization ξ_{tot} is mainly determined by the transport rate of co-ions in an ion-exchange membrane. The co-ion transport rates are generally quite low with the exception of the protons and hydroxide ions. Their mobility is an order of magnitude higher than that of other ions. The exceptionally high mobility of the H^+-ion can be explained by the transport mechanism of protons. The molecular interaction of water dipoles with the proton results in the formation of the hydronium-ion and the so-called tunneling transport of the proton from one water molecule to the next as has been discussed earlier. The transport tunneling mechanism explains not only the extraordinary high mobility of protons but is also one of the reasons for the high permeability of anion-exchange membranes for protons while they are nearly impermeable for other salt cations. The permselectivity of cation-exchange membranes for hydroxide ions is similarly poor as the permselectivity of anion-exchange membranes for protons. This leads to a leakage of protons and hydroxide ions into the salt solution as indicated in Figure 4.15. The loss of protons and hydroxide ions due to their high permeability results in a low current utilization especially when high acid and base concentrations are to be achieved.

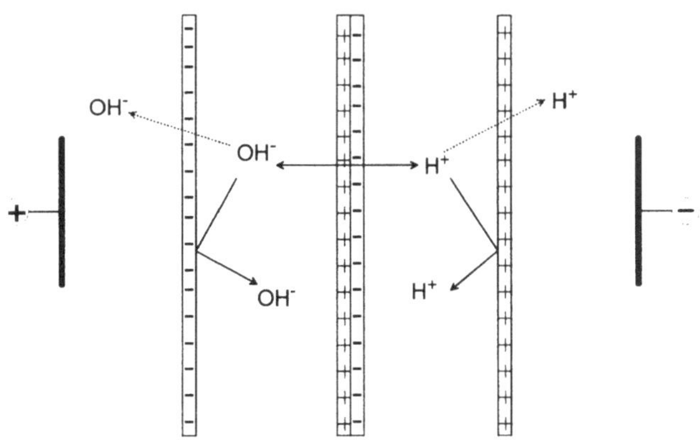

Fig. 4.15 Schematic diagram illustrating the leakage of protons and hydroxide ions through the anion- and cation-exchange membranes from the acid and base flow stream, respectively.

4 THE PRINCIPLE OF DIFFUSION DIALYSIS

Diffusion dialysis with ion-exchange membranes is mainly used today to recover acids or bases from a mixture with salts. The process is based on the fact that ion-exchange membranes generally show a high permeability for counterions while being more or less impermeable for co-ions. An exception are protons and hydroxide ions which can easily permeate both cation- and anion-exchange membranes for reasons that have been discussed earlier.

4.1 The Mass Transport in Diffusion Dialysis

The principle of the process is illustrated in Figure 4.16. In diffusion dialysis the driving force for the transport of ions through the ion-exchange membrane is their gradient in the electrochemical potential, and in contrast to electrodialysis no external electrical potential is applied.

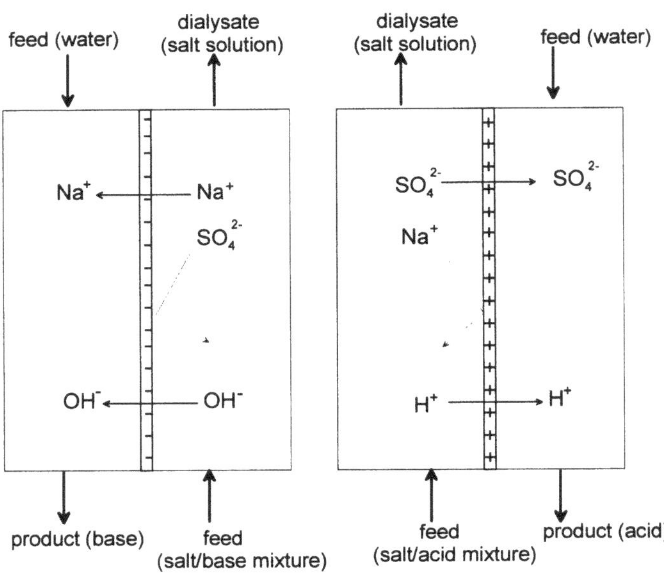

Fig.4.16 Schematic drawing illustrating the principle of diffusion dialysis utilizing a cation-exchange membrane to recover a base and an anion-exchange membrane to recover an acid from a mixture with salts.

The mass transport in diffusion dialysis is determined by the transport of ions through the membranes, and the fluxes of the feed and product solution through the cells separated by the membranes. It can be determined by a complete mass balance which considers the convective transport in the bulk solutions and the transport of ions and water across the membrane. The overall mass transport and the concentration changes in the bulk solutions are illustrated in Figure 4.17. The schematic drawing indicates the convective flux of the components in x-direction, i.e. parallel to the membrane surface in the bulk solutions in opposite direction and the fluxes of the components and the center of mass in z-direction, i.e. across the membrane.

To describe the mass balance in diffusion dialysis the entire system is divided in two subsystems, i.e. the two bulk solutions and the mass balance is applied separately to each system. Furthermore it is assumed that the membrane is in equilibrium with the two adjacent bulk solutions.

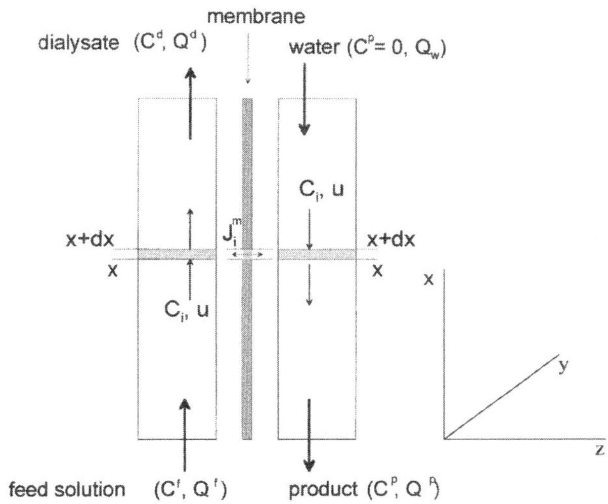

Fig. 4.17 Schematic diagram illustrating the mass transport in a diffusion dialysis cell system due to convection in the direction of the x-coordinate and diffusion in z-direction (C, Q, u, and J refer to concentration, convective flow rate, linear velocity, and flux across the membrane, respectively, the superscripts d, f, and p refer to feed, dialysate, and product, respectively, the subscripts i and w refer to a component and water.

The change in concentration in the bulk solutions is obtained by the following material balance:

$$\frac{\partial(A_{cell}\, dx\, C_i)}{\partial t} = (u\, A_{cell}\, C_i)_x - (u\, A_{cell}\, C_i)_{x+dx} + A_m\, J_i^m \qquad (4.128)$$

Here C_i and u are the concentration of a component i and the linear velocity in the bulk solution, A_{cell} is the cross-section of the bulk solution containing cell, A_m is the surface area of the membrane, and x is the directional coordinate parallel to the membrane surface, J_i^m is the flux of a component i across the membrane.

The surface of the membrane A_m is given by the length of the cell denoted as X_{cell} and its width denoted as Y_{cell}. The cross-section A_{cell} of the cell is given by the width and the thickness of the cell, denoted as Z_{cell}.

Differentiation of Equation 4.128 leads to:

$$\frac{dC_i}{dt} = -u\frac{dC_i}{dx} - C_i\frac{du}{dx} + \frac{X_{cell}\, J_i^m}{Z_{cell}\, dx} \qquad (4.129)$$

Here X_{cell} and Y_{cell} are the length and the width of the cell.

The continuity equation which describes the center of mass change in the bulk solution is given by:

$$\frac{d\rho}{dt} = -\rho\frac{du}{dx} - u\frac{d\rho}{dx} + \frac{A_m \sum_i J_i^m M_i}{A_{cell}\, dx} \qquad (4.130)$$

Here M_i is the molecular mass of the component i.

The mass transport across the membrane is determined by the electrochemical potential gradient of the different components across the membrane. It is given according to Equation 2.144 by:

$$J_i^m = -Đ_i^m \left[\left(1 + C_i^m \frac{d\ln\gamma_i^m}{dC_i^m}\right)\frac{dC_i^m}{dz} + z_i C_i^m \frac{F}{RT}\frac{d\varphi}{dz}\right] + C_i^m v^m \qquad (4.131)$$

Here J_i^m is the total flux of a cation or anion through the membrane, $Đ_i^m$ is the Stefan-Maxwell diffusion coefficient, C_i^m is the concentration, and γ_i^m the activity coefficient of the component i in the membrane, v^m is the convective flux through the membrane, F is the Faraday constant, φ is the electrical potential, and R and T have their general meaning.

Equation 4.131 describes the flux through the membrane as the sum of three terms, i.e. diffusion due to a concentration gradient, migration due to a potential gradient, and convection due to a center of mass flux.

The concentration of the various ions in the solution at the membrane surface is determined by the Donnan potential which describes the electrochemical equilibrium between the membrane and the adjacent bulk solutions. The Donnan potential can according to Equation 2.91 to a first approximation be expressed by:

$$\varphi_{don} = \varphi^m - \varphi^s = \frac{RT}{z_i F}\ln\frac{a_i^s}{a_i^m} = \frac{RT}{z_i F}\ln\frac{C_i^s}{C_i^m}\frac{\gamma_i^s}{\gamma_i^m} \qquad (4.132)$$

Here φ_{don} is the Donnan potential, φ^m and φ^s are the potentials in the membrane and the solution, a is the activity, C is the concentration, γ is the activity coefficient, the subscript i refers to cation or anion, and the superscripts s and m refer to membrane and solution.

In diffusion dialysis no external electrical potential is applied. However, due to the differences in the transport numbers of co- and counterions in an ion-exchange membrane and the required electroneutrality the potential gradient is established which causes additional migration of the coins.

In diffusion dialysis there is generally no hydrostatic pressure difference between the two solutions separated by the membrane, and the center of mass velocity can be expressed by a hydrodynamic coefficient and osmotic and electroosmotic difference across the membrane.

Furthermore, there are no excess charges in the system, i.e.:

$$\sum z_i C_i^m + C_{fix} = 0 \qquad (4.133)$$

and there is no electrical current through the membrane

$$\sum_i J_i^m z_i = 0 \qquad (4.134)$$

By solving the coupled differential Equations 4.29 and 4.30 for the mass and momentum balance under consideration of Equations 4.131 to 4.134 numerically with a appropriate computer program [W. Neubrand, 1999] the diffusion dialysis process can be mathematically modeled. The various constants such as the diffusion coefficients and the transport numbers of the different components in the membrane, or the hydrodynamic permeability and fixed ion concentration of the membrane can be obtained from the literature or from independent measurements. The concentrations of the dialysate and the product can be expressed as a function of the various membrane and solution properties such as the total membrane area, the fixed charge density of the membrane, the flow velocities and flow directions of the bulk solutions in the cells and the cell geometry, i.e. cell length, width, and thickness.

The model also shows that the acid or base concentration in the product can exceed its concentration in the feed solution when the counterion concentration in the feed solution is higher than the proton or hydroxide ion concentration. This is always the case because the salt in the feed solution increases the counterion concentration according to the salt concentration. The high counterion concentration results in a diffusion potential and thus in an additional driving force for the transport for the co-ions.

For the practical dialysis process and equipment design the above described mathematical model has a serious disadvantage. All boundary layer effects in the bulk solution are disregarded. Concentration polarization in the laminar boundary layer at the membrane surface can drastically change the membrane permselectivity and the actual driving force for the transport of the different components.

For the practical design of the diffusion dialysis process a simplified description based on a mass balance and the mass transport between the two solutions separated by the membrane is suggested in the literature [Y. Kobuchi et al., 1984]. The mass transport is determined by a mass transfer coefficient and the concentration difference between the two solutions. The mass transfer coefficient includes both the transfer through the membrane and the two boundary layers at the membrane surface. It is, therefore, a function of the boundary layer thickness which is determined by the cell design and the flow velocities in the cell.

Disregarding any volume changes due to osmosis or convection and disregarding the membrane potential which is caused by the different transport numbers of the co- and counterions in the ion-exchange membrane the mass balance in a dialysis stack is given by:

$$N_i = J_i^m A = Q^f \left(C_i^f - C_i^d\right) = Q^p C_i^p \qquad (4.135)$$

Here N_i is the flux of a component from a feed solution to a receiving, i.e. the product solution, J_i^m is the flux rate through the membrane, A is the membrane area, Q is the volume flux through the dialysis cell, and the superscripts f, d, and p refer to feed at the entrance of the cell, to dialysate, i.e. the feed at the exit of the cell, and to product, i.e. the product in the receiving solution at the exit.

The flux of a component from the feed solution into the receiving solution is given by:

$$N_i = J_i^m A^m = k^x A^m \Delta C_i \qquad (4.136)$$

Here k^x is a mass transfer coefficient describing the mass transport through the membrane and the adjacent laminar boundary layers into the bulk solution at the membrane surface, A^m

is the membrane area in the diffusion cell, and ΔC_i is the concentration difference between the two solutions.

Since the concentrations in the bulk solutions change during the passage through the cell ΔC_i also changes with the length of the cell. To calculate the mass transport between the two solutions an average $\Delta \overline{C}_i$ must be introduced which is given by the integral average of the concentration in the solutions on both sides of the membrane. The average concentration across the membrane is given by:

$$\Delta \overline{C} = \frac{\left(C_i^f - C_i^p\right) - C_i^d}{\ln \frac{\left(C_i^f - C_i^p\right)}{C_i^d}} \tag{4.137}$$

Disregarding volume changes and concentration changes due to osmotic and electroosmotic effects the acid recovery rate can be calculated [Y. Kobuchi et al., 1984] by:

$$\Delta = \frac{C_i^p}{C_i^f}\frac{Q^p}{Q^f} = \frac{Q^p}{Q^f}\frac{\left[e^{\frac{k^x A}{Q^f}\left(1-\frac{Q^f}{Q^p}\right)} - 1\right]}{\left[\frac{Q^p}{Q^f} e^{\frac{k^x A}{Q^f}\left(1-\frac{Q^f}{Q^p}\right)} - 1\right]} \tag{4.138}$$

Here Δ is the recovery rate of the desired product, i.e. the acid or the base, C_i is the concentration of a component i, k^x is the mass transfer coefficient of the component i through the membrane and adjacent boundary layers, A is the membrane area, Q is the volume flow through the dialysis cell, and the superscripts f, p, and d refer to feed, product, and dialysate, respectively..

The recovery rate Δ determines how much of the acid or base, which is in the mixture with a salt, can be recovered. Δ has a value between 0 and 1. It is related to the concentration of the recovered product and the concentration in the depleted solution at the cell exit by:

$$C_i^p = \frac{Q^f}{Q^p} \Delta C_i^f \tag{4.139}$$

and

$$C_i^d = (1 - \Delta) C_i^f \tag{4.140}$$

According to the Equations 4.138 to 4.140 the concentration of a component such as an acid or base in the final product can be determined as a function of the overall mass transfer coefficient, the total membrane area, and the volume flow rates of the feed and the product.

If it is assumed that the volume flow rates of the feed solution and the product recovery solution are identical and counter-current, a very simple relation can be derived for the product recovery rate. Because of equal flow rates the concentration changes in the solutions with respect to the distance along the flow path is linear and the concentration difference across the membrane is equal to the arithmetic mean concentration difference. Thus is:

$$\Delta\overline{C} = \frac{\left(C_i^f - C_i^p - C_i^d\right)}{2} \tag{4.141}$$

Combination of Equation 4.141 with the mass balance Equations 4.135 and 4.136 leads to:

$$\Delta\overline{C} = C_i^f - C_i^p \tag{4.142}$$

and

$$k^{\times}A\left(C_i^f - C_i^p\right) = Q^p C_i^p \tag{4.143}$$

Rearranging of Equation 4.143 gives the recovery rate of the product recovered from the feed solution:

$$\frac{C_i^p}{C_i^f} = \frac{k^{\times}A}{k^{\times}A - Q^p} \tag{4.144}$$

Here C and Q are the concentration and the flow rate, $k^{\times}$ is the mass transfer coefficient for the diffusion of a component through the membrane, the subscript i refers to a component, and the superscripts f and p refer to feed and product.

The modeling of diffusion dialysis on the basis of concentration differences only as suggested in the literature [Y. Kobuchi et al., 1984] does not consider electrochemical interactions of the different ions, i.e. a diffusion potential or osmotic and electroosmotic effects, and can therefore only be considered as a first approximation for the design of the diffusion dialysis process.

4.2 Complicating Factors in Dialysis Mass Transport Modeling

All mass transport models described so far do not take osmotic effects into account. In practical applications of diffusion dialysis the water from the stripping solution to the feed solution due to osmosis is generally quite significant because of the concentration difference between the two solutions. In a typical industrial size dialyzer used for instance to recover acids from a mixture with metal salts the volume of the stripping solution decreases as much as 20 %. The relatively high concentration of the salts in the feed solution has an dramatic effect on the activity of the individual ions. Of special relevance is the activity of the anion which can differ from their concentration by more than 30%. Therefore, a reliable mathematical model should take osmotic water transport and ion activity coefficient into account. Since data concerning the ion activity in the salt mixture and the osmotic water transport through ion-exchange membranes are often not readily available mathematical models have only limited value for the design of the dialysis processes and equipment.

The simple model which considers only the diffusive flux of the component to be recovered through the membrane does not take into account the fluxes caused by the electrical potential which is the result of the difference in the overall ion concentration in the two solutions separated by the membrane. Therefore, the simple model cannot provide an explanation for the experimentally observed fact that in countercurrent diffusion dialysis of a salt and acid mixture the concentration of the acid in the product solution is significantly higher than in the feed solution. For a more accurate modeling of diffusion dialysis the numerical solution with an appropriate computer program of the comprehensive mass and momentum balance as expressed in the Equations 4.128 to 4.131 is certainly a better suited to interpret the experimental results and to design the process for a given application.

5 THE PRINCIPLE OF CONTINUOUS ELECTRODEIONIZATION

In the production of deionized water the conventionally used mixed bed ion-exchange resin is more and more replaced today by continuous electrodeionization. The process is similar to conventional electrodialysis. Anion-and cation-exchange membranes are arranged in alternating series and form individual cells between two electrodes. One of two adjacent cells, i.e. the diluate cell is filled with a mixed bed ion-exchange resin which absorbs the ions from a feed stream and enhances their migration into the adjacent cell, i.e. the concentrate cell. The principle of the process which is illustrated in Figure 4.18 has been suggested more than 30 years ago [O. Kedem et al.,1976], [G.J. Gittens et al., 1965], [Z. Matejka, 1971], [E. Korngold, 1975]. However, only recently the process has become available on a large commercial scale.

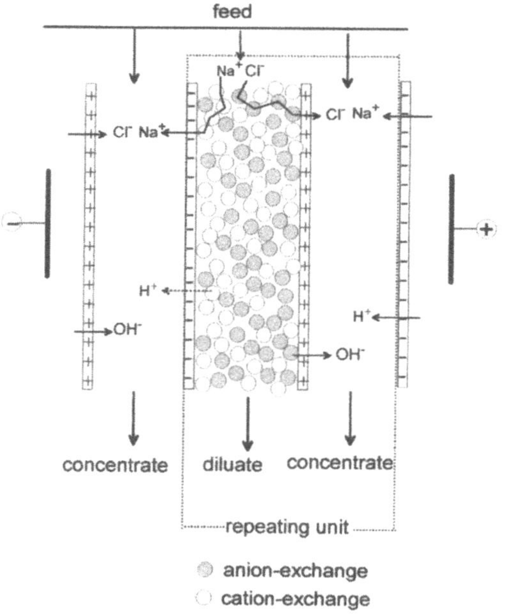

Fig. 4.18 Schematic drawing illustrating the principle of the ion transport in the ion-exchange resin filled diluate cell of a continuous electrodeionization apparatus.

Compared to water deionization by mixed bed ion-exchange resin the continuous electrodeionization has the advantage of being a continuous process, and no regeneration of the ion-exchange resin is required which is generally labor intensive and costly. Compared to conventional electrodialysis the continuous electrodeionization has the advantage that the conductivity in the resin filled diluate cell is increased by more than two orders of magnitude. This effects the limiting current density and thus the required membrane area and the energy consumption. Although the continuous electrodeionization process yields high quality deionized water at reasonable costs it is not without technical problems. High pressure losses and non-uniform flow distribution of the feed stream in the resin filled cell often result in poor process efficiency. Also the removal of weakly dissociated components such as boric and carbonic acid or silica is often not satisfactory. Several modifications of the original continuous electrodeionization process including the use of bipolar membranes which have increased the overall efficiency of the process are described in the literature [S. Thate, 2002], [G. C. Ganzi et al., 1994].

5.1 Mass Transfer in Continuous Electrodeionization

The design of the continuous electrodeionization system is closely related to the design of a conventional electrodialysis stack. However, it is largely based on experimentally determined parameters, and a comprehensive description of the mass transfer in the resin filled cell that is based on theoretical considerations is difficult. Most studies concerning the continuous electrodeionization process, referred to in the literature, are based on experimental investigations carried out with a given stack design, feed solution composition and required product water quality without relating the different experimental parameters in more detail to the overall performance of the process [A.J Giuffrida et al., 1986]. The transport mode of ions in the resin filled cell under the driving force of an electrical potential gradient and the possible water dissociation at the contact points of the cation and anion resin beads is still subject of extensive studies mostly carried out in industry to improve the current utilization and overall process efficiency. A comprehensive study of the mass transfer in a resin filled electrodialysis cell is given by Glueckauf [E. Glueckauf, 1965]. In his analysis Glueckauf assumes that the transport of ions in an electrodialysis cell filled with a mixed bed ion-exchange resin and very dilute solutions takes place mainly by migration through the ion-exchange resin and not through the solution. Thus the transfer of ions in the deionization compartment proceeds in two stages. In the first stage the ions are transferred from the solution to the ion-exchange resin particles. In the second stage the ions are transported through the resin particle or along its surface by migration to the surface of the ion-exchange membrane forming the compartment wall through which they will leave the compartment and pass into the concentrate solution. The first stage in the electrodeionization process, i.e. the transport from the solution to the ion-exchange particle surface is diffusion controlled in the aqueous phase since the ions have to diffuse through a thin film of unstirred solution to the surface of the ion-exchange particle.

As a result of the two stage removal process of the ions from the deionisation compartment, involving ion fluxes in two dimensions, it is necessary to set up two mass balance equations to model the entire process. The first equation concerns the diffusive transfer of ions from the solution into the ion-exchange particles and the second concerns the migration of ions along the particles from the interior of the compartment to the appropriate membrane which permits the passage of the ions out of the compartment. The combination of the two processes then leads to the final relation which describes the entire process as a function of the various design and process parameters.

5.2 Energy Requirements in Continuous Electrodeionization

The energy requirements can be estimated by applying the same general relations as in conventional electrodialysis with the exception that the electrical resistance in the ion-exchange resin filled diluate cell is difficult to calculate on a purely theoretical basis. It must generally be determined experimentally for a given set of boundary conditions. Furthermore, the current utilization is also lower than in conventional electrodialysis. One of the reasons for the low current utilization is the electrodialytic water dissociation which occurs at the membrane surfaces and at the contact points of cation- and anion-exchange resin beads as indicated in Figure 4.18. Since the concentration difference between the dilute and the concentrate solution is generally rather high, i.e. generally 2 to 3 orders of magnitude, and the membranes are not strictly permselective there is a substantial back diffusion of the salt from the concentrate to the diluate solution through the ion-exchange membranes. This not only reduces the current utilization it also effects the quality of the product water, i.e. the lowest achievable salt concentration in the diluate solution. The current utilization is a function of the current density and the overall mass transfer coefficient and the compartment geometry [G.J. Gittens et al., 1965]. The current utilization is furthermore a function of the type of ion-exchange resin and for low feed solution concentration, i.e. for a diffusion controlled mass transport, it is also a function of the linear flow velocity of the solution in the ion-exchange bed. Typical values for the concentration of the product water that can be obtained in practical application of the continuous electrodeionization process are in the range of 0.05 to 0.5 $\mu S\ cm^{-1}$.

For the technical application the concentration factor, i.e. the ratio of the concentrate to the diluate solution that can be achieved is of importance. The achievable concentration factor depends on the permselectivity of the ion-exchange membranes and is limited by the back diffusion of the electrolyte from the concentrate into the diluate compartment. This back diffusion is proportional to the concentration difference between the concentrate and diluate solution. Concentration factors obtained in practical applications are in the range between 100 and 300.

5.3 Complicating Factors in Continuous Electrodeionization Modeling

The mass transport in electrodeionization is the result of a series of complex parallel and consecutive steps. Some of them are difficult to model such as the transport of the ions through the laminar boundary layer around the individual ion-exchange particles which can vary significantly in size and shape. The transport mechanism of the ions adsorbed by the ion-exchange particle is not completely clear. The ions can migrate through the ion-exchange particle or along its surface. Which of the different pathways is the dominating one will depend on the cross-linking density of the ion-exchange material and on the particle surface to volume ratio. The contact points between the ion-exchange particles carrying the same fixed charges is also important for the overall ion transport in a cell filled with a mixed bed ion-exchange resin.. Furthermore, water splitting at the contact points of the cation- and anion-exchange particles can obscure the current utilization and often leads to local pH-value changes in the solution which effect the degree of dissociation of weak acids and bases.

Thus, presently the design of an industrial type continuous electrodeionization system is based mainly on experimental results obtained in laboratory or pilot plant tests. A comprehensive and reliable mathematical model which explains all experimental results and is suited for the construction of continuous electrodeionization systems for various practical applications must still be developed.

6 THE PRINCIPLE OF DONNAN DIALYSIS

Donnan dialysis is an ion-exchange membrane separation process in which ions of the same electrical charge are exchanged between two solutions through an ion-exchange membrane. The fluxes of the ions are coupled, i.e. the flux of a counterion through an ion-exchange membrane caused by a concentration difference between the two solutions separated by the membrane is leading to a counterion transport in opposite direction. The reason for the coupling of the two fluxes is the electroneutrality requirement. Since in an ion-exchange membrane coions are more or less excluded from permeating the membrane, electroneutrality in the two solutions can only be achieved when identical numbers of counterions are transported in opposite direction through the membrane. The principle of the transport is illustrated in Figure 4.19 which shows two electrolyte solutions, i.e. a NaCl solution, and a $CaCl_2$ solution separated by a cation-exchange membrane. The NaCl solution which has a much higher concentration than the $CaCl_2$ solution is referred to as stripping solution. The $CaCl_2$ solution has a lower concentration and is referred to as feed solution. The difference in the Na^+-ion concentration between the stripping and the feed solution acts as driving force for the transport of Na^+-ions from the stripping into the feed solution. Since the Cl^--ions cannot permeate the cation-exchange membrane Na^+-ions can only leave the stripping solution when the equivalent number of charges, i.e. Ca^{2+}-ions are transported in opposite direction from the feed solution into the stripping solution.

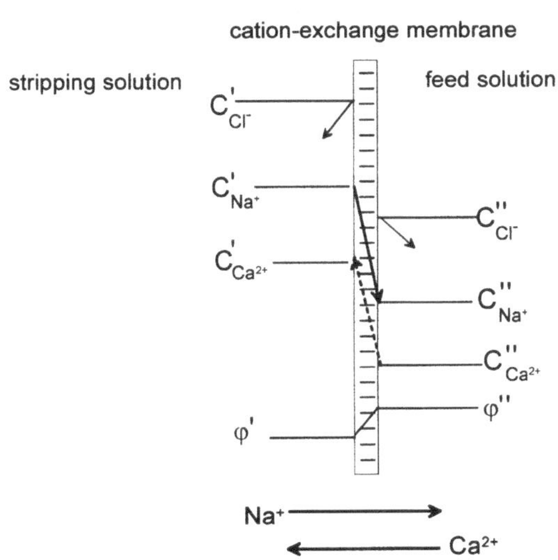

Fig. 19 Schematic drawing illustrating the principle of Donnan dialysis with NaCl in the stripping and $CaCl_2$ in the feed solution.

To maintain electroneutrality in both solutions the transport of x moles Ca^{2+}-ions through the membrane requires the transport of 2 x moles of Na^{+}-ions. Equilibrium between the stripping and the feed solution is referred to as Donnan equilibrium. It is achieved when the electrochemical potential of the Na^{+}-ions and the Ca^{2+}-ions is equal in both solutions. The exchange of monovalent and divalent ions through an ion-exchange membrane for water softening is one of the envisioned applications of Donnan dialysis.

6.1 Mass Transport in Donnan Dialysis

The mass transport in Donnan dialysis is determined by the transport of ions through the membranes, and the flow velocities of the feed and stripping solutions through the compartments separated by the membranes. The overall mass transport can be calculated by the same relations used for determining the mass transport in diffusion dialysis, i.e. by a complete mass balance which considers the convective transport in the bulk solutions and the transport of ions and water across the membrane as has been discussed earlier.

In Donnan dialysis there is no applied external electrical potential and no net flux of electrical charges, i.e. no electric current through the membrane. For a strictly permselective ion-exchange membrane the fluxes of coions are 0 and because of the electroneutrality requirement the sum of the electrical charges carried by the counterion flux is also 0. Thus, there is no net transport of electrical charges across the membrane and the conservation of charges requires:

$$\sum_i z_i J_i = 0 \tag{4.145}$$

Here J_i is the flux of counterions through the membrane and z_i is their valence.

The fluxes of ions across the membrane is the result of an electrochemical potential gradient expressed by activity differences of the different ions in the two solutions separated by the membrane. Equilibrium between the two phases is achieved when the electrochemical potential of all ions in the two solutions are equated and when there are no net ion flux across the membrane. Thus, in equilibrium is:

$$\eta_i^1 = \eta_i^2 = RT \ln a_i^1 + z_i F\varphi^1 = RT \ln a_i^2 + z_i F\varphi^2 \tag{4.146}$$

Here η is the electrochemical potential, a is the activity, j is the electrical potential, z is the valence, R is the gas constant, T the absolute temperature, the subscript i refers to an ion, and the superscripts 1 and 2 refer to the two solutions separated by the membrane.

Equation 4.146 shows that an activity difference of an ion in two solutions separated by a membrane will result in an electrical potential difference when the two solutions are in equilibrium. This potential difference is referred to as Donnan potential. It is obtained by rearranging Equation 4.146:

$$\varphi_{Don} = \varphi_2 - \varphi_1 = \frac{RT}{F} \ln \left(\frac{a_i^1}{a_i^2} \right)^{\frac{1}{z_i}} \tag{4.147}$$

The Donnan potential act on all ions that can permeate the membrane and in equilibrium there is no potential difference between the two solutions. Thus is:

$$\sum_i \left(\frac{a_i^1}{a_i^2} \right)^{\frac{1}{z_i}} = 0 \tag{4.148}$$

Equation 4.148 describes the equilibrium state between all mobile ions in the system. If the system is composed of two electrolytes as indicated in the schematic drawing of Figure 18 which shows a high NaCl solution in the stripping solution and a low $CaCl_2$ solution in the feed the equilibrium activity distribution is given by:

$$\frac{a_{Na^+}^{str}}{a_{Na^+}^{f}} = \left(\frac{a_{Ca^{2+}}^{str}}{a_{Ca^{2+}}^{f}} \right)^{\frac{1}{2}} \tag{4.149}$$

If for dilute solutions it is assumed that to a first approximation the activities can be expressed by the concentrations the Donnan equilibrium is given by:

$$\frac{[Na^+]^{str}}{[Na^+]^{f}} = \left(\frac{[Ca^{2+}]^{str}}{[Ca^{2+}]^{f}} \right)^{\frac{1}{2}} \tag{4.150}$$

Here the superscripts f and str refer to feed and stripping solution.

Equation 4.150 shows that the transport of x moles of Ca^{2+}-ions through the membrane requires the transport of 2 times x moles of Na^+-ions in opposite direction to maintain electroneutrality in the system.

The equilibrium concentrations can also be expressed by the initial concentrations of the Na^+-ions in the stripping and the Ca^{2+}-ions in the feed solution, by the number of moles transported through the membrane, and by the volumes of the stripping and feed solution.

$$\left(\frac{\left(C_{Ca^{2+}}^{f} - \frac{x}{V^f} \right)}{\frac{x}{V^f}} \right)^{\frac{1}{2}} = \frac{\frac{2n}{V^{str}}}{C_{Na^+}^{str} - \frac{2n}{V^{str}}} \tag{4.151}$$

Here $C_{Ca^{2+}}^{f}$ and $C_{Na^+}^{str}$ are the initial concentrations of the Ca^{2+}-ions in the feed solution and the Na^+-ions in the stripping solution, n is the number of moles Ca^{2+}-ions transferred through the membrane, V^f and V^{str} are the volumes of the feed and the stripping solution.

From Equation 4.151 the of moles of Ca^{2+}-ions that can be removed from a feed solution for a given initial concentration of Na^+-ions in the feed solution can be calculated..

Assuming equal volumes of feed and stripping solution and an initial Na^+-ion concentration of 1 mol L^{-1} and an initial Ca^{2+}-ion concentration of 0.1 mol L^{-1} x will be ca. 0.0948. That means that 95% of the Ca^{2+}-ions have been removed from the feed solution when the equilibrium is reached.

Equation 4.150 determines the maximum amount of a component that can be removed from a feed solution for given initial concentrations of the stripping and the feed solution. To design a Donnan dialysis system for a given application it is necessary to determine the required membrane area, the feed and stripping solution flow rate, and process path length for a given cell geometry, and initial and final concentrations in the feed and stripping solution.

The process modeling is based on a complete mass balance which is determined by applying the same basic procedure as used in modeling the mass transport in diffusion dialysis which has been described in detail by Equations 4.128 to 4.134. In Donnan dialysis the fluxes of the counter ions through the ion-exchange membrane are also equal, but in opposite direction. For a system consisting of a NaCl stripping and $CaCl_2$ feed solution as described in Figure 4.19 the fluxes of the Na^+-ions and the Ca^{2+}-ions are given by:

$$N_{Na^+} = A\,J_{Na^+} = 2N_{Ca^{2+}} = 2A\,J_{Ca^{2+}} = -2A\left(\frac{Đ_{Na^+}C_{Na^+}Đ_{Ca^{2+}}C_{Ca^{2+}}}{Đ_{Na^+}C_{Na^+} + Đ_{Ca^{2+}}C_{Ca^{2+}}}\right)\frac{d\ln a_{Na^+}}{dz} \qquad (4.152)$$

Here Đ and C are the diffusion coefficient and the concentration of the Na^+-and Ca^{2+}-ion in the membrane, J and N are the flux rate and the total flux of the ions through the membrane, A is the membrane area, a is the activity, and z the directional coordinate perpendicular to the membrane surface.

Equation 4.152 indicates that due to the electroneutrality requirement the flux of the Na^+-ion is not only determined by its diffusion coefficient and concentration in the membrane, but it depends also on the diffusion coefficient of the Ca^{2+}-ions and their concentration in the membrane.

In the steady state the fluxes of ions in the Donnan dialysis stack does not change with time but it is changing along the process path in direction parallel to the membrane surface. To calculate the required membrane area and process path length the average flux over the entire path length must be determined from the complete mass balance.

Donnan dialysis is used today in water softening and in waste water treatment, but it has not reached the technical and commercial relevance of other electromembrane processes.

7 OTHER ION-EXCHANGE SEPARATION PROCESSES

There are several other ion-exchange membrane separation processes such as piezo-dialysis or the use of ion-exchange membranes in combination with chemical reactions or in energy storage and conversion devices. Ion-exchange membranes are also used or control release systems. Membranes carried positive or negative electric charges are used in ultrafiltration, nanofiltration, reverse osmosis, and in pervaporation. A discussion of these processes, however, is beyond the scope of this book.

8 LIST OF SYMBOLS

A	area	$[m^2]$
a	constant	$[A\ s\ mol^{-1}]$
a	activity	$[mol\ m^{-3}]$
a	constant in limiting current density	$[A\ s\ mol^{-1}]$
B	pre-exponential factor	$[s^{-1}]$

b	constant	[-]
C	concentration	[mol m^{-3}]
$\overline{C}$	average concentration	[mol m^{-3}]
$^{\Delta}C_s$	salt concentration change due to transport through membrane	[mol m^{-3}]
D	diffusion coefficient (Fick)	[m^2 s^{-1}]
Đ	diffusion coefficient (Stefan-Maxwell)	[m^2 s^{-1}]
$\overline{D}$	average value of diffusion coefficient	[m^2 s^{-1}]
E	electric field	[V m^{-1}]
E	electrical energy	[A V s]
E_A	activation energy	[[J mol^{-1} $^oK^{-1}$]
G	Gibbs free energy	[J]
F	Faraday constant	[A s $equivalent^{-1}$]
I	current	[A]
i	current density	[A m^{-2}]
$\overline{i}$	average current density	[A m^{-2}]
J	flux	[mole m^{-2} s^{-1}]
K_a	acid dissociation equilibrium constant	[mol^2 m^{-6}]
K_b	base dissociation equilibrium constant	[mol^2 m^{-6}]
K_w	water dissociation equilibrium constant	[mol^2 m^{-6}]
k	reaction rate constant	[s^{-1}]
k_d	water dissociation rate constant	[s^{-1}]
k_d^{cr}	water dissociation rate constant due to catalytic reaction	[s^{-1}]
k_s	mass transfer coefficient	[m s^{-1}]
$k^{\times}$	mass transfer coefficient	[m s^{-1}]
k_{eff}	efficiency coefficient of pump	[-]
k_d	water dissociation rate	[s]
L_{pp}	process path length	[m]
N	flux	[mol s^{-1}]
N	number	[-]
n	number of moles	[-]
Q	volume flux	[m^3 s^{-1}]
P	power	[V A]
Q	volume flux	[m^3 s^{-1}]
R	resistance	[Ω]
$\overline{R}$	average resistance	[Ω]
r	area resistance	[Ω cm^2]
R	gas constant	[J $^oK^{-1}$ mol^{-1}]
T	transport number	[-]
$\overline{T}$	average transport number	[-]
T	temperature	[^oK]
t	time	[s]
U	voltage	[V]
u	ion mobility	[m^2 s^{-1} V^{-1}]
u	velocity	[m s^{-1}]
V	volume	[m^3]
$\overline{V}$	molar volume	[m^3 mol^{-1}]

x	directional coordinate	[m]
y	directional coordinate	[m]
z	directional coordinate	[m]
z	valence	[eq mol^{-1}]
X	cell length	[m]
X	molar fraction	[-]
Y	cell width	[m]

Greek symbols

α	coefficient	[m]
Δ	cell thickness	[m]
Δ	difference	[-]
ε	permittivity	[C^2 J^{-2} m^{-1}]
η	electrochemical potential	[J mol^{-1}]
γ	activity coefficient	[-]
η	current efficiency	[-]
φ	electrical potential	[V]
λ	transition region thickness	[m]
ν	stoichiometric coefficient	[-]
ξ	current utilization	[-]
Δ	cell thickness	[m]
μ	chemical potential	[J mol^{-1}]
Λ	equivalent conductivity	[cm^2 $\Omega^{-1}$$eq^{-1}$]
ξ	current utilization	[-]
Ψ	membrane permselectivity	[-]

Superscripts

am	anion-exchange membrane
as	acid
b	bulk solution
bm	bipolar membrane
bs	base
cm	cation-exchange membrane
diff	diffusion
f	feed
b	bulk solution
b	constant
c	concentrate
d	diluate
e	electrode rinse solution
fc	feed concentrate
fd	feed diluate
i	inlet

m	membrane surface
m	membrane matrix
mig	migration
diff	diffusion
p	product
str	stripping solution
tr	transition region

Subscripts

a	anion
ac	acid
c	cation
cell	cell
cag	catalytic active group
d	dissociation
des	desalination
eq	equivalent
F	Faraday efficiency
fix	fixed ion
s	salt
eq	equivalent
lim	limiting current density
m	membrane
ms	membrane selectivity
n	number of moles
p	pressure
prod	production
Don	Donnan potential
tot	total
st	stack
sol	solution
w	water

LITERATURE

Aritomi, T., van den Boomgard T., Strathmann, H., 1996. Current-voltage curve of a bipolar membrane at high current density, Desalination, 104: 13-18.

Atkins, P.W., 1990. Physical chemistry, Oxford University Press, Oxford, U.K.

Bockris, J., Reddy, A.K.N., 1970. Modern electrochemistry, Plenum Press, New York.

Bird, R.B., Stewart, W.E., Lightfoot, E.N., 1965. Transport phenomena, John Wiley & Sons New York.

Choi, J.H., Park, J.S., Moon, S.H., 2002. Direct measurement of concentration distribution within the boundary layer of an ion-exchange membrane, J. Colloid Interface Sci., 251: 311-317.

Cowan, D.A., Brown, J.H., 1959. Effect of turbulence on limiting current density in electrodialysis, Ind. Eng. Chem., 51: 1445-1448.

El Moussanoui, R., Pourcelly, G., Maeck, M., Hurwitz H.D., Gavach, C.,1994. Co-ion leakage through bipolar membranes, influence on I-V response and water splitting efficiency, J. Membrane Sci., 90: 283-292.

Ganzi, G.C., Guiffrida, A.J., Griffin, C., 1994. Electrodeionization apparatus, US Patent 5,316,637 May 31.

Gittens, G.J., Glueckauf, E., 1965. The application of electrodialysis to demineralization, A.I.Ch.E-I.Chem.E. Symposium Series, 9: 79-83.

Giuffrida, A.J., Jha, A.D., Ganzi, G.C., 1986. Electrodeionization apparatus, US Patent 4,632,745, December 30.

Glueckauf, E., 1959. Electrodeionization through a packed bed, British Chemical Engineering, 4: 646-651.

Helfferich, F., 1962. Ion-Exchange,McGraw-Hill, London.

Huffmann, E.L., Lacey, R.E., 1972. Engineering and economic considerations in electromembrane processing, in: Industrial processing with membranes, edts.:Lacey, R.E., Loeb, S., pp. 39-55, John Wiley & Sons, New York.

Hurwitz, H.D., El Moussanoui, R., 1998. Single film membrane, process for obtaining it and use thereof, US patent 5,840,192.

Kedem, O., Maoz, Y., 1976. Ion-conducting spacer for improved electrodialysis, Desalination 19: 465-470.

Kobuchi, Y., Matsunaga, Y., Noma, Y., 1987. Application of ion-exchange membranes to the recovery of acids by diffusion dialysis and electrodialysis, in: Synthetic polymer membranes, edts.: Sedlacek, B., Kahovec, J., W. de Gruyter & Co. Berlin.

Korngold, E., 1975. Electrodialysis process using ion-exchange resins between membranes, Desalination, 16: 225-233.

Krol, J.J., Wessling, M., Strathmann, H., 1999. Concentration polarization with monopolar ion-exchange membranes, J. Membrane Sci., 162: 145-154.

Krol, J.J., Jansink, M., Wessling, M., Strathmann, H., 1998. Behavior of bipolar membranes at high current density, water diffusion limitation, Separation& Purification Technology 14: 41-52.

Krol, J. J., 1997. Monopolar and bipolar ion-exchange membrane mass transport limitation, Ph.D.-Thesis, University of Twente, The Netherlands, ISBN 9036509866.

Kortüm, G., 1957. Lehrbuch der Elektrochemie, Verlag Chemie, Weinheim, Germany.

Lee, H.-J., Safert, F., Strathmann, H., Moon, S.-H., 2002. Design of an electrodialysis desalination plant, Desalination, 142: 267-286.

Mafe, S., Ramirez, P., Alcaraz, A., Aguilella, V., 2000. Ion transport and water splitting in bipolar membranes: Theoretical background, in: Bipolar membrane technology, edt.: Kemperman, A.J.B., Twente University Press, Enschede, The Netherlands.

Mafe, S., Ramirez, P., 1997. Electrochemical characterization of polymer ion-exchange bipolar membranes, Acta Polym., 48: 234-250.

Maletzki, F., Rösler, H.W., Staude, E., 1992. Ion transfer across electrodialysis membranes in the overlimiting current range: stationary voltage current characteristics and current noise power spectra under different conditions of free convection, J. Membrane Sci., 71: 105-115.

Manzanares, J.A., Kontturi, K., Mafe, S. Aguilella, V.M., Pellicer J., 1991. Polarization effects at cation-exchange membrane-solution interface, Acta Chem. Scand., 45: 115-121.

Mauro, A., 1962. Space charge regions in fixed charge membranes and the associated property of capacitance, Biophys. J., 2: 179-198.

Mani, K.N., 1991. Electrodialysis water-splitting technology, J. Membrane Sci., 58: 117-138.

Matejka, Z., 1971. Continuous Production of high purity water by electro-deionisation, J. Appl. Chem. Biotechnol. 21: 117-120.

McRae, W.A., 1983. Electrodialysis, in: Desalination technology developments and practice, pp. 249-264, edt.: Porteous, A., Applied Science, New York.

Neubrand, W., 1999. Modellbildung und Simulation von Elektromembranverfahren, Ph.D.-Thesis, University of Stuttgart, Germany, ISBN 3-89722-216-7.

Onsager, L., 1934. Deviation from Ohm's law in weak electrolytes, J. Chem. Phys., 2: 559-615.

Ramirez, P. ,Aguilella, V.M., Manzanarees, J.A., Mafe, S., 1992. Effecs of temperature and ion transport on water splitting in bipolar membranes, J. Membrane Sci., 73: 191-201.

Ramirez, P., Rapp, H.J., Reichle, S., Strathmann, H., Mafe, S., 1992. Current voltage-curves of bipolar membranes, J. Appl. Phys., 72: 259-264.

Ramirez, P., Manzanares, J.A., Mafe, S., 1991. Water dissociation effects in ion transport through anion-exchange membranes with thin cation-exchange surface film, Ber. Bunsenges. Phys. Chem.95: 499-507.

Rapp, H.J., 1995. Die Elektrodialyse mit bipolaren Membranen, Theorie und Anwendungen, Ph.D.-Thesis, University of Stuttgart, Stuttgart Germany.

Rubinstein, I., Shtilman, L., 1979. Voltage against current curves of cation-exchange membranes, J. Chem. Soc., Faraday Trans. II, 75: 231-246.

Rubinstein, I., Warshawsky, A., Schechtman, L., Kedem, O., 1984. Elimination of acid-base generation in electrodialysis, Desalination, 51: 55-60.

Rubinstein, I., 1991. Electroconvection at an electrically inhomogeneous permselective interface, Phys. Fluids, 3: 2301-2309.

Sata, T., Yamane, R., Mitzutani, Y., 1969. Concentration polarization phenomena in ion-exchange membrane electrodialysis, studies of the diffusion boundary layer by six different measurements, Bulletin of the Chem. Society of Japan, 42: 279-284.

Schaffer, L.H., Mintz, M.S., 1966. Electrodialysis, in: Principles of desalination, edt.: Spiegler, K.S., pp. 3-20. Academic Press New York.

Shockley, W., 1949. The theory of p-n junctions in semiconductors and p-n junction transistors, Bell Syst. Techn. J., 29: 435-489.

Simons, R., Khanarian, G., 1978. Water dissociation in bipolar membranes: Experiment and Theory, J. Membrane Biol., 38: 11-30.

Simons, R., 1979. The origin and elimination of water splitting in ion-exchange membranes during water demineralization by electrodialysis, Desalination , 28: 41-42.

Simons, R., 1984. Electric field effects on proton transfer between ionizable groups and water in ion-exchange membranes, Electrochemica Acta, 29:151-158.

Simons, R., 1985. Water splitting in ion-exchange membranes, Electrochim. Acta, 30: 275-285.

Simons, R., 1993. Preparation of high performance bipolar membrane, J. Membrane Sci., 78: 13-23.

Simons, R., 1993. A mechanism for water flow in bipolar membranes, J. Membrane Sci., 82: 65-73.

Siwak, L.R., 1992. Here is how electrodialysis reverses and why DER works, Desalination &Water Reuse, Vol. 2/4, pp. 26-30.

Spiegler, K.S., 1958. Transport process in ionic membranes, Trans. Faraday Soc., 54:1408-1428.

Spiegler, K.S., 1956. Electrochemical operations, in: Ion exchange technology, edts.: Nachod, F.C., Schubert, J., pp. 118-181, Academic Press, New York.

Spiegler, K.S., 1971. Polarization at ion exchange membrane-solution interface, Desalination, 9: 367-385.

Strathmann, H., Krol, J.J., Rapp, H.J., Eigenberger, G., 1997. Limiting current density and water dissociation in bipolar membranes, J. Membrane Sci., 125: 123-142.

Thate, S., 2002. Untersuchung der elektrochemischen Deionisation zur Reinstwasser herstellung, Ph.D.-Thesis, University of Stuttgart, Stuttgart Germany, ISBN 3-89722-911-0.

Taky, M., Pourcelly, G., Lebon, F., Gavach, C., 1992. Polarization phenomena at the interface between an electrolyte solution and an ion-exchange membrane, J. Electroanal. Chem., 336: 171-194.

Timashev, S.F., Kirganov, E.V., 1981. Mechanism of the electrolytic decomposition of water in bipolar ion-exchange membranes, Translated from Elektrokhimiya, 17: 440-443.

Wilhelm, F.G., Pünt, I., van der Vegt, N.F.A., Wessling, M., Strathmann, H., 2001. Optimisation strategies for the preparation of bipolar membranes with reduced salt ion leakage in acid-base electrodialysis, J. Membrane Sci., 182: 13-28.

Wilson, J.R., 1960. Design and operation of electrodialysis plants, in: Demineralization by Electrodialysis, edt. J.R. Wilson, Butterworth Publishers, London .

Zabolotsky, V.I., Sheldeshov, N.V., Gnusin, N.P., 1988. Dissociation of water molecules in systems with ion-exchange membranes, Russ. Chem. Revs., 57: 801.

Zabolotsky, V.I., Gnusin, N.P., Shel'deshov, N.V., Pismenskaya, N.D., 1986. Investigation of the catalytic activity of secondary and tertiary amine groups in the dissociation of water in bipolar MB-2 membrane, translated from Elektrokhimiya, 21: 1059-1069.

Zabolotsky, V.I., Gnusin, N.P., Shel'deshov, 1984. The current-voltage characteristic of the transition region in MB-2 bipolar membranes, translated from Elektrokhimiya, 20: 1340-1345.

Zabolotsky, V.I., Shel'deshov, N.V., Gnusin, N.P., 1986. Influence of the ionogenic groups on the dissociation constants of water in bipolar ion-exchange membranes, translated from Elektrokhimiya, 22: 1676-1679.

Chapter 5

Ion-Exchange Membrane Process and Equipment Design

1 INTRODUCTION

The performance of ion-exchange membrane separation processes in practical applications is not only a function of membrane properties and feed solution composition but is also determined by several process and equipment design parameters such as the stack construction, i.e. the cell geometry and the spacer configuration, the feed flow velocities and mode of operation, i.e. batch or continuous operation with co- or counter-current flow streams. These parameters effect the cost of the processes by determining the investment as well as operating costs. Equipment and process design requirements are quite different in the different ion-exchange membrane separation processes and their application. For electrodialysis and its application in brackish water desalination most of the data necessary for the process and equipment design can be obtained from the literature or from theoretical considerations. In other processes such as acid or base recovery by diffusion dialysis very little data are available in the literature or can be obtained by theoretical considerations, and the design of industrial size equipment relies largely on data obtained from laboratory and pilot plant experiments.

2 DESIGN OF ELECTRODIALYSIS PROCESSES AND EQUIPMENT

The basic concept of electrodialysis has been discussed earlier and is shown in Figure 2.1. This figure depicts the transport of ionic species in a cell arrangement consisting of cation- and anion-exchange membranes in alternating series forming an array of individual cells between two electrodes. Although there are many different components necessary for the proper and efficient operation of an electrodialysis plant such as the electrical power supply, pumps and control and monitoring devices, the stack is a key component in an electrodialysis unit. For designing and operating an electrodialysis stack certain process parameters have to be taken into account and be controlled such as the feed solution concentration and the desired product and brine concentrations. Various modes of operation are possible such as batch type or continuous. In many cases the desired desalination or concentration of a feed solution can not be obtained in a single pass through the electrodialysis cells. To achieve a higher degree of desalination or concentration two or more stacks are placed in series. When high product recovery rates are required part of the brine may be recycled. The flow streams through a stack can be counter- or co-current. The velocities of the flow streams will effect the concentration polarization and thus the limiting current density which determines the required membrane area for a given capacity desalination plant. The flow velocities, on the other hand, determine the degree of desalination and concentration that can be achieved in one pass in a stack of a given geometry and the pressure loss which has to be overcome by the flow stream circulation pumps. Certain process parameters are fixed by the feed and product solution properties. Other parameters may be varied in a certain range and can thus be utilized to optimize the process such as the current density, the applied voltage, the flow stream velocities, etc. Many parameters are interdependent and often counteracting in terms of overall process efficiency. They all must be taken into account for constructing the various hardware components of an electrodialysis unit as well as designing the process. The key component in electrodialysis is the stack, its design, and the mode of operation which determine the technical feasibility and the economics of the process in the various applications.

2.1 The Electrodialysis Stack Design and Mode of Operation

A cell arrangement of an electrodialysis stack is shown schematically in Figure 4.1.

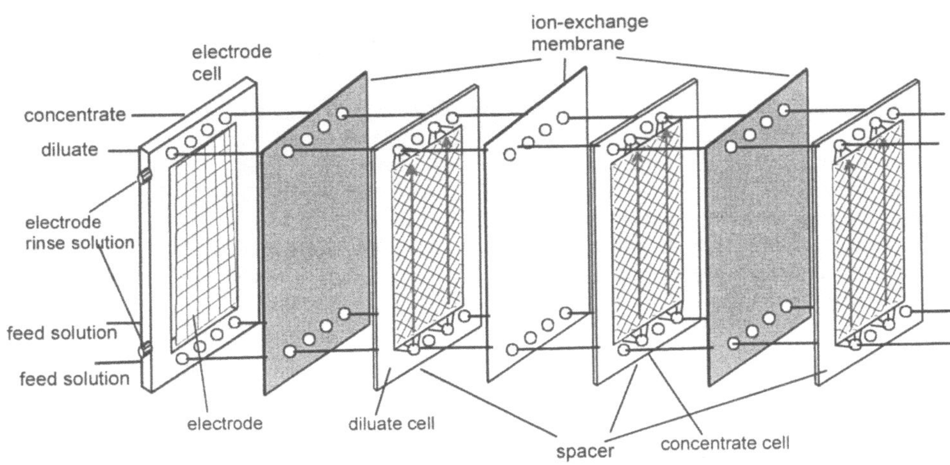

Fig. 5.1 Schematic drawing illustrating the construction of a sheet-flow electrodialysis stack.

It consists of a multitude of cells placed in parallel between two electrodes. The different cells are separated by ion-exchange membranes. Gaskets around the perimeter of the cell confine the solutions within the cells. In alternating cells a feed solution is concentrated and desalinated, respectively. The hydraulic circuits in two adjacent cells are referred to as diluate and concentrate flow streams and the two adjacent cells are referred to as a cell pair. The gaskets not only separate and seal the membranes but also contain manifolds to distribute the process fluids in the different compartments. The supply ducts for the diluate and the concentrate are formed by matching holes in the gaskets, the membranes, and the electrode cells. The ion-exchange membranes are separated by a spacer screen which supports the membranes and also provides turbulent mixing of the solutions.

In designing an electrodialysis stack, several general criteria concerning the mechanical, the hydrodynamic, and the electrical properties have to be considered. Since some of these criteria have opposing effects the final stack construction is generally a compromise between several conflicting parameters. A proper electrodialysis stack design should provide a maximum effective membrane area per unit stack volume and ensure equal and uniform flow distribution in each compartment. Any leakage between the diluate, concentrate, and the electrode cells should be prevented. The spacer screen should provide a maximum of mixing of the solutions at the membrane surfaces and should cause a minimum in pressure loss. There are various stack design concepts described in the literature [E.L. Huffmann et al., 1972], [L.H. Schaffer et al., 1966]. In praxis, however, only two different stack types are used on a large scale. One is the so-called sheet-flow and the other the tortuous path-flow stack.

2.1.1 The Sheet-Flow Electrodialysis Stack

In a sheet-flow electrodialysis stack the solution flow is approximately in a straight path from the entrance to the exit ports, which are located on opposite sides in the gasket as indicated in Figure 5.1. A typical sheet-flow spacer gasket with the manifolds for the distribution of the flow streams and two different spacer screens, i.e. a) a so-called Vexar screen and b) a woven screen are shown in Figure 5.2.

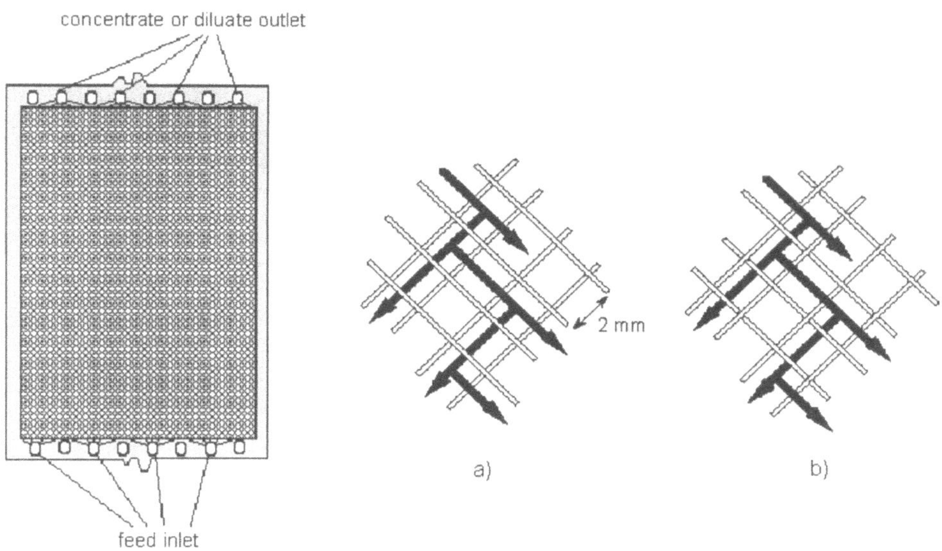

Fig. 5.2 Concept of a sheet-flow electrodialysis spacer gasket and the two different types of screens used in commercial stacks.

The distance between the membrane sheets, i.e. the cell thickness, should be as small as possible to keep the energy consumption due to electrical resistance of the solutions as low as possible. In industrial-size electrodialysis stacks, membrane distances are typically between 0.3 to 2 mm. A spacer is introduced between the individual membrane sheets to support the membrane and to help to control the feed solution flow distribution.

The most serious design problem for an electrodialysis stack is that of assuring uniform flow distribution in the various compartments and good mixing of the solutions in the cell at low pressure losses. Different types of spacer screens have been studied for their use in a sheet-flow stack. Two of them are mainly used today. One is a woven and the other a non-woven netting as also indicated in Figure 5.2. The main difference between the non-woven and the woven spacer is the pressure loss when the solutions are pumped through the stack. The non-woven spacer has a significantly lower pressure loss than a woven spacer of comparable mesh size and flow distribution. In a practical electrodialysis system, 200 to 1,000 cation- and anion-exchange membranes are installed in parallel to form an electrodialysis stack providing 100 to 500 cell pairs. At each end of a stack there is an end-plate that contains the electrodes and the connections that rinse the electrodes.

2.1.2 The Tortuous Path Flow Stack

In the tortuous path flow stack, the membrane spacer gaskets have a long serpentine cut-out which defines a long narrow channel for the fluid path. The objective is to be able to operate the stack with a substantially higher current density due to the high flow velocities in the channels in comparison to the sheet flow design. The high flow velocities require higher residence times of the solution in the cells. This is achieved by the long serpentine flow path of the system. A tortuous path flow spacer gasket is shown schematically in Figure 5.3.

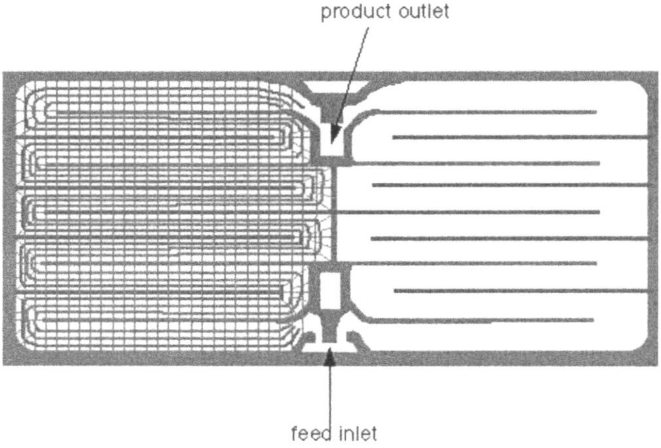

Fig. 5.3 Schematic drawing illustrating the geometry of a tortuous path spacer.

Solution flow velocities in sheet-flow stacks are typically 3 to 10 cm s^{-1}, whereas in tortuous path flow stacks the solution flow velocities are 15 to 50 cm s^{-1}. Because of higher flow velocities and longer flow paths, higher pressure drops in the order of 2 to 3 bars are obtained in stacks with tortuous path flow spacers than in sheet-flow systems where pressure drops are between 0.5 and 2 bars. However, higher velocities help to reduce the deposition of suspended solids such as polyelectrolytes, humic acids, surfactants, and biological materials.

2.2 Electrodialysis Process Design

In addition to the actual stack and the power supply, an electrodialysis plant consists of several other components essential for a proper operation, such as pumps, process monitoring and control devices, and feed solution pretreatment equipment. Again, as in the design of the electrodialysis stack, there are several operating modes for the electrodialysis process. Depending on the feed solution composition and the product requirements electrodialysis units may be operated in a batch-type, in a continuous, or in a feed-and-bleed mode with partial recycle of the diluate and concentrate streams. Independent of the mode of operation there are two basic concepts of using electrodialysis in large scale applications. The first concept is referred to as the unidirectional electrodialysis[K.S. Spiegler, 1956] and the second as electrodialysis reversal [W.E. Katz, 1979].

2.2.1 Unidirectional Electrodialysis System

In an unidirectional operated electrodialysis stack an electrical potential gradient is established between the anode and cathode across the stack which is composed of diluate and concentrate cells in alternating series. The electric field is permanently applied in one direction and the diluate and concentrate cells are also permanently fixed over the period of operation. A flow diagram of a typical unidirectional operated electrodialysis plant is shown in the simplified flow scheme of Figure 5.4.

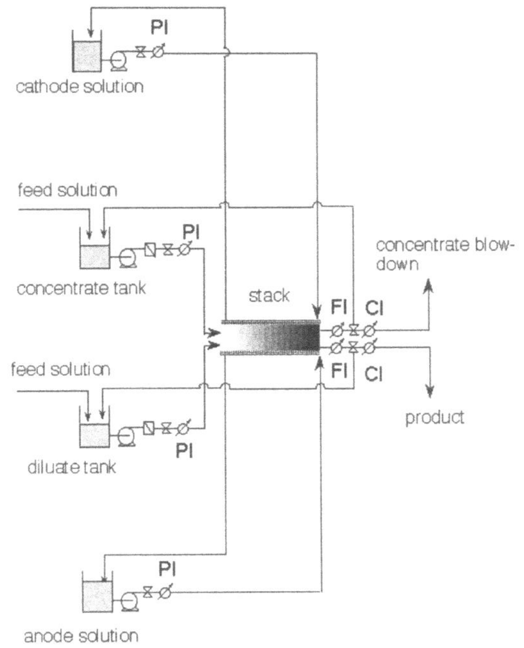

Fig. 5.4 Flow scheme of an unidirectionally operated electrodialysis plant with partial recycling of the diluate and concentrate solutions.

After the proper pre-treatment, the feed solution is pumped through the actual electrodialysis unit, which generally consists of one or more stacks in series or parallel. A deionized solution and a concentrated brine are obtained. The depleted and the concentrated process streams leaving the last stack are collected in storage tanks when the desired degree of concentration or depletion is achieved, or they are recycled if further concentration or depletion is desired. In some cases an acid is added to the concentrated stream to prevent scaling of carbonates and hydroxides. To prevent the formation of free chlorine by anodic oxidation, the electrode cells are sometimes rinsed with a separate solution that does not contain chloride ions and generally consists of 0.5 n sodium sulfate solution. In many cases, however, the feed or brine solution is also used in the electrode cells.

Unidirectionally operated electrodialysis plants are rather sensitive to membrane fouling and scaling and often require a substantial feed solution pre-treatment and stack cleaning procedures in form of periodical rinsing of the stack with acid or detergent solutions. The unidirectional operating concept is mainly used today for applications in the food and drug industry where solutions often contain valuable components which must be recovered in the concentrate or in the diluate. Also, in the treatment of industrial waste waters which contain toxic components such as heavy metal ions mostly unidirectional electrodialysis plants are used to minimize the loss of toxic or valuable materials. In desalination of brackish or surface waters generally electrodialysis reversal is applied which always results in some loss of the product water.

2.2.2 Membrane Fouling and its Control by Electrodialysis Reversal

Suspended and colloidal matters, polyelectrolytes, organic anions, and multivalent salts near the saturation level can cause severe problems in electrodialysis due to precipitation on the membrane surfaces or by partial penetration into the membranes. Precipitation of suspended matter, silicates, and salts with low solubility such as calcium carbonates or iron hydroxides may occur within the actual flow channels resulting in high hydrodynamic pressure losses and non-uniform flow distribution in the stack. Precipitation on the surfaces of the membranes also causes an increase of the electrical resistance of the stack and may lead to physical damage of the membranes. Especially organic anions such as humates can precipitate on the anion-exchange membranes as humic acid and cause a sharp increase in the electric resistance. Mechanical cleaning and treatment with dilute bases and acids can generally restore the original properties of the membranes. More severe is the poisoning of membranes by organic anions that are small enough to penetrate the membranes but whose electromobility is so low that they virtually remain inside the membrane causing a drastic increase in the membrane resistance. Certain detergents are also the cause of this type of poisoning which is very difficult to deal with and can best be avoided by a proper pre-treatment of the feed solution.

The various pretreatment procedures such as precipitation, flocculation, or ion-exchange and rinsing cycles can be substantially reduced by a simple but very effective operating mode which is referred to as electrodialysis reversal. In the electrodialysis reversal operating mode which has been developed by Ionics Incorporated [W.E. Katz, 1979], [F.H. Meller, 1984] the polarity of the electric field applied to the electrodialysis stack as the driving force for the transport of ions is reversed in certain time intervals. Simultaneously the flow streams are reversed, i.e. the diluate cell becomes the concentrate cell and vice versa with the result that matter that has been precipitated at the membrane surface will be redisolved and removed with the flow stream passing through the cell [L.R. Siwak, 1992]. The principle the electrodialysis reversal operating mode is illustrated in Figure 5.5. This figure shows a typical electrodialysis cell formed by a cation- and anion-exchange membrane between two electrodes. If an electric field is applied to a feed solution containing negatively charged particles or large organic anions these components will migrate to the anion-exchange membrane and be deposited on its surface. If the polarity is reversed the negatively charged components will now migrate away from the anion-exchange membrane back into the feed stream and the membrane properties are restored. This procedure has been very effective not only for the removal of precipitated colloidal materials but also for removing precipitated inorganing salts and is used today in almost all electrodialysis water desalination systems.

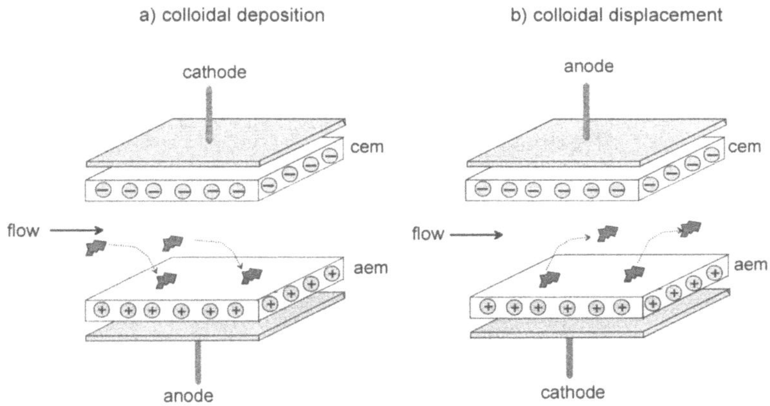

Fig. 5.5 Schematic drawing illustrating the removal of deposited negatively charged colloidal components from the surface of an anion-exchange membrane by reversing the electric field.

Electrodialysis reversal is illustrated in Figure 5.6. However, reversing the polarity of a stack has to be accompanied with a reversal of the flow streams. This always leads to some loss of product and requires a more sophisticated flow control. The flow scheme of an electrodialysis plant operated with reversed polarity is shown in Figure 4.9. In this operating mode, the polarity of the current is changed at specific time intervals ranging from a few minutes to several hours. In the reverse polarity operating mode, the hydraulic flow streams are reversed simultaneously, i.e., the diluate cell will become the brine cell and vice versa.

The advantage of the reversed polarity operating mode is that precipitation in the brine cells will be redesolved when the brine cell becomes the diluate cell in the reverse operating mode. During the reversal of the polarity and the flow streams there is a brief period when the concentration of the desalted product exceeds the product quality specification. The product water outlet has a concentration sensor which controls an additional three-way valve. This valve diverts high concentrated product to waste and then, when the concentration returns to the specified quality, directs the flow to the product outlet. Thus, in electrodialysis reversal there is always a certain amount of the product lost to the waste stream. This is generally no problem in desalination of brackish water. It might, however, be not acceptable in certain applications in the food and drug industry when feed solutions with high value products are processed.

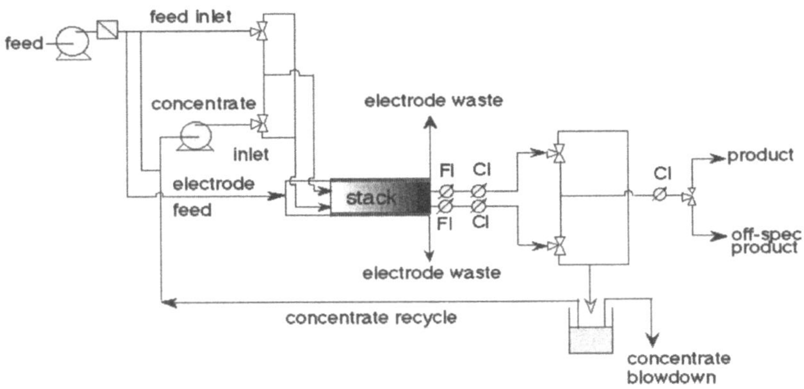

Fig. 5.6 Flow scheme of electrodialysis reversal in a continuous operating mode with the feed solution also used as electrode rinse.

2.3 Continuous and Batch Type Operation in Electrodialysis

The degree of desalination that can be achieved in passing the feed solution through a stack is a function of the solution concentration, the applied current density, and the residence time of the solution in the stack. If the flow rates of diluate and concentrate through the stack are relatively high the degree of desalination or concentration which can be achieved in a single path is quite low and often not sufficient to meet the required product qualities. If this is the case the electrodialysis can either be operated as a batch process, a multistage process, or a process with feed and bleed in which the diluate or the concentrate or both are partially recycled. A pure batch-type process is hardly used in the production of potable or high quality process water. Most commonly the desired degree of desalination is obtained by passing the solutions through a second or third stack, i.e. by staging a required number of stacks. The staging can be done by complete physical separation of the stacks with each having its own set of electrodes, i.e. its own power supply and feed pumps, or the required process path length can be achieved by internal staging as illustrated in Figure 5.7.

In internal staging only one pump and one set of electrodes is needed. The disadvantage of the internal staging is that the total number of stages is limited by the pressure losses of the solution in the stack. These pressure losses can lead to differences in the hydrostatic pressure between the different flow streams with the consequence of leakages between the diluate and the concentrate flow streams. Furthermore, the average current density over the stack is lower in internal staging than the average current density that is obtained when the stages are provided in separated stacks.

A disadvantage of simple staging as illustrated in Figure 5.7 is a constant recovery rate fixed at 50% of the feed solution. If variable recovery rates are desired the system must be operated in a feed and bleed mode as illustrated in Figure 5.8.

In the feed and bleed mode both the brine and the product concentration can be determined independently and very high recovery rates can be obtained.

There is a large variety in the feed and bleed operating mode in conjunction with the staging of stacks. Sometimes it is convenient to apply the feed and bleed operating mode to all stacks in series, and in other cases it is more economical to use the feed and bleed operation mode only in a single stack of stage system.

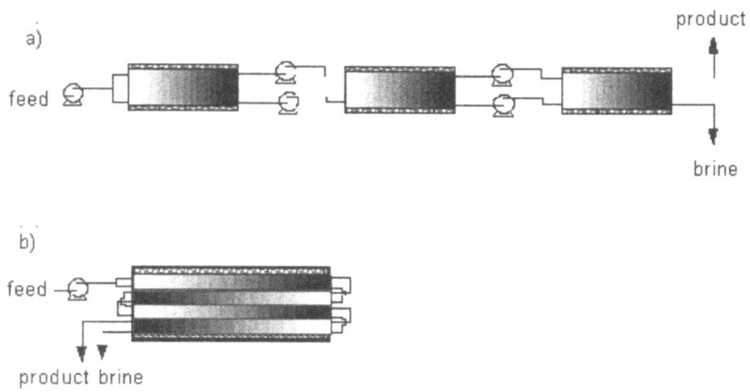

Fig. 5.7. a) Staging of individual stacks with independent power supply and diluate and concentrate pumps, b) internal staging using one pair of electrodes and one pump.

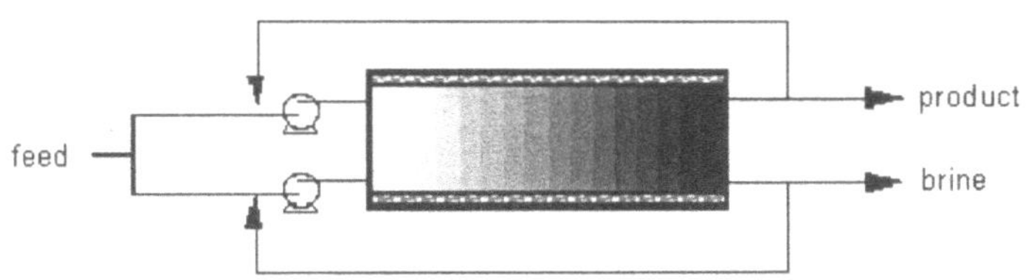

Fig. 5.8 Flow scheme of an electrodialysis stack operated in a feed and bleed mode, i.e. with partial recycling of the diluate and concentrate solutions.

2.4 Mass Transport in an Electrodialysis Stack

In an electrodialysis stack ions are removed from one compartment, which is referred to as the diluate cell, and concentrated in the other, i.e. the concentrate cell. The degree of desalination can be expressed by a material balance between the feed, the concentrate, and the diluate solutions. It is a function of the residence time of the different solutions in the cells, i.e. the flow velocities of the solutions, the membrane area in a cell pair, and the applied current density. The degree of desalination and concentration of a feed solution is given by the concentration difference between the feed and the product solution which is identical to the difference between the feed and the brine solution.

The mass transport in an electrodialysis stack is the result of an electrochemical potential gradient across the membrane stack and can be determined by a mass balance between the flow streams in the two compartments of a cell pair as has been discussed in detail earlier. Calculation of the complete mass balance between the diluate and concentrate flow stream can be rather complex for multi-component solutions and high ion concentrations, which means high osmotic pressures and activity coefficients that deviate largely from 1. The calculation gets even more difficult when the two flow streams are counter-current or have different flow velocities, and a hydrostatic pressure difference between the two compartments is obtained. However, in many electrodialysis applications, e.g. in the desalination of brackish water the concentrate and diluate cells have identical geometries and flow conditions, there is an equal flow rate in both cells and the membrane stack is operated in a co-current flow as indicated in Figure 5.9.

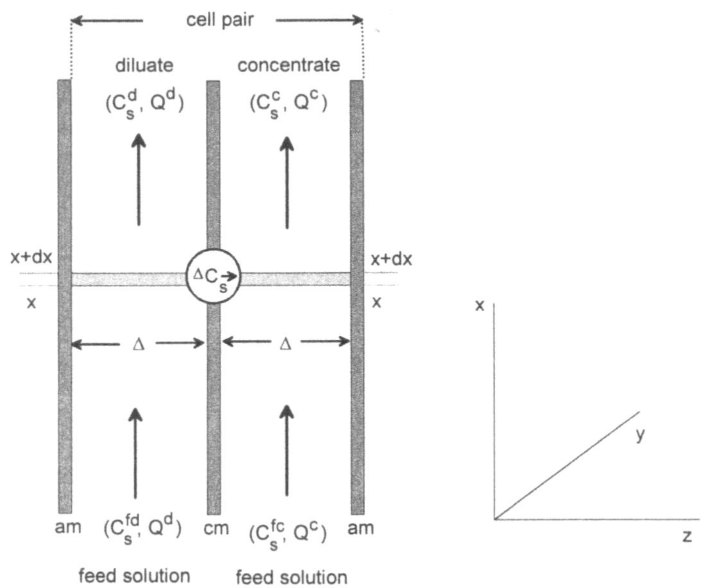

Fig 5.9 Schematic drawing illustrating the flow of a diluate and concentrate stream parallel to the membrane surface in a cell pair of an electrodialysis stack and the concentration change between diluate and concentrate solution $^{\Delta}C_s$ due to the ion flux through the membrane.

This figure shows schematically a cell pair of an electrodialysis stack containing a diluate and concentrate flow stream parallel to the membrane surface in x-direction. The thickness of both cells is Δ. Therefore, no hydrostatic pressure difference exists between the two flow streams. Furthermore, it can be assumed that the concentration potential between the two flow streams due to different concentrations in the diluate and concentrate solutions is negligibly low compared with voltage drops due to the resistance of the solutions, and changes in the resistance of the solutions due to the boundary layer effects are negligible. Finally it is assumed that all diffusion processes between the two flow streams can be neglected and the only driving force for the transport is the electrical potential gradient which is constant over the entire cell length.

Under the above stated assumption the degree of desalination of a given feed solution, i.e. the changes in the ion concentration in the diluate and concentrate cells in an area element can be calculated according to Equation 4.5 to:

$$dC_s^d = -dC_s^c = d^\Delta C_s = \frac{i\xi}{FQ_{cell}} dA_{cell} = \frac{U_{cell}}{\frac{\Delta(C_s^c + C_s^d)}{\Lambda_s C_s^c C_s^d} + r^{am} + r^{cm}} \frac{\xi Y dx}{FQ_{cell}} \tag{5.1}$$

Here i is the current density, U_{cell} is the applied voltage across the cell pair, dA_{cell} is a cell pair area element, Δ is the thickness of the cell, Λ_S is the equivalent conductivity of the solution, r^{am} and r^{cm} are the resistances of the anion- and cation-exchange membranes, and dC_S^d and dC_S^c are the concentration changes in a volume element of the diluate and concentrate cell, $d^\Delta C_s$ is the concentration change in a volume element of the diluate and concentrate cell due to the ion flux through the membrane, Q_{cell} is the flow stream in the diluate or concentrate cell, Y is the cell width, and dx is a distance change in direction of the flow stream parallel to the membrane surface as depicted in Figure 5.9. Thus is dA=Ydx.

In Equation 5.1 the concentrations are given in equivalents per volume instead of salt or ion concentration. The equivalent concentration is the salt concentration multiplied by the valence and stoichiometric coefficient.

The equivalent conductivity Λ_s is an average value for the concentration range considered. The concentration change in the volume element of the diluate and concentrate solution $d^\Delta C_s$ is given by the ion flux across the membrane from the diluate to the concentrate cell. In steady state this flux is constant but it changes from the cell entrance to the cell exit.

The rate of the salt transfer between the cell entrance to the cell exit is given by the mass balance applied to the salt:

$$C_s^c - C_s^{fc} = C_s^{fd} - C_s^d = {}^\Delta C_s \tag{5.2}$$

Here C_S^d and C_S^c are the concentrations of the diluate and the concentrate at the exit, and C_s^{fc} and C_s^{fd} are the feed concentrations of the diluate and concentrate solution at the entrance of the desalination stack, ${}^\Delta C_s$ is the concentration difference between a feed and concentrate solution achieved during the desalination process, i.e. the residence time of the solutions in the stack.

The change in concentration of the diluate and concentrate solutions from the entrance to the exit of the cell is obtained by integration over the cell length according to Equation 4.15:

$$\ln\frac{C_s^c C_s^{fd}}{C_s^d C_s^{fc}} + \frac{\Lambda_s(r^{am} + r^{cm})(C_s^{fd} - C_s^d)}{\Delta} = \frac{\Lambda_s \xi U_{cell} YX}{FQ_{cell}\Delta} \tag{5.3}$$

Here C_s^{fd} and C_s^{fc} are the concentrations of the diluate and the concentrate at the cell inlet, C_s^d and C_s^c are the concentrations of the diluate and the concentrate at the cell outlet, Λ_s is equivalent conductivity of the salt solution, r^{am} and r^{cm} are the area resistances of the anion- and cation-exchange membrane, Δ is the cell thickness, i.e. the distance between the cation- and anion-exchange membrane, Q_{cell} is the flow velocity of the diluate or concentrate stream through a cell, U_{cell} is the voltage drop across a cell pair, ξ is the current utilization, F the

Faraday constant, Y and X are the width and the length of a cell, i.e. Y times X = A_{cell}, i.e. the area of the cation- or anion-exchange membrane in the cell.

The total flow of the diluate or concentrate solution through a stack is given by:

$$Q_{st} = N\,Q_{cell} = N\Delta Yu \tag{5.4}$$

Here Q_{st} is the total solution flow of the diluate, respectively, the concentrate through the entire stack, Q_{cell} is the flow through one cell, N is the number of cells in the stack, D is the thickness and Y the width of a cell, and u is the linear flow velocity in the cell.

The total voltage across the stack U_{st} is given by:

$$U_{st} = NU_{cell} \tag{5.5}$$

It should also be noticed that all concentrations of specific ion species are calculated using equivalents instead of moles.

Equations 5.3 to 5.5 give the relations between the various stack construction data such as cell length, cell width, and cell thickness, the solution properties such as the concentrations of the feed, the concentrate and the product, and their conductivity, the flow velocity in the cells and the applied voltage. But to design an electrodialysis stack for a certain plant capacity on the basis of Equation 5.3 several boundary conditions must be taken into consideration. One of these boundary conditions is that the current density cannot exceed a certain value.

The applied voltage U is related to the current density according to Equation 4.9 by:

$$i = \frac{U_{cell}}{\dfrac{\Delta\left(C_s^c + C_s^d\right)}{\Lambda_s C_s^c C_s^d} + r^{am} + r^{cm}} \tag{5.6}$$

The current density and thus the voltage that can be applied in a practical electrodialysis, however, is limited due to concentration polarization and ion depletion at the membrane surfaces in the diluate cell. The limiting current density is determined by the stack geometry and the solution flow velocity.

2.5 Effect of Process Design on Limiting Current Density

The causes for the limiting current density and their consequences have been discussed earlier. The relation between the limiting current density and various process parameters such as membrane properties and diluate concentration are given by Equation 4.29. The limiting current density for the counterion in an ion-exchange membrane is given by:

$$i_{lim} = \frac{F\,k_s^s C_s^d}{T^m - T^s} \tag{5.7}$$

Here i_{lim} is the limiting current density, F is the Faraday constant, k_s^s is the mass transfer coefficient, and C_s^d the concentration in the diluate solution; T^m and T^s are the transport numbers of the counterion in the membrane and in the solution.

Introducing Equations 5.7 and 5.6 into Equation 5.5 gives the maximum voltage $U_{st,lim}$ that should be exceeded in electrodialysis:

$$U_{st,lim} = \frac{F\,k_s^s}{\left(T^m - T^s\right)} \frac{N\Delta}{\Lambda_s} \left[\frac{C_s^d}{C_s^c} + 1 + \frac{\Lambda_s C_s^d}{\Delta} \left(r^{am} + r^{cm} \right) \right] \tag{5.8}$$

Here $U_{st,lim}$ is the limiting applied voltage across a stack, F is the Faraday constant, k_s^s is the mass transfer coefficient, C_s^d is the concentration in the diluate solution; T^m and T^s are the transport numbers of the counterion in the membrane and in the solution, Λ_s is the equivalent conductivity of the salt solution, r^{am} and r^{cm} are the area resistances of the anion- and cation-exchange membrane, Δ is the cell thickness, and N is the number of cell pairs in a stack.

Introducing Equation 5.8 into Equation 5.3 and rearranging gives the required membrane area of a cell pair and when multiplied with the number of cell pair in the stack it gives the membrane area of the complete stack as function of the applied voltage and the feed, the diluate and the concentrate solution, the cell geometry, the solution flow velocities, and mass transfer coefficient in the laminar boundary layer.

$$A_{cell} = \frac{\left[\ln \frac{C_s^c C_s^{fd}}{C_s^d C_s^{fc}} + \frac{\Lambda_s \left(r^{am} + r^{cm} \right) \left(C_s^{fd} - C_s^d \right)}{\Delta} \right]}{\left[\frac{C_s^d}{C_s^c} + 1 + \frac{\Lambda_s C_s^d}{\Delta} \left(r^{am} + r^{cm} \right) \right]} \frac{Q_{cell}\,F\,C_s^d}{\xi\; i_{lim}} \tag{5.9}$$

and

$$A_{st} = N_{cell} A_{cell} = N_{cell} \frac{\left[\ln \frac{C_s^c C_s^{fd}}{C_s^d C_s^{fc}} + \frac{\Lambda_s \left(r^{am} + r^{cm} \right) \left(C_s^{fd} - C_s^d \right)}{\Delta} \right]}{\left[\frac{C_s^d}{C_s^c} + 1 + \frac{\Lambda_s C_s^d}{\Delta} \left(r^{am} + r^{cm} \right) \right]} \frac{Y \Delta u \left(T^m - T^s \right)}{\xi\; k_s^s} \tag{5.10}$$

Here C_s^{fd} and C_s^{fc} are the concentrations of the diluate and the concentrate at the cell inlet, C_s^d and C_s^c are the concentrations of the diluate and the concentrate at the cell outlet, Λ_s is the equivalent conductivity of the salt solution, r^{am} and r^{cm} are the area resistances of the anion- and cation-exchange membrane, Δ is the cell thickness, i.e. the distance between the cation- and anion-exchange membrane, Y is the cell width, A_{cell} is the cell pair area, A_{st} is the total cell pair area installed in the stack, and N_{cell} is the number of cell pairs installed in the stack, Q_{cell} is volume flow rate through a cell, u is the linear flow velocity of the diluate or concentrate stream in the cell, k_s^s is the mass transfer coefficient, and ξ is the current utilization.

From the cell pair area required for a certain capacity plant and the cell dimensions the process path length can be determined by:

$$L_{pp} = \frac{A_{cell}}{Y} \tag{5.11}$$

Here is L_{pp} the process path length for a given plant capacity and cell dimensions. It becomes identical with the cell length X when desalination is achieved in one path through the cell.

Introducing Equation 5.11 into 5.9 leads to:

$$L_{pp} = \frac{\left[\ln\frac{C_s^c C_s^{fd}}{C_s^d C_s^{fc}} + \frac{\Lambda_s\left(r^{am}+r^{cm}\right)\left(C_s^{fd}-C_s^d\right)}{\Delta}\right]}{\left[\frac{C_s^d}{C_s^c}+1+\frac{\Lambda_s C_s^d}{\Delta}\left(r^{am}+r^{cm}\right)\right]} \frac{\Delta u\left(T^m-T^s\right)}{\xi\ k_s^s} \tag{5.12}$$

Here L_{pp} is the process path length, C_s^{fd} and C_s^{fc} are the concentrations of the diluate and the concentrate at the cell inlet, C_s^d and C_s^c are the concentrations of the diluate and the concentrate at the cell outlet, Λ_s is the equivalent conductivity of the salt solution, r^{am} and r^{cm} are the area resistances of the anion- and cation-exchange membrane, Δ is the cell thickness, T^m and T^s are the transport numbers of the counterions in the membrane and in the solution, k_s^s is the mass transfer coefficient, ξ is the current utilization, and u the linear flow velocity.

2.6 Energy Required for the Transfer of Ions in Electrodialysis

The required electrical power for a certain desalination process is given by the total current passing through a stack multiplied by the applied voltage. The total current passing through a cell pair is given by:

$$I = \frac{Q_{cell} F\left(C_s^{fd}-C_s^d\right)}{\xi} \tag{5.13}$$

Here I is the total electric current passing through the cell pair, Q_{cell} is the flow rate of the solution in the diluate cell, C_s^{fd} and C_s^d are the diluate feed and product equivalent concentrations.

The electric power P_{des} required for the actual desalination process in a cell pair is given by:

$$P_{des} = U_{cell} I = \frac{F Q_{cell}\left(C_s^{fd}-C_s^d\right)}{\xi} U_{cell} \tag{5.14}$$

Introducing Equation 5.3 and 5.6 into 5.14 and rearranging gives the power requirements for the desalination process in a cell pair:

$$P_{des} = U_{cell}\, I = \frac{1}{A_{cell}}\left[\frac{\Delta \ln \dfrac{C_s^{fd}}{C_s^{fc}}\dfrac{C_s^{c}}{C_s^{d}}}{\Lambda_s\left(C_s^{fd} - C_s^{d}\right)} + r^{am} + r^{cm}\right]\left[\frac{Q_{cell}^{d} F\left(C_s^{fd} - C_s^{d}\right)}{\xi}\right]^2 \qquad (5.15)$$

Multiplying the power requirements for one cell pair with the time of operation and the number of cell pairs in the stack gives the energy required for the amount of product obtained in this time:

$$E_{des} = N_{cell} U_{cell}\, I\, t = \frac{N_{cell} t}{A_{cell}}\left[\frac{\Delta \ln \dfrac{C_s^{fd}}{C_s^{fc}}\dfrac{C_s^{c}}{C_s^{d}}}{\Lambda_s\left(C_s^{fd} - C_s^{d}\right)} + r^{am} + r^{cm}\right]\left[\frac{Q_{cell}^{d} F\left(C_s^{fd} - C_s^{d}\right)}{\xi}\right]^2 \qquad (5.16)$$

Thus, the specific energy required for an unit volume product is:

$$E_{des/spec.} = \frac{N_{cell} U_{cell}\, I}{Q_{st}} = \frac{N_{cell}}{A_{cell} Q_{st}}\left[\frac{\Delta \ln \dfrac{C_s^{fd}}{C_s^{fc}}\dfrac{C_s^{c}}{C_s^{d}}}{\Lambda_s\left(C_s^{fd} - C_s^{d}\right)} + r^{am} + r^{cm}\right]\left[\frac{Q_{cell}^{d} F\left(C_s^{fd} - C_s^{d}\right)}{\xi}\right]^2 \qquad (5.17)$$

Here E_{des} and $E_{des/spec}$ are the desalination costs for the desalination in electrodialysis and the cost per unit volume product, I is the total current and U_{cell} the voltage per cell, N_{cell} is the number of cell pairs in a stack, t is the time of operation; C_s^{fd} and C_s^{fc} are the equivalent concentrations of the diluate and the concentrate at the cell inlet, C_s^{d} and C_s^{c} are the concentrations of the diluate and the concentrate at the cell outlet, Λ_s is equivalent conductivity of the salt solution, r^{am} and r^{cm} are the area resistances of the anion- and cation-exchange membrane, Δ is the cell thickness, ξ is the current utilization, and Q_{cell} and Q_{st} are the diluate flow velocities in a cell and in the entire stack, respectively, A_{cell} is the cell pair area.

2.7 Practically used Relations for the Design of an Electrodialysis Plant

The above listed equations provide the theoretical base for the design of an electrodialysis plant. For a practical application these equations, however, must be modified to meet the specific stack construction used in the plant. One major problem is the determination of the mass transfer coefficient for the transport of ions in the laminar boundary layer at the membrane surface k_s^s. This coefficient determines the limiting current density and is a function of the solution flow velocity, the spacer configuration used in the stack, etc. Therefore, it is difficult to calculate the limiting current density theoretically. In practice the limiting current density is determined experimentally for a specific stack construction and spacer geometry as a function of the solution flow velocity and concentration. It can be described according to Equation 4.30 by the following relation:

$$i_{lim} = a\, C_s^d\, u^b \tag{5.18}$$

Here i_{lim} is the empirically determined limiting current density, u is the linear flow velocity in the cell, and a and b are constants which are determined by measurements of the limiting current density at various linear flow velocities [H.J. Lee et al., 2002].

The empirical Equation 5.18 is widely accepted for the design of a water desalination plant. In practical applications of electrodialysis the flow distribution in a cell may not be completely uniform and in certain areas less efficient mixing of the solution might be experienced. Therefore, a safety factor is introduced and the limiting current density used in praxis is given by:

$$i_{prac} = s\, i_{lim} = s\, a\, C_s^d\, u^b \tag{5.19}$$

Here i_{prac} is the current density applied in a practical application, and s is a safety factor which always is $s<1$.

Furthermore, in a practically used electrodialysis stack the cells contain a spacer which supports the membranes and promotes the mixing of the solutions. This spacer, however, decreases the volume of a cell available for the flow of the solutions, thus increasing the linear flow velocity. Therefore, the practical flow rate can be written as:

$$Q_{prac} = \frac{N_{cell} A_{prac} \Delta u \alpha}{L_{prac}} = N_{cell}\, Y\, \Delta\, u\, \alpha \tag{5.20}$$

Here is Q_{prac} the flow in a desalination stack used in a practical application, A_{prac} and L_{prac} are the cell pair area and process path length used in a practical application, and α is a correction factor taking the effect of spacer on the volume flow rate into account. It is $Q_{prac} = Q \times \alpha$, where Q is the flow rate calculated from the cell dimensions and the linear flow velocity.

The spacer also effects the current in the cell. Due to a shadow effect part of the membrane is not accessible for the current. Therefore, in a practical application the required membrane area is always larger than the calculated cell pair area. It is:

$$A_{prac} = \frac{A}{\beta} \tag{5.21}$$

Here A and A_{prac} are the calculated and the in a practical application required cell pair area, β is a correction factor which accounts for the shadow effect. It is $\beta \le 1$.

Introducing the practical current density as given by Equation 5.19 and the correction factor β for the spacer shadow effect into Equation 5.10 gives the practically required membrane area:

$$A_{st} = N_{cell} A_{prac} = \frac{\left[\ln \frac{C_s^c C_s^{fd}}{C_s^d C_s^{fc}} + \frac{\Lambda_s (r^{am} + r^{cm})(C_s^{fd} - C_s^d)}{\Delta}\right] F\, C_s^d\, Q_{prac}}{N_{cell}\left[\frac{C_s^d}{C_s^c} + 1 + \frac{\Lambda_s C_s^d}{\Delta}(r^{am} + r^{cm})\right] s\, i_{lim} \beta\, \xi} \tag{5.22}$$

Introducing Equaitions 5.19 and 5.20 into Equation 5.22 leads to:

$$A_{st} = N_{cell} A_{prac} = \frac{\left[\ln \frac{C_s^c C_s^{fd}}{C_s^d C_s^{fc}} + \frac{\Lambda_s (r^{am} + r^{cm})(C_s^{fd} - C_s^d)}{\Delta}\right] F\, u^{(1-b)}\, Y\, \Delta\, \alpha}{\left[\frac{C_s^d}{C_s^c} + 1 + \frac{\Lambda_s C_s^d}{\Delta}(r^{am} + r^{cm})\right] s\, a\, \beta\, \xi} \qquad (5.23)$$

Here A_{prac} and A_{st} are the area of one cell pair and the area of all cell pairs in a stack required in a practical electrodialysis desalination process, Q_{prac} is the flow rate of the diluate as defined by Equation 5.20, β and α are correction factors for the shadow effect and cell volume due to the spacer, s is a safety factor for uncertainties in the experimentally determined limiting current density, i_{lim} is the limiting current density as defined by Equation 5.19; N_{cell} is the number of cell pairs in a stack, C_s^{fd} and C_s^{fc} are the equivalent concentrations of the diluate and the concentrate at the cell inlet, C_s^d and C_s^c are the concentrations of the diluate and the concentrate at the cell outlet, Λ_s is the equivalent conductivity of the salt solution, r^{am} and r^{cm} are the area resistances of the anion- and cation-exchange membrane, Δ is the cell thickness, Y is the cell width, u is the linear flow velocity in the cell, and ξ is the current utilization.

The practically required path length is obtained by combination of Equations 5.20 to 5.23:

$$L_{prac} = \frac{\left[\ln \frac{C_s^c C_s^{fd}}{C_s^d C_s^{fc}} + \frac{\Lambda_s (r^{am} + r^{cm})(C_s^{fd} - C_s^d)}{\Delta}\right] F\, u^{(1-b)}\, \Delta\, \alpha}{\left[\frac{C_s^d}{C_s^c} + 1 + \frac{\Lambda_s C_s^d}{\Delta}(r^{am} + r^{cm})\right] s\, a\, \beta\, \xi} \qquad (5.24)$$

Here L_{prac} is the process path length in a practical electrodialysis desalination process, a is the experimentally determined factor in the limiting current density, and u is the linear flow velocity in the cell. All other symbols have the same meaning as defined in Equations 5.22 and 5.23.

2.7.1 Energy Requirements in a Practical Electrodialysis Desalination Process

The total energy requirements in a practical electrodialysis process is the additive of four terms:

- the energy required for the transport of ions from a feed to a concentration solution,
- the energy necessary to pump the solutions through the stack from a feed solution reservoir to the product tank and brine disposal,
- the energy consumed by the electrode reactions, and
- the energy required to operate various process control and measuring instruments.

The last two terms can generally be neglected in larger capacity plants. However, they can be significant in specific applications of small capacity plants used in the chemical industry or in the treatment of certain waste water streams.

The specific energy for the desalination process is given by Equation 5.17. Introducing in this equation the practically required cell pair area, the empirically determined limiting current density, the safety factor, and the shadow effect of the spacer gives the specific energy required in a practical electrodialysis desalination process to:

$$E_{prac/spec} = \frac{N_{cell}}{A_{prac} Q_{prac}} \left[\frac{\Delta \ln \frac{C_s^{fd}}{C_s^{fc}} \frac{C_s^c}{C_s^d}}{\Lambda_s \left(C_s^{fd} - C_s^d \right)} + r^{am} + r^{cm} \right] \left[\frac{Q_{prac} F \left(C_s^{fd} - C_s^d \right)}{N_{cell} \xi} \right]^2 \qquad (5.25)$$

Here $E_{prac/spec}$ is the desalination cost for the desalination in electrodialysis per unit volume product, N_{cell} is the number of cell pairs in a stack, C_s^{fd} and C_s^{fc} are the equivalent concentrations of the diluate and the concentrate at the cell inlet, C_s^d and C_s^c are the concentrations of the diluate and the concentrate at the cell outlet, Λ_s is equivalent conductivity of the salt solution, r^{am} and r^{cm} are the area resistances of the anion- and cation-exchange membrane, Δ is the cell thickness, ξ is the current utilization, Q_{prac} is the diluate flow stream of the entire stack, and A_{prac} is the practically used cell pair area.

The energy needed to pump the various flow streams through the stack has been discussed earlier in Chapter 4 and is given by:

$$E_{p/spec} = \frac{E_p}{Q^d t} = k_{eff} \frac{\left(Q^d \Delta p^d + Q^c \Delta p^c + Q^e \Delta p^e \right)}{Q^d} \qquad (5.26)$$

Here $E_{p/spec}$ is the total energy per unit diluate water for pumping the diluate, the concentrate, and the electrode rinse solution through the stack, k_{eff} is an efficiency term for the pumps, and Q^d, Q^c, and Q^e are the volume flow rates of the diluate, the concentrate, and the electrode rinse solution through the stack.

The energy consumed by the electrode reaction and required for the operation of the control and measuring instruments are generally neglected or taken as a fixed item which is between 1 and 3% of the energy used for the ion transfer and pumping of the solutions.

2.8 Electrodialysis Power Supply and Instrumentation

The driving force for the desalination in electrodialysis is an electrical potential gradient. Therefore, electrodialysis requires a direct current of a certain amperage depending on the cell pair area and the required current density. The voltage necessary to obtain the required current depends on the resistance of a cell pair and the number of cell pairs placed between two electrodes as discussed earlier. The inability to alter the direct current voltage readily makes it desirable to operate an electrodialysis unit at constant voltage. Since the voltage across the stack is constant and the resistance of the solutions flowing through a stack is changing along the flow path from the entrance to the exit of the cell due to the change in concentration in the solutions the current density is also changing along the flow path. In certain cases it is economically advantageous to operate the different stacks of a staged system at different voltages. This requires, however, multiple rectifiers and the additional investment cost must be balanced against incurred additional membrane area. In most applications the objective in electrodialysis is the production of a product water of a predetermined quality regardless of variations in the feed solution composition or temperature. Constant product water salinity can be achieved by varying the applied voltage and thus the current density in a stack or by adjusting the feed and bleed streams accordingly by changing the re-circulation rates of the diluate and concentrate flow streams.

The electrodialysis process is determined by a number of process variables which must be measured and controlled. Several of these variables are pre-determined by the feed solution composition and the required product quality or recovery rates. Other process variables such as feed flow velocities, the applied voltage, the current density, the process path length, or the feed and bleed rates can be varied in a certain range taking into account technical feasibility and economic considerations. However, many process variables are interdependent. For instance, an increase in the flow stream velocities results in shorter residence time of the solutions in the stack with the consequence of lower desalination rates in a single path. On the other hand the flow velocity in the diluate cell determines also the limiting current density, and high flow velocities mean high current densities and a high degree of desalination per unit membrane area. However, high flow velocities cause higher pressure losses in the stack which do not only increase the pumping power consumption, but may cause internal and external leakages.

To control the different process variables the electrodialysis unit must be equipped with a number of monitoring devices that are coupled with the power supply and hydraulic system. Usually the feed solution, the brine, and the product water are continuously monitored in terms of their salinities, their flow velocity, their conductivity, and their temperature. Flow meters and pressure gauges are installed at the feed solution inlet and brine and product water outlet points of the stack. Generally, pH-meters are also installed at the product water outlet to control eventually occurring pH-changes due to water splitting caused by excessive concentration polarization. Data generated by the monitoring instruments are fed back to control devices which then adjust different operating parameters such as the flow velocities or the applied voltage according to a predetermined program to guaranty the desired product capacity and quality.

The instrumentation especially for multi-stack units with independent power supplies and hydraulic systems can be quite complex and are a significant fraction of the total investment costs.

2.9 Pretreatment Procedures and Safety Requirements

In electrodialysis many feed solutions used for the production of potable or industrial process water contain suspended and colloidal matter or soluble salts close to their saturation level. These components cause severe membrane fouling and poisoning. They can even lead to blocking the flow channels in the stack. A pretreatment of the feed solution which guarantees the removal of these materials is therefore mandatory for a trouble free operation of an electrodialysis plant.

Filtration of the feed solution is applied in almost all electrodialysis systems as a pretreatment step to remove particulate materials. Often, however, additional procedures such as flocculation and precipitation of metal hydroxides or ion-exchange may be necessary as pretreatment of the feed solution. Feed waters containing significant quantities of calcium carbonates and bicarbonates are likely to cause precipitation in the concentrate stream at the anion-exchange membrane surface due to pH-changes resulting from water splitting which is obtained when the limiting current density is exceeded locally. To some extent this problem can be avoided by acidification of the concentrate stream. More difficult is the control of calcium sulfate precipitation. Acidification has no effect and in severe cases, i.e. when in the feed solution the calcium sulfate concentration is close to the saturation point, a softening step is required prior to the electrodialysis desalination process.

Before the product water of an electrodialysis unit can be used as potable or industrial process water certain post treatment procedures may be necessary. Since micro-organisms are not removed from a feed stream by electrodialysis most commonly a chlorination of the product water is required. When the concentrate stream has been acidified the product water pH-value must be readjusted. Other measurements may be necessary depending on the feed water composition and required product water quality.

2.10 Process Economics

Process design and economics are closely related in electrodialysis. The total costs are the sum of fixed charges associated with amortization of the plant investment costs and of operating costs such as energy and maintenance costs. Membrane replacement costs are sometimes regarded as a separate item because of the shorter life of the membranes. In electrodialysis water desalination the useful life of the membranes is in the order of 5 to 7 years. Other components such as the power supply, the valves, piping, pumps, instrumentation, stack spacers etc. usually last 10 to 20 years. Membrane and other equipment costs and their useful life under operating conditions vary widely depending on the manufacturer, plant size and plant location. Thus, for a final cost calculation the plant size and location must be considered.

Investment costs

Investment costs include non-depreciable items such as land and depreciable items such as the electrodialyis stacks, pumps, electrical equipment, monitoring, and control devices and membranes. The total investment costs of an electrodialysis plant are the costs of the membrane and the costs of all other components required for a certain plant capacity. The membrane area required for a certain plant capacity can be calculated by Equation 5.23. It is related to the feed, the diluate, and concentrate concentration. It is furthermore a function of the linear flow velocity and the cell thickness as will be shown later in a sample calculation.

Operating costs

The operating costs are mainly determined by the energy consumption and by maintenance costs. The energy consumption in electrodialysis is determined by the electrical energy required for the actual desalting process and the energy for pumping the solution through the stacks. The energy for the actual desalting process, i.e. the ion transfer from the feed solution to the concentrate, is directly proportional to the amount of ionic species to be removed, i.e. the concentration of the feed solution and the product. But it also depends on the applied cell pair area and on the linear flow velocity in the stack. It can be calculated by

Equation 5.25. The pumping energy is independent of the concentration of the feed solution and given by Equation 5.26.

2.10.1 Calculation of an Electrodialysis Plant

The calculation procedure for designing an electrodialysis plant is based on a number of input parameters which consist of values related to a given stack geometry, solution properties, and operating strategies. The total process path length as calculated by Equation 5.24 is related to the path length of a given stack by:

$$L_{prac} = L_{st} N_{st} \tag{5.27}$$

Here are L_{prac} and L_{st} the practical process path length and the length of a stack, and N_{st} is the number of stacks in series in a plant.

The length of a stack is given by the manufacturer. The number of stacks in series N_{st} is 1 or a positive integer. This means that Equation 5.27 generally does not balance when the practical process path length is used as calculated by Equation 5.24. Therefore, the practical process path length must be adjusted to meet the requirements of Equation 5.27. The practical process path length is related to the product concentration, the feed concentration, the concentrate concentrations, the linear flow velocity, the limiting current density, etc. as expressed by Equation 5.24. However, this equation cannot be solved for a desired product concentration and process path length explicitly, and an iteration must be executed based on an assumed value of the product concentration and the given length of the stack or the stacks in series.

Another important parameter for the design of a plant is the recovery rate applied in the process. The recovery rate determines which fraction of a feed solution is obtained as product, i.e.:

$$\Delta = \frac{Q^{p}_{prac}}{Q^{f}_{prac}} \tag{5.28}$$

Here Δ is the recovery rate, and Q^{p}_{prac} and Q^{f}_{prac} are the flow rates of the product and the two feed solutions of the diluate and concentrate cell.

The recovery Δ rate has a value between 0 and 1. If in a stack, where the diluate and concentrate cells have identical geometries the recovery rate is 0.5, the product flow rate is identical to that of the concentrate. For any other recovery rate product and concentrate flow rates are different, and one flow stream must be operated in a feed and bleed mode to obtain identical linear flow velocities in the diluate and concentrate cells.

The concentrations of diluate and concentrate are related to the recovery rate by a mass balance. If identical flow velocities of the diluate and concentrate streams are assumed the required concentrations of the concentrate feed solution and the final concentration of the concentrate for a given diluate feed concentration and required product concentration are given by:

$$C^{c}_{s} = \frac{C^{fd}_{s} - \Delta C^{d}_{s}}{1 - \Delta} \quad \text{and} \quad C^{fc}_{s} = \frac{C^{fd}_{s}(1-\Delta)}{\Delta} + \frac{C^{c}_{s}(2\Delta - 1)}{\Delta} \tag{5.29}$$

The calculation procedure for the construction of a typical brackish water electrodialysis desalination plant are described in the following example. The plant shall have a product water capacity of 350 m^3 d^{-1}. The feed water has a salinity of 3500 g m^{-3} which is according to the assumed salt composition a concentration of 53.3 eq m^{-3}. The product salinity should be less than 350 g m^{-3} which is a concentration of 5.3 eq m^{-3} for the assumed salt composition. The product recovery should be higher than 75 %.

The calculation procedure

The calculation procedure is illustrated in the flow chart of Figure 5.10.

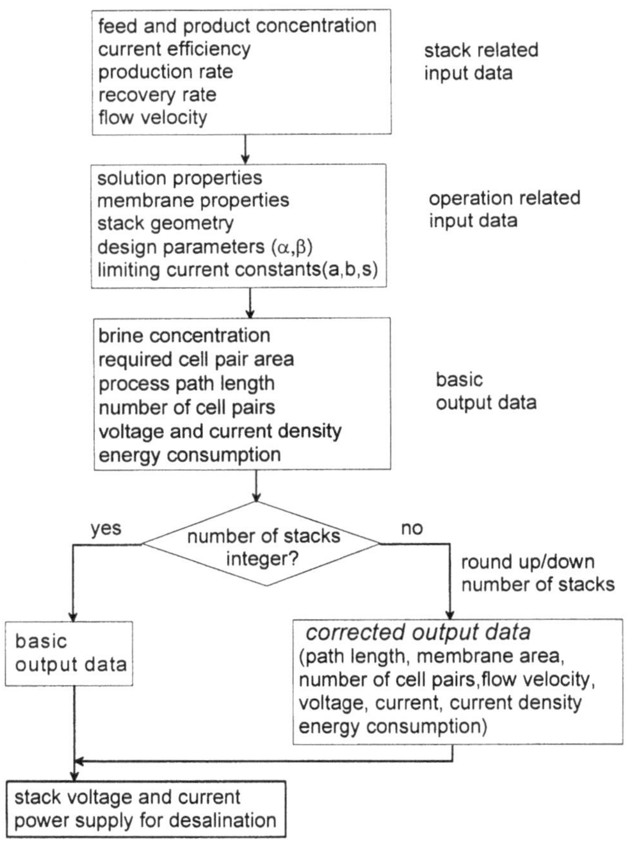

Fig. 5.10 Flow sheet illustrating the calculation procedure for an electrodialysis desalination plant.

The input data consist of values related to the plant hardware components such as the membrane properties, the electrodialysis stack geometry including safety factors taking into account the current utilization and the effect of the spacer on the cell volume or the flow distribution, the feed and product solution concentrations, and the operating strategies such as feed flow velocities and recovery rate. Output data are the process path length, the number of stacks in series or in parallel, the total required membrane area, the limiting current density, the applied voltage, and the energy consumption. For the cost analysis energy required for pumping the solutions through the stack must be calculated and charges related to the total investment and operation must be estimated.

For the model calculation the following assumptions have been made:

- diluate and concentrate cells have identical geometries,
- the flow streams of diluate and concentrate are co-current and of equal velocity,
- the activity coefficients of the salt in the diluate and concentrate are 1,
- concentration potentials and boundary layer effects are neglected.

Input data such as plant capacity, feed and product water salinity or stack design are freely selected, other data such as membrane properties, limited current density constants, current efficiency are taken from the literature.

Input Data:

concentration feed inlet	C_s^{fd}	0.0583 keq m^{-3}
concentration diluate outlet	C_s^{d}	0.00583 eq m^{-3}
recovery rate	Δ	0.75
equivalent conductance of solution at 20°C	Λ	10.5 S m^2 keq^{-1}
thickness of cell	Δ	0.00065 m
linear flow velocity	u	0.075 m s^{-1}
constant a (limited current density calculation)	a	25 000 A s^b m^{1-b} keq^{-1}
constant b (limited current density calculation)	b	0.5
Faraday constant	F	9.65×10^7 As keq^{-1}
volume factor of spacer	α	0.8
area factor accounting for shadow effect	β	0.7
total area resistance of membranes	r^{am}, r^{cm}	0.0007 Ω m^2
safety factor	s	0.7
current efficiency	ξ	0.9
production capacity of the plant		350.0 m^3 d^{-1}
effective width of cell	Y	0.42 m
effective length of flow path per cell	X	0.725 m

Output Data

concentration concentrate inlet	C_s^{c}	0.1634 keq m^{-3}
concentration concentrate outlet	C_s^{fc}	0.2159 keq m^{-3}
theoretical limiting current density	i_{lim}	39.92 A m^{-2}
experimentally determined limiting current density	i_{lim}	27.95 A m^{-2}
total length of flow path	L_{pp}	3.622 m
cell pair voltage drop	U_{cell}	0.325 V
total membrane area	2 N_{cell} A_{st}	752.34 m^2
total number of cell pairs	N_{cell}	247.31
total current through the stack	I	92.206 A
potential drop per stack	U_{st}	80.20
number of stacks in series	L_{pp}/X	4.995

Since the number of stacks must be an integer, the actual stack number is chosen to be 5 and the resulting parameters are recalculated based on the actual stack number.

The recalculated data		
number of stacks in series	N_{st}	5
length of flow path	$L_{pp}\, N_{st}$	3.625 m
number of cell pairs	N_{cell}	247
limiting current density	i_{imp}	27.967 Am^{-2}
membrane area	$2\, N\, A_{st}$	753.771 m^2
power consumption	$E_{des/spec}$	0.406 $kWhm^{-3}$
potential drop across cell pair	U_{cell}	0.3246 V
current	I	92.3807 A
potential drop per stack	U_{st}	80.115 V

The results of the calculations are shown in Figures 5.11 and 5.12 which show the flow diagram of the calculated electrodialysis unit consisting of 5 stacks in series and the concentration changes as a function of the process path length for the calculated example.

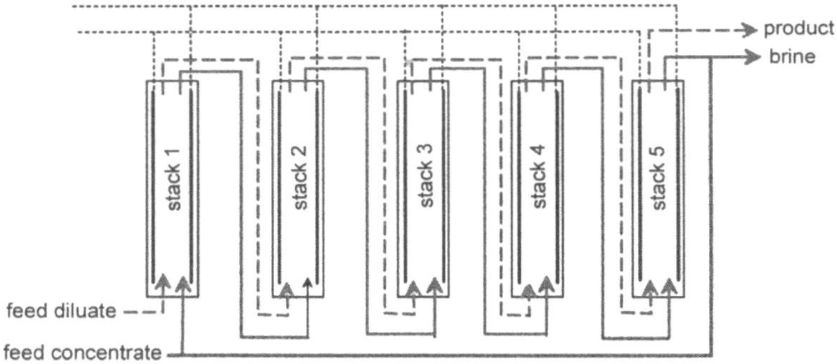

Fig.5.11 Flow diagram illustrating the stack arrangement of the calculated electrodialysis plant consisting of 5 stacks with 247 cell pairs each.

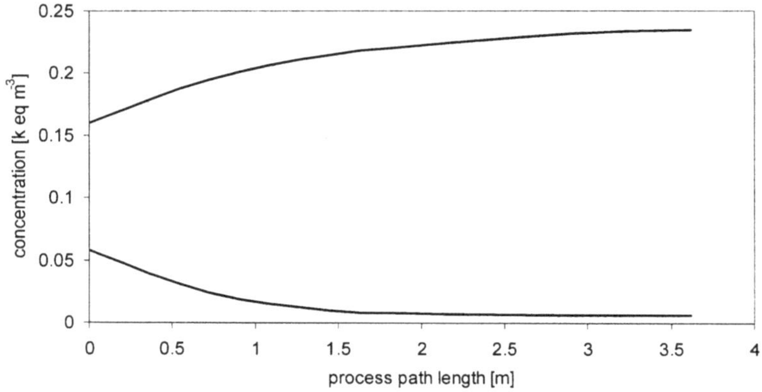

Fig. 5.12 Concentration of diluate and concentrate as function of the process path length.

While the diluate and concentrate solutions pass through the stacks their concentration is changing as shown in Figure 5.12 which shows the concentration changes as a function of the total process path length for the calculated example indicating that the desalination rate is decreasing with the process path length due to the decrease in current density with increasing process path length. The rate of desalination expressed by the slope of the curves is much higher at the beginning of the process path than at the end.

2.10.2 Estimation of Desalination Costs

The estimation of the product water costs is based on a number of assumptions. Especially the investment costs are difficult to estimate since membrane and other equipment costs and their useful life under operating conditions vary widely depending on the manufacturer, plant size and plant location. Thus, for a final cost calculation the plant size and location must be considered. In this example calculation it is assumed that investment and maintenance costs are proportional to the required membrane area and can be expressed as constant factor while the operating costs are determined by energy consumption. The basic parameters for the calculations of the product water costs are:

General plant equipment and energy costs:	
plant capacity	350 m^3 d^{-1}
operating time	24 h d^{-1}, 330 d y^{-1}
membranes and stack costs	150.- US \$ m^{-2}
power supply, pumps, piping, and instrumentation	20% of stack costs
membrane and equipment life	5 years
operating and maintenance costs	1% of investment costs y^{-1}
electric energy costs	0.10 US \$ $(kWh)^{-1}$
land and infrastructure	50 000.- US \$
interest on invested capital	4% p.a.
Capital related and operating costs:	
investment costs (total)	135,720.- US \$
stack costs (754 × 150)	113,100.- US \$
power supply and equipment (20% of 113,100.-)	22,620.- US \$
capital costs (amortization and interest on capital)	34572.80 US \$ y^{-1}
operating and maintenance costs:	1357.20 US \$ y^{-1}
energy costs (total):	5889.30 US \$ y^{-1}
desalination energy	4689.30 US \$ y^{-1}
operating energy	1200.- US \$ y^{-1}
total costs of product water	418193.30 US \$ y^{-1}
costs of unit product water	0.36 US \$ m^{-3}

Product water costs:

The product water costs in the above described example calculation are made under certain assumptions in terms of feed and product water quality, and equipment and energy costs. These items may vary substantially depending on feed water composition, plant capacity and location, etc. However, some general relations are true for all electrodialysis applications. The water production costs are increasing with the feed water salinity assuming a given product water quality as indicated in Equation 5.23 which relates the required membrane area to the feed solution concentration and Equation 5.25 which gives a relation between the desalination energy and the feed solution concentration. Pumping energy requirements and investment for equipment such as pumps, piping and control devices are independent of the feed solution concentration. The product water costs as function of the feed solution concentration assuming a constant product concentration are illustrated in Figure 5.12 which shows schematically the operating costs which also include the energy costs and the membrane area and other investment related costs as well as total desalination costs as function of the feed water concentration.

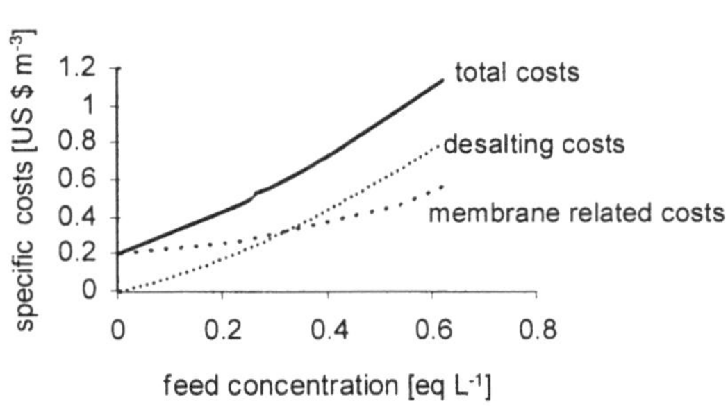

Fig. 5.13 Schematic diagram illustrating the electrodialysis desalination costs as function of the feed solution concentration assuming constant product concentration.

Figure 5.13 shows that both the operating and the investment related costs and thus the total product costs are strongly increasing with the feed solution concentration when the product concentration and all other parameters are kept constant. The absolute cost values are subject to plant capacity and location, costs of hardware components, labour costs, etc. Costs of water produced in a commercial electrodialysis desalination plant can vary significant, i.e. than 50% from that given in the example calculation.

2.10.3 Discussion of Parameter Values and Model Assumptions

In the calculation of the above described example a number of assumptions have been made which may or may not effect the reliability of calculation. The effect of the different assumptions will be discussed.

The equivalent conductivity

One of the assumptions for the derivation of the model was that the conductivity is proportional to the salt concentration. However, the equivalent conductivity of electrolytes Λ_s is concentration dependent and is decreasing as the square root of the ion concentration. In this case the concentration dependence of Λ_s can be fitted over a wide range of concentration by a linear regression curve [H.J. Lee et al., 2002]. Using the given linear regression curve the equivalent conductance can be calculated from a given concentration value and it can be shown that although the concentration may change by an order of magnitude the equivalent conductivity change is less than 10%. Therefore, it seems to be justified to use an average value of 10.63 S m² keq^{-1} in the above described example of brackish water desalination.

The membrane resistance

In the above example the membrane area resistances r^{am} and r^{cm} of 7 Ω cm² are assumed for the anion- plus cation-exchange membrane together. This value is based on the area resistance of commercially available membranes as listed in various company brochures.

The effect of osmosis and electroosmosis water transport on the current utilization

In the above example the water transport due to osmosis and electroosmosis has been neglected. This can be justified by calculating the osmotic and the electroosmotic water flux. The osmotic water flux is the result of the osmotic pressure driving force and the membrane water permeability. The osmotic pressure difference between the concentrate and the product solution in the above described example is less than 6 bar. The water permeance of the ion exchange membranes is in the order of 1.5×10^{-9} m s^{-1} bar^{-1}. This results in an osmotic water flux from the diluate to the concentrate solution of 0.5 m^3 d^{-1} which can be neglected compared to the water production of 350 m^3 d^{-1}. The electroosmotic water flux can be estimated by the number of moles of water removed from the diluate per mole of migrated ion which is in the order of 2-10 [V.V. Nikonenko et al., 1999]. For the above described example the electroosmotic water flux is in the order of 1 to 2 m^3 d^{-1}. Thus, the total water transport from the diluate to the concentrate due to osmosis and electroosmosis is less than 1 % of the total water production and can be neglected.

The desalination energy consumption

The total desalination energy consumption is the product of the total current and the applied voltage. The current is directly proportional to the number of ions transferred from the diluate to the concentrate solution. However, the current through a stack is not completely utilized for the transport of ions. It is effected by the membrane permselectivity, the water transport and current leakage through the manifold as has been discussed earlier. In the above described example a 90% current utilization has been assumed. This value seems reasonable since in a well constructed stack the current utilization is mainly caused by the incomplete permselectivity of the ion-exchange membranes, which is in most commercial ion-exchange membranes higher than 92% as shown in the Table of Appendix I.

The potential drop in the electrodialysis stack is the result of the electric resistances of the solutions and the membranes, and of the concentration potential between the solutions separated by the membranes. In addition there are potential drops at the electrodes and in the electrode rinse solution. For conditions specified in the example the concentration potential is in the order of 0.02 V which is less than 2% of the applied cell voltage. In a stack which contains 100 cell pairs the voltage drop due to the resistance of the electrode rinse solution and the electrode reactions can also be neglected. Therefore, the assumption made in the example calculation that the voltage drop can be expressed to a first approximation by the electric resistance of the solutions and the membrane in the stack is justified.

The limiting current density

The limiting current density is very difficult to estimate theoretically since the mass transfer coefficient in the laminar boundary layer at the membrane surface in the diluate solution is a function of the linear flow velocity and several other sometimes ill defined parameters such as the spacer geometry, membrane surface roughness, etc. Therefore, the experimentally derived relation given in Equation 5.18 is generally used. The key parameters in this equation are the constants a and b. The experimental procedure of deriving these constants for a sheet-flow stack design is described in the literature [H.J. Lee et al., 2002]. The experimentally derived data indicate that for a linear diluate flow rate between 2 and 6 cm s^{-1} a and b may be considered as constant.

The effect of the spacer on current density and cell volume
The so-called shadow effect of the spacer on current density can be estimated by considering the spacer construction or be determined experimentally by comparing the operation of a cell pair with and without a spacer. Generally, the effect of the spacer on the current density can be estimated within an error range of about 5%.

Assumptions made in designing an electrodialysis process for a practical application
The performance of electrodialysis in practical applications depends on membrane properties, feed solution composition and equipment design parameters such as the feed flow velocities, operation mode, the cell geometry, and the spacer configuration in an electrodialysis stack. In the above described example, the basic relations for the design of a practical electrodialysis brackish water desalination plant using a sheet flow stack design were derived. In the determination of the different process and equipment design parameters a number of approximations have been made which are specific for the discussed application and anticipated plant components and cannot be generalized. In other applications with other hardware components and feed and product solution properties the described calculations must be modified to meet the specific conditions of the application and used system components.

3 BIPOLAR MEMBRANE ELECTRODIALYSIS PROCESS DESIGN

The basic concept of electrodialytic water dissociation with bipolar membranes has been discussed in detail earlier. The main application of this process is the production of an acid and a base from the corresponding salt in a cell arrangement consisting of cation- and/or anion-exchange membranes and bipolar membranes in alternating series between two electrodes forming an array of individual cells as shown earlier in Figure 1.3. A repeating unit may be composed of a three cell arrangement as indicated in Figure 5. 14 a) or a two cell arrangement as indicated in Figures 5.14 b) and c).

In the three compartment cell design one cation-exchange, one anion-exchange, and one bipolar membrane with three flow streams in between them, i.e. the salt solution, the acid, and the base represent a cell unit. In the two compartment cell design a cation-exchange and a bipolar membrane, or an anion-exchange membrane and a bipolar membrane with two flow streams in between are arranged in parallel. One of the flow streams contains the salt solution in a mixture with the produced acid or base. A device in which these cell units are "stacked" with electrodes on both ends is referred to as an electrodialysis stack. Although there are many different components necessary for the proper and efficient operation of an electrodialytic water dissociation plant such as the electrical power supply, pumps and control and monitoring devices, the stack is the key component. For designing and operating an electrodialysis stack certain process parameters have to be taken into account and must be controlled such as the desired acid and base concentrations, and the concentration of the original salt solution.

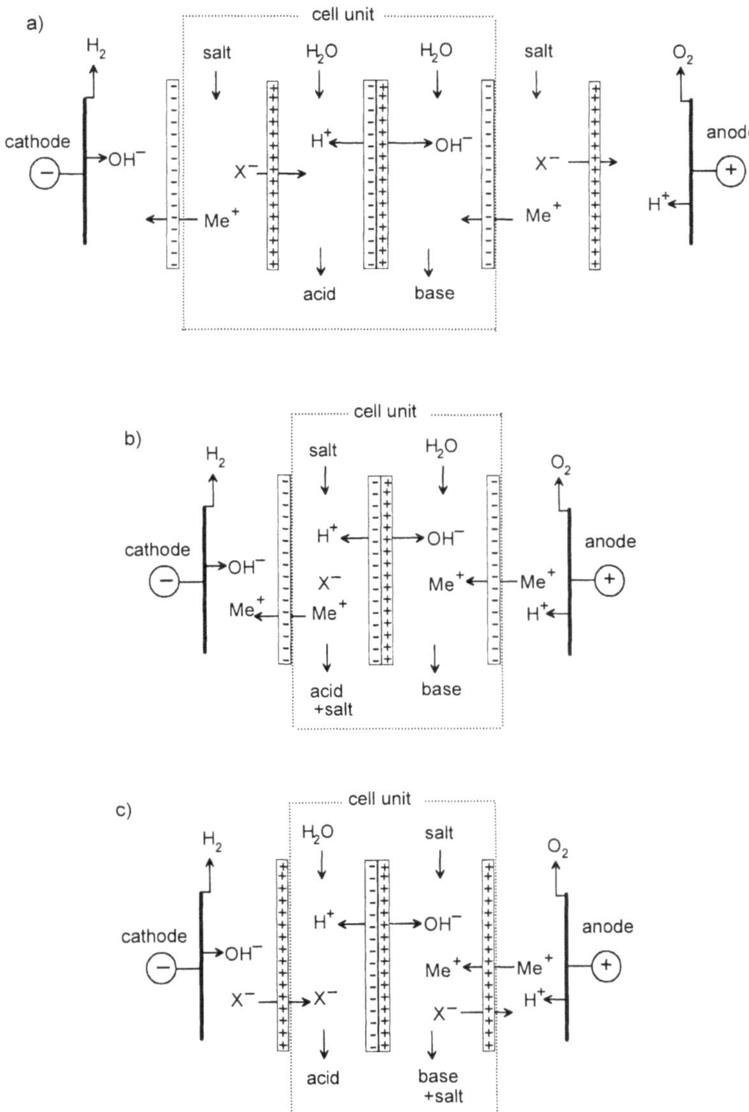

Fig. 5.14 Schematic diagram illustrating the cell arrangement in an electrodialytic water dissociation stack indicating a) a three cell unit composed of an anion-exchange, a cation-exchange and a bipolar membrane, b) a two cell unit composed of a cation-exchange and a bipolar membrane, and c) a two cell unit composed of an anion-exchange and a bipolar membrane.

Various modes of operation are possible such as batch, feed and bleed or continuous process. The flow velocities of the different solutions in the stack determine the degree of acid and base concentration that can be achieved in one pass in a stack of a given geometry and the pressure loss which has to be overcome by the flow stream circulation pumps. Certain process parameters are fixed by the feed and the required product properties. Other parameters may be varied in a certain range and can thus be utilized to optimize the process, such as the current density, the applied voltage, the flow stream velocities, etc. Most parameters are interdependent and often counteracting in terms of overall process costs [K.J. Liu et al., 1977].

In a three cell arrangement relative pure acids and bases are obtained. However, an additional monopolar membrane is needed which will effect the investment cost. The two cell arrangement will result in an acid or a base in a mixture with the salt. The two cell arrangement is often applied when the product has a relatively low conductivity. This is, for instance, the case when certain organic acids which have a low degree of dissociation such as lactic acid are recovered from a fermentation broth. Without additional salt the poor conductivity in the acid compartment would result in excessive power consumption.

3.1 The Design of an Electrodialysis Stack with Bipolar Membranes

Just as in conventional electrodialysis the membrane stack is also the key component in electrodialytic water dissociation with bipolar membranes. The stack design used mainly in electrodialysis with bipolar membranes is basically the same as that of the sheet flow stack used in conventional electrodialysis with the exception, that the basic unit of an electrodialysis stack with bipolar membranes consists generally of three cells. Therefore, three separate flow streams have to be fed through the manifold in the individual cells that form the repeating unit of the stack. A further major difference between conventional electrodialysis and an electrodialysis with bipolar membranes is the applied current density and the voltage drop over a cell unit. The current density used in electrodialysis with bipolar membranes for the production of acids and bases is generally an order of magnitude higher than in most conventional electrodialysis applications. The voltage drop across a cell unit is also higher than in conventional electrodialysis because of the additional resistance of the bipolar membrane and the concentration potential between the reaction layer of the membrane and the outer phases. Furthermore, the current density is not limited due to boundary layer effects which can generally be neglected, but is caused by limitations of the water transport into the transition region of the bipolar membrane. To operate an electrodialysis stack of given geometry efficiently in a certain capacity plant the required membrane area, the applied voltage, and the process path length must be calculated. As a consequence of the high current density the cell unit area is usually significantly lower in a bipolar membrane electrodialysis than that of the cell pair area in conventional electrodialysis. Because of the higher voltage drop over a cell unit in electrodialysis with bipolar membranes the number of repeating cell units is also much lower than in conventional electrodialysis. Assuming a current density of 1000 A m^{-2} and a voltage drop per repeating cell unit of 2 V a stack with 1 m^2 unit area and 100 cell units would have a total power input of 200 kW. This energy must be dissipated as heat during the operation and therefore limits the size of a stack in electrodialysis with bipolar membranes.

3.1.1 The Applied Voltage and Current Density in a Bipolar Membrane Stack

The required membrane area can be expressed as the cell unit area which consists of a cation-exchange, an anion-exchange membrane, and a bipolar membrane. The cell unit area is related to the total current required for the production of a certain amount of an acid and a base by the current density.

$$A_{cell} = \frac{I}{i} \tag{5.30}$$

Here A_{cell} is the cell unit area, I is the total current passing through the stack or stacks in series, and i is the average current density.

The total current to be used for a certain production capacity is given by:

$$I = \frac{Q_p F \Delta C_p}{N_{cell} \xi} \tag{5.31}$$

Here Q_p is the total amount of product, ΔC_p is the concentration difference between the product solution in the feed at the entrance of the first stack and at the exit of the final stack, N is the number of cell units in a stack, F is the Faraday constant, and ξ is the current utilization.

The average current density can be determined from the applied voltage and the resistance of the cell unit. The voltage drop obtained across a cell unit is illustrated in Figure 5.15.

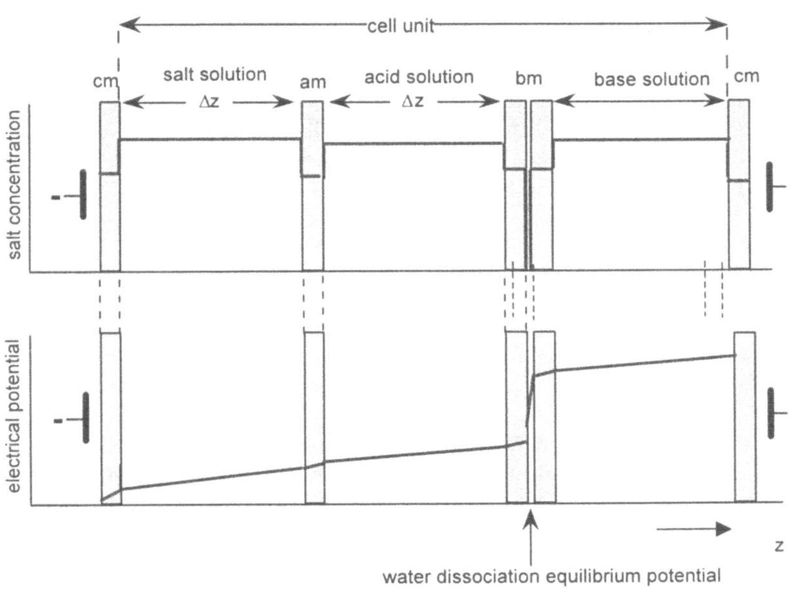

Fig. 5.15 Schematic drawing illustrating the ion concentration profiles and the voltage drop across a cell unit at a certain point along the flow channel of an electrodialysis stack with bipolar membrane.

This Figure 5.15 shows the concentration profiles at a certain point along the flow channel in direction perpendicular to the membrane surfaces that are established at steady state in a cell unit when a current flows through an electrodialysis stack. It has been assumed that the concentrations of the salt solution and the acid and the base are about equal.

When concentration potentials between the three solutions are neglected the total voltage drop consists of three major contributions:

- the potential drop due to the water dissociation equilibrium,
- the potential drop due to the resistance of the solutions, and
- the potential drop due to the resistance of the membranes.

A cell unit contains the salt, the acid and base solutions, and the three membranes between these solutions. Concentration changes due to boundary layers effects are neglected in the schematic drawing. The concentration profiles across a cell unit at a certain point depend on various process and cell design parameters. The total voltage drop is the sum of the water dissociation potential resembling an electromotive force and the voltage drop due to the resistance of the solutions and membranes.

In practical applications of electrodialysis the stacks consist of geometrically identical cells which are operated in co-current flow. Generally, the following approximations can be made:

- the equivalent conductivity is independent of concentration over the range of interest,
- the concentration potentials are negligibly low, compared to the potential drops caused by the resistance of the solutions and the membranes,
- the flow of the solutions is co-current and of equal velocity, and
- changes in the resistance of the solutions due to boundary layer effects are also neglected or averaged over the cross-section of the cell,
- the concentrations are expressed in equivalent per unit volume.

With these assumptions the voltage drop across a cell unit is given according to Equations 4.123 by:

$$\Delta U_{cell} = i\left[\Delta\left(\frac{1}{\Lambda_s C_s}+\frac{1}{\Lambda_a C_a}+\frac{1}{\Lambda_b C_b}\right)+r^{tr}+r^{am}+r^{cm}+r^{bml}\right]+\frac{RT}{F}\ln\frac{C^{tr}_{H^+}C^{tr}_{OH^-}}{C^{b}_{H^+}C^{b}_{OH^-}} \qquad (5.32)$$

Here ΔU_{cell} is the voltage drop across a cell unit, i is the current density at any point along the cell in direction of the flow velocity parallel to the membrane surface, Δ is the thickness of the individual cells, C is the concentration, Λ is the equivalent conductivity, r the area resistance, R the gas constant, T the absolute temperature, F the Faraday constant, and the superscripts tr, am, cm, bml, and b refer to transition region, cation-exchange membrane, anion-exchange membrane, the two layers of the bipolar membrane, and the bulk solution, respectively; the subscripts s, a, and b refer to salt, acid, and base, respectively.

Although Equation 5.32 contains the approximation of some terms, it predicts the total potential drop across a cell unit reasonably accurately.

The concentration in the solutions is changing during the process from the feed solution concentration to the product solution concentration. To obtain the average current density at a constant applied voltage the concentrations expressed in Equation 5.32 must be replaced by average concentrations. As shown earlier the average concentration is given by the integral average over the process path length. The average concentration for each ion is:

$$\overline{C}_i = \frac{C_i^{fi} - C_i^{pi}}{\ln\left(\frac{C_i^{fi}}{C_i^{pi}}\right)} \qquad (5.33)$$

Here $\overline{C}_i$ is an average concentration of the component i in a cell, C_i^{fi} and C_i^{pi} are the concentrations in the bulk solution at the cell entrance, i.e. the beginning of the process path length and the cell exit, i.e. the exit of the process path length.

Introducing Equation 5.33 into Equation 5.32 leads to:

$$\Delta U_{cell} = i\left[\Delta\left(\sum_i \frac{\ln\left(\frac{C_i^{fi}}{C_i^{pi}}\right)}{\Lambda_i\left(C_i^{fi} - C_i^{pi}\right)}\right) + r^{tr} + r^{am} + r^{cm} + r^{bml} + \frac{RT}{iF}\ln\frac{C_{H^+}^{tr} C_{OH^-}^{tr}}{\frac{C_{H^+}^{fH^+} - C_{H^+}^{pH^+}}{\ln\frac{C_{H^+}^{fH^+}}{C_{H^+}^{pH^+}}}\,\frac{C_{OH^-}^{fOH^-} - C_{OH^-}^{pOH^-}}{\ln\frac{C_{OH^-}^{fOH^-}}{C_{OH^-}^{pOH^-}}}}\right] \qquad (5.34)$$

Here ΔU_{cell} is the voltage drop across a cell unit, i is the average current density in the cell unit, the subscript i refers to salt, acid, and base. All other symbols have the same meaning as described in Equation 5.32.

3.1.2 Membrane Area and Process Path Length in a Bipolar Membrane Stack

To determine the required membrane area and the process path length for a given capacity plant the current density must be related to the total current. The required membrane area is given by:

$$A = \frac{I}{i} \qquad (5.35)$$

Here A is the required membrane area and I the current required for a given capacity plant.

The current is related to the volume flux of the product and the concentrations of the feed and the product by:

$$I = \frac{Q_p F\left(C_p^{fp} - C_p^{p}\right)}{\xi} \qquad (5.36)$$

Here Q_p is the product volume flow, F is the Faraday constant, C_p^{fp} and C_p^{p} are the concentrations of the product at the cell entrance and the product at the cell exit.

By combining Equations 5.35 and 5.36 the required membrane area for a certain plant capacity can be determined as a function of the current density and the current utilization by:

$$A = \frac{Q_p F \left(C_p^{fp} - C_p^{p}\right)}{i\,\xi} \tag{5.37}$$

For a given stack design, i.e. a given cell width and length and the number of cell units in the stack the required process path length and the number of stacks can be calculated by:

$$L_{pp} = \frac{A}{YN_{cell}} = \frac{Q_p F \left(C_p^{fp} - C_p^{p}\right)}{i\,\xi\, Y\, N_{cell}} \tag{5.38}$$

and

$$N_{st} = \frac{Q_p F \left(C_p^{fp} - C_p^{p}\right)}{i\,\xi\, Y\, N_{cell} X_{cell}} \tag{5.39}$$

Here L_{pp} is the process path length, A is the cell unit area, Y is the cell width, Q_p is the product volume flow, F is the Faraday constant, C_p^{fp} and C_p^{p} are the concentrations of the salt at the beginning and the end of the process path length; i is the average current density, ξ is the current utilization, N_{cell} is the number of cell units in a stack, and N_{st} is the number of stacks in series.

For a given stack construction the number of stacks must be an integer and must eventually be recalculated by a round-up or round-down procedure by changing the current density and the flow velocities.

An important design criteria in electrodialysis with bipolar membranes is the total energy dissipation in the stack. Due to the relatively high current density and voltage drop per cell unit both the cell unit area and the number of cells in a stack are limited by the maximum energy that can be dissipated the stack as heat.

3.2 Process Design Aspects and Hardware Components

In addition to the actual stack and the power supply, an electrodialysis plant with bipolar membranes consists of several other components essential for a proper operation, such as pumps, process monitoring and control devices, etc. Again, as in the design of the stack, there are several operating modes for the bipolar membrane stack. Depending on the feed solution concentration and the product requirements the process may be operated in a batch mode or continuously in one or more stages or in a feed-and-bleed mode as illustrated in Figure 5.16.

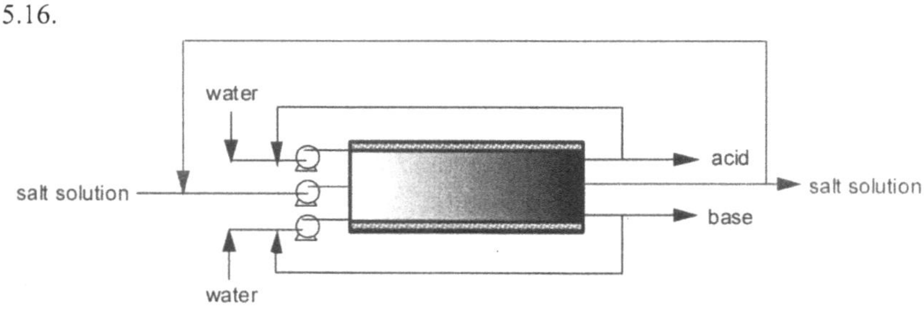

Fig. 5.16 Flow scheme of a bipolar membrane electrodialysis stack operated in a feed and bleed mode, i.e. with partial recycling of the acid, the base, and the salt solution.

3.2.1 Power Supply and Instrumentation

Bipolar membrane electrodialysis requires a direct current of a certain amperage depending on the cell unit area and the applied current density. The voltage necessary to obtain the required current depends on the resistance of a cell unit and the number of cell units placed between two electrodes. The inability to alter direct current voltage readily makes it desirable to operate an electrodialysis unit at constant voltage. In most applications the objective in electrodialytic water dissociation with bipolar membranes is the production of an acid and a base of a predetermined quality. Constant product concentration can be achieved by adjusting the feed and bleed streams accordingly by changing the recirculation rates of the acid, the base, and the salt solution flow streams.

To guarantee a certain capacity and quality of product the electrodialysis unit must be equipped with a certain number of monitoring and control devices such as conductivity, temperature, flow and pH-meters. Data generated by the monitoring instruments are fed back to control devices which then adjust different operating parameters such as the flow velocities or the applied voltage.

3.2.2 Membrane Fouling Control and Pretreatment Procedures

Salt solutions used in electrodialytic water dissociation with bipolar membranes for the production of acids and bases may contain suspended and colloidal matter, polyelectrolytes, large organic anions or multivalent ions. These components can cause severe membrane fouling due to precipitation on the membrane surfaces or by partial penetration into the membranes. As the result of a pH-value increase in the base containing cell multivalent ions will precipitate. This can lead to a blocking of the flow channels in the stack and can cause an increase of the electrical resistance of the stack. A pretreatment of the feed solution which guarantees the removal of all multivalent ions is therefore mandatory for a trouble-free operation of the process. The various pretreatment procedures such as filtration and especially ion-exchange to remove multivalent ions are generally applied.

3.3 Process Costs

The economics of electrodialytic water dissociation with bipolar membranes is determined by different process variables. Some of these variables are predetermined by the feed solution and the required product quality. Others, such as the applied voltage, the current density, the feed flow velocities, or the feed and bleed rates can be varied in a certain range taking into account the stack geometry and membrane properties. However, many process variables are interdependent. Variations in the velocities of the flow streams result in shorter residence time of the solutions in the stack with the consequence of lower concentration changes in a single path. Higher current densities require a lower membrane area for a certain capacity plant but consume more energy. To optimize the process for a given feed and product requirement all process variables and their interdependency must be taken into account. A complete cost calculation of the electrodialytic water dissociation with bipolar membranes should also include all pre- and post-treatment procedures and peripheral equipment necessary for a proper operation.

The total production costs in electrodialysis with bipolar membranes are the sum of fixed charges associated with amortization of the plant investment costs and of operating costs such as energy and maintenance requirements. They are a function of the mono- as well as bipolar membrane properties, of the feed solution composition, and of process and equipment design parameters such as stack construction and operating current density.

3.3.1 Investment Costs in Electrodialysis with Bipolar Membranes

Investment costs include non-depreciable items such as land and depreciable items such as the electrodialysis stacks, pumps, electrical equipment, monitoring and control devices, and membranes. The total investment costs of an electrodialysis plant used for the production of acids and bases are a function of the required membrane area which is directly proportional to the plant capacity. The useful life of the membranes is in the order of 3 to 5 years. Other components such as the power supply, valves, piping, pumps, instrumentation, stack spacers etc. usually last 10 to 15 years.

The required cell unit area for a given capacity plant is inverse proportional to the current density and is direct proportional to the amount of acid and base produced from a given feed solution as shown in Equation 5.37. The cell unit area is the area of one cell unit which consists of a cation-exchange, an anion-exchange, and a bipolar membrane.

For an electrodialysis stack operated at constant applied voltage the current density should not exceed a certain value which is given by the diffusion rate of water from the bulk solutions into the transition region of the bipolar membrane [H. Strathmann et al., 1997]. The limitation of the current due to insufficient water diffusion is illustrated in Figure 5.17 which shows the potential drop across a bipolar membrane as a function of the current density determined with a 1 mol L^{-1} Na_2SO_4 solution. The results indicate a limiting current density of ca. 450 mA cm^{-2} due to water transport limitations [J.J. Krol et al., 1998].

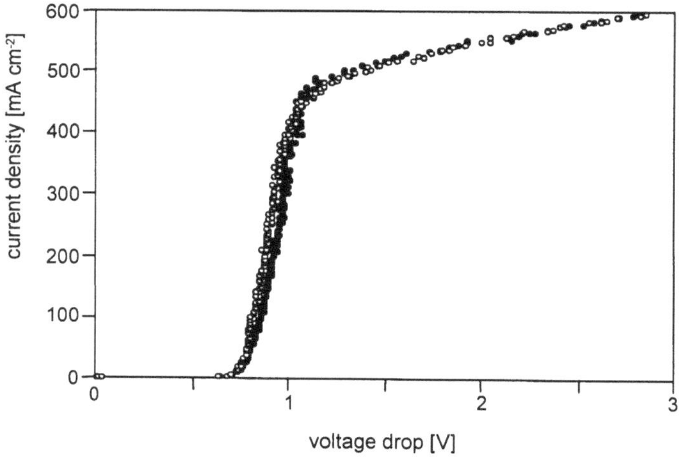

Fig. 5.17 Potential drop as a function of the current density of a commercial bipolar membrane (BP- 1 of Tokuyama Soda) determined with a 1 mol L Na_2SO_4 test solution [J.J. Krol et al., 1998].

The schematic drawing in Figure 5.17 indicates three characteristic areas for the current versus voltage curve of a bipolar membrane. First, a very small current is observed at low potential drop. This current is caused by salt leakage through the bipolar membrane. Then the potential drop across a bipolar membrane is increasing with the current being nearly constant. When the voltage drop approaches 0.8 V at current densities it increases drastically until it reaches ca. 450 mA cm^{-2}. Then the resistance of the membrane increases drastically again because of the limitation of the water transport into the bipolar membrane. The membrane is dried out and will be destroyed when this value is exceeded for a longer time period. In practical applications of the electrodialytic water dissociation the current density will for technical and economical reasons be significantly lower than the water transport limited current density as will be discussed later. Generally, the current density in a practical application will not exceed 100 mA cm^{-2}.

For a constant cell geometry, membrane properties, and applied voltage the current density varies from the cell entrance to the cell exit due to concentration changes in the product and feed solutions, and an average current density is defined as discussed earlier. The required membrane area is given by:

$$A_{am} = A_{cm} = A_{bm} = \frac{Q_p F\left(C_p^{fp} - C_p^{p}\right)}{i\,\xi} \tag{5.40}$$

The required membrane area is also significantly effected by the current utilization ξ, which can be quite low when high product concentrations are required. The poor current utilization is caused mainly by the proton and hydroxide ion leakage through the anion- and cation-exchange membranes.

3.3.2 Energy Requirements in Electrodialytic Water Dissociation

The energy required in an electrodialytic water dissociation with bipolar membranes is an additive of three terms:

- the electrical energy required to transfer the salt ions from the feed solution, and protons and hydroxide ions from the transition region of the bipolar membrane into the acid and base solutions,
- the energy required for the water dissociation in the bipolar membrane, and
- the energy consumption due to the pumping of the solutions through the stack.

Energy consumption due to electrode reactions can generally be neglected when more than 50 to 100 cell units are stacked between the two electrodes [K.N. Mani, 1991]. The energy required for the dissociation of the water is reversible while the energy required for the transfer of ions is irreversible and determined by the electrical resistance of the solutions and the membranes. This part of the electrical energy is converted to heat dissipated to the surrounding.

3.3.3 The Reversible Water Dissociation Energy

The reversible, i.e. the minimum energy that is required to dissociate water into protons and hydroxide ions, i.e. to produce an acid and a base according to the following reaction:

$$2\,H_2O \rightarrow H_3O^+ + OH^-$$

at constant temperature and pressure can be calculated by the equation for a concentration chain of solutions with different H^+-ion activities, i.e. pH-values. The reversible Gibbs free energy required for the production of acids and bases in a bipolar membrane is:

$$\Delta G = -\,RT\ln\frac{C^{tr}_{H^+}\cdot C^{tr}_{OH^-}}{C^{b}_{H^+}\cdot C^{b}_{OH^-}} = -2.3\ RT\,\Delta pH \qquad (5.41)$$

Here ΔG is the Gibbs free energy, R is the gas constant, T is the absolute temperature, F is the Faraday constant, $C^{tr}_{H^+}$, $C^{tr}_{OH^-}$, $C^{b}_{H^+}$, and $C^{b}_{OH^-}$ are the proton and hydroxide ion concentrations in the transition region of the bipolar membrane and the bulk solutions in contact with the bipolar membrane, and ΔpH is the difference between the pH-values of the two solutions separated by the bipolar membrane.

For 1n acid and base solutions in the two phases separated by the membrane ΔG is 0.0222 kWh mol^{-1} at 25°C. It is the reversible energy to transfer one mole of water into one mole of protons and hydroxide ions, i.e. 1n acid and base. The reversible energy is related to the electrical energy by:

$$\Delta G = \Delta U_{rev}\,F \qquad (5.42)$$

Here is ΔU_{rev} the reversible voltage drop between the two solutions separated by the bipolar membrane which is 0.828 Volt for the 1n acid and base solutions.

3.3.4 Practical Energy Requirements in Bipolar Membrane Electrodialysis

The production of an acid and a base requires, in addition to the reversible energy for the water dissociation, energy for the transport of salt ions from the salt solution through the cat- and anion-exchange membranes, and protons and hydroxide ions form the bipolar membrane transition region into the acid and base solutions. This additional energy is dissipated in form of heat as a result to the friction encountered by the ions when migrating through the solutions and membranes. The total energy required in electrodialytic water dissociation in a practical process is given be the current passing through the stack multiplied with the total voltage drop encountered between the electrodes.

$$E_{pro} = I\,\Delta U\,t \qquad (5.43)$$

Here E_{pro} is the energy consumed in a stack for the production of an acid and a base, I is the current passing through a stack or a series of stacks, ΔU is the voltage applied across the stack, i.e. between the electrodes, and t is the time of operation.

The voltage drop in a stack arrangement is caused by the electrical resistances of the solutions and the membranes in the stack and by the potentials which are established between solutions of different ion concentrations. The electrical resistance is caused by the friction of the ions with the membrane matrix and the water molecules while being transferred from one solution to another, resulting in an irreversible energy dissipation in the form of heat generation. The voltage drop per cell unit under the assumption that the stacks consist of geometrically identical cells which are operated in cocurrent flow with equal velocity is given by Equation 5.32. The required current for the production of a given amount of acid and base is given by Equation 5.36. Introducing Equations 5.32 and 5.36 into Equation 5.43, and considering the voltage drop across the entire stack, the total energy for the production of a given amount of acids and bases from the corresponding salt solution is given by:

$$E_{pro} = I\, N_{cell}\, \Delta U_{cell}\, t = N_{st}\, i\left(\frac{\Delta}{\sum_i \Lambda_i \overline{C}_i} + r^{tr} + r^{am} + r^{cm} + r^{bml} + \frac{2.3\, RT\, \Delta pH}{i\, F}\right)\left(\frac{Q_p F\left(C_p^{fp} - C_p^{p}\right)}{N_{cell}\, \xi}\right) t \quad (5.44)$$

Here E_{pro} is the energy for the production of a certain amount of acid and base, I is the current passing through the stack, N_{cell} is the number of cell units in a stack, ΔU_{cell} is the voltage drop across a cell unit, i is the current density, Δ is the thickness of the individual cells, $\overline{C}_i$ is the average concentration, and Λ_i is the equivalent conductivity, r is the area resistance, ξ is the current utilization, R is the gas constant, T the absolute temperature, F the Faraday constant, and ΔpH is the difference in the pH-value between the solution in the transition region of the bipolar membrane and the two adjacent bulk solutions, the subscript i refers to salt, acid, and base, The superscripts tr, am, cm, and bml refer to the transition region, the cation-exchange membrane, the anion-exchange membrane, and the two layers of the bipolar membrane, C_p^{fp} and C_p^{p} refer to the concentrations of the acid or base at the cell inlet, and to the concentration of the acid or base at the cell outlet, Q_p is the total flow of the acid or base through the stack, and t is the time.

The total current I passing through the stack which is identical with the current through a cell unit is related to the average current density i and the area of a cell A_{cell} unit by:

$$I = \frac{Q_p F\left(C_p^{fp} - C_p^{p}\right)}{N_{st}\, \xi} = i\, A_{cell} \quad (5.45)$$

Introducing Equation 5.45 into 5.44 gives the energy of the production of a certain amount of acid and base:

$$E_{pro} = N_{st}\, A_{cell}\left(\frac{\Delta}{\sum_i \Lambda_i \overline{C}_i} + r^{tr} + r^{am} + r^{cm} + r^{bml} + \frac{2.3\, RT\, \Delta pH}{i\, F}\right)\left(\frac{Q_p F\left(C_p^{fp} - C_p^{p}\right)}{A_{cell}\, N_{st}\, \xi}\right)^2 t \quad (5.46)$$

The meaning of the different symbols is identical to that described in Equation 5.44.

The term $\frac{Q_p F(C_p^{fp} - C_p^p)}{A_{cell} N_{st} \xi}$ is identical to the current density. This means that for a given stack design the acid and base production energy E_{pro} is proportional to the i^2.

The average concentrations of the acid, the base, and the salt in the bulk solutions $\overline{C}_i$ can be expressed by an integral average of the solutions at the cell inlet and the cell outlet C_i^{fi} and C_i^{pi}, respectively, as indicated by Equation 5.33.

The average differences in the Δ pH -values of proton and hydroxide ion concentrations in the transition region of the bipolar membrane and the acid and base solutions in the cells adjacent to the bipolar membrane is given by:

$$\Delta\,pH = 2.3\ln\frac{C_{H^+}^{tr}\,C_{OH^-}^{tr}}{\dfrac{C_{H^+}^{fH^+} - C_{H^+}^{pH^+}}{\ln\dfrac{C_{H^+}^{fH^+}}{C_{H^+}^{pH^+}}}\;\dfrac{C_{OH^-}^{fOH^-} - C_{OH^-}^{pOH^-}}{\ln\dfrac{C_{OH^-}^{fOH^-}}{C_{OH^-}^{pOH^-}}}} \tag{5.47}$$

Here $\Delta\,pH$ is an average pH-difference between the cells, C_i^{fi} and C_i^{pi} are the concentrations in the bulk solution at the cell entrance and the cell exit, respectively.

By introducing Equations 5.33 and 5.47 into Equation 5.46 the electrical energy consumption for the production of an acid and a base in electrodialytic water dissociation with bipolar membranes can by calculated as a function of the applied current density, the produced acid and base concentrations, and the volume of acid and base, and by the current utilization which is effected by hydraulic leaks between the cells, or by current flow through the manifold system and coion transport, especially protons and hydroxide ions through not strictly permselective membranes. All these effects are expressed in the current utilization ξ which is defined as the ratio of current required theoretically to the current required in practice.

$$\xi = \frac{I_{theo}}{I_{prac}} \tag{5.48}$$

3.3.5 The Current Utilization

Current utilization is a problem in all electromembrane processes. In the practical electrodialytic water dissociation with bipolar membranes its effect is especially severe. When current leakages through the manifold are avoided by a proper stack design the incomplete current utilization is caused mainly by incomplete membrane selectivity and water transport due to osmosis and electroosmosis.

The total current utilization is then given by:

$$\xi_{tot} = \xi_F\,\eta_w = N\,\eta_m\,\eta_w \tag{5.49}$$

Here ξ is the current utilization, η is an efficiency term, N is the number of cell units in a stack, and the subscripts tot, F, m, and w refer to total efficiency, Faraday efficiency, membrane selectivity efficiency, and water transport through the membranes. The efficiency terms η_m and η_w are all defined in such a way that under ideal conditions they approach 1.

The so-called Faraday current utilization ξ_F is the most important one and can be rather low due to the poor membrane selectivity at high electrolyte concentrations and the poor retention of anion-exchange membranes for protons as will be discussed later [H. Strathmann et al., 2000]. The current efficiency loss due to water transport η_W can generally be neglected in electrodialytic water dissociation.

3.3.6 Membrane Co- and Counterion Selectivity

The transport rate of coions in an ion-exchange membrane is generally quite low because of the Donnan exclusion, and the transport number is close to zero with the exception of the transport rate of protons and hydroxide ions [F.G. Donnan et al., 1932]. The transport number of counter-ions in an ion-exchange membrane is generally quite high and not very different for different ions. The transport rates of different ions through a membrane are proportional to their permeability. The permeability of ions is the product of their concentration and mobility in the membrane matrix. In a solution the concentration of the ions is a given process parameter and their mobility is determined by the ion radius according to the Stokes-Einstein relation. In an ion-exchange membrane the concentration of coions is determined by the Donnan exclusion. The mobility of the ions in the membrane depends on their radius and on membrane properties. The mobility of H^+- and OH^--ions is an order of magnitude higher than that of salt ions because of the transport mechanism of protons in water which also is the reason for the high permeability of anion-exchange membranes for protons while they are nearly impermeable for other salt cations. The same mechanism also holds for the transport of hydroxide ions and thus the permeability of hydroxide ions in a cation-exchange membrane is much higher than that of salt anions.

The high permeability of protons in anion-exchange membranes and hydroxide ions in cation-exchange membranes results in poor current utilization especially at high acid and base concentrations. In the practical application of electrodialysis production of acids and bases with bipolar membranes the current utilization is generally between 0.7 and 0.8 depending on the anion-exchange membrane.

3.3.7 Processes Effecting the Electrodialysis with Bipolar Membranes

The actual energy required to operate the electrodialytic water dissociation process exceeds the energy theoretically required for the dissociation of the water and the transport of ions between the different solutions. The process efficiency is effected in several ways. Concentration polarization at the membrane and electrode surfaces will increase the voltage drop across the stack. In a well designed stack these effects, however, are generally of minor relevance. The main sources for poor overall efficiency is the low current utilization caused by the salt permeation through the bipolar membrane and by the proton and hydroxide leakage through the anion-exchange and cation-exchange membrane, respectively [R. Simons, 1990]. The relatively short useful life of the membranes under operating conditions in strong acids and bases also effects the economics of the process.

3.3.8 Operating and Investment Costs

The operating costs are mainly determined by the energy consumption required for the water dissociation process and the transfer of ions which is proportional to the amount of produced acid and base as shown in Equation 5.46 and by maintenance costs. The energy for pumping the solution through the stacks is relatively low compared to the energy consumption for the water dissociation. The maintenance costs are generally expressed as a percentage of the investment costs. The energy dissipation due to the electrical resistance of the cell is increasing with the current density, since the electric energy for a given resistance is proportional to the square of the current, whereas the ion transfer is directly proportional to the current. Hence, the energy consumption for the production of a given amount of product increases linearly with the current density as shown in Equation 5.46. In electrodialysis with bipolar membranes the current density is limited by the water diffusion into the transition region of the bipolar membrane. As shown in Figure 5.17 the limitation of current density due to the water diffusion in bipolar membranes is in the order of 3000 to 5000 A m^{-2}. In practical applications, however, the current density is generally less than 1000 A m^{-2} since high current densities result in high energy consumption and high heat generation in the stack which can damage the membrane.

The investment costs are proportional to the membrane area required for a given amount of products and the useful life of the membranes and other plant components under operating conditions. The required membrane area for a given plant capacity depends on the applied current density. The required membrane area is according to Equation 5.40 inverse proportional to the applied current density. The life of the membranes and especially that of the bipolar membrane is quite short when the process is used for the production of acids and bases because of the exposure of the membranes to strong acids and bases and due to the heat generation in the process. This leads to rather high investment costs in many applications of bipolar membranes compared to conventional electrodialysis.

3.3.9 The Total Electrodialytic Water Dissociation Process Costs

The total costs of the electrodialytic water dissociation with bipolar membranes are the sum of fixed charges associated with the amortization of the plant investment costs and of the operating costs which include energy and maintenance costs and all pre- and post-treatment procedures. The total costs are a function of the membrane properties, of the feed solution composition, the required acid and base concentrations, and several process and equipment design parameters such as stack construction and operating current density. A dominant role in the total water dissociation process costs plays the applied current density since it effects the investment costs as well as the energy costs directly. The required membrane area is decreasing with increasing current density while the consumed energy is increasing with the current density. Thus, the total product costs which are the sum of energy costs, amortization and maintenance costs are a function of the current density and will reach a minimum at a certain current density. This is schematically shown in Figure 5.18. Here, the total process costs, the energy costs, the amortization on membrane and hardware investment, and the maintenance costs are shown as function of the applied current density.

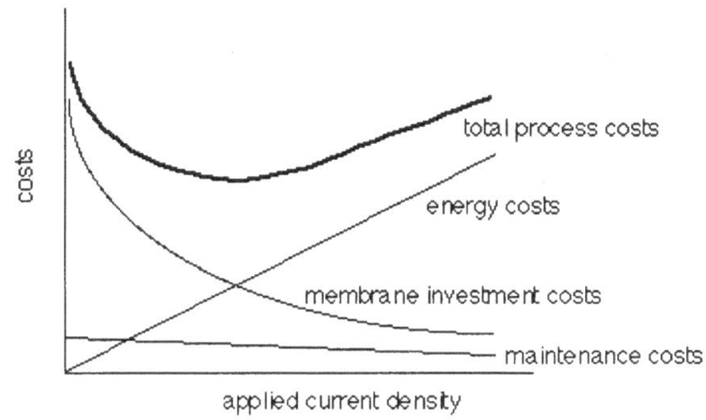

Fig. 5.18 Schematic diagram illustrating the process costs in electrodialytic water dissociation with bipolar membranes as a function of the applied current density.

Figure 5.18 shows that for a given plant capacity the energy costs increase linearly with the current density, while the membrane investment costs and, therefore, also the total investment related costs decrease in a hyperbolic function with the current density. The maintenance costs are related to the total installed membrane area and will decrease also with the decreasing membrane area, i.e. with increasing current density. The total costs which are the sum of operating and investment related costs will reach a minimum at a certain current density representing the optimum operation as far as the applied current density is concerned.

The optimum operating current density in an electrodialytic water dissociation process depends to a large extent on the equipment, and especially the membrane costs and life, and on the cost of energy. The current density to be applied in electrodialytic water dissociation with bipolar membranes is determined mainly by economical considerations. However, it should not exceed the limiting current density caused by water transport limitations as pointed out earlier. In practical applications of the electrodialytic water dissociation process the applied current density usually does not exceed 1000 A m^{-2}.

3.4 Costs Calculation for a Bipolar Electrodialysis Process

The total costs of the production of acids or bases with bipolar membrane electrodialysis depends on the application and a number of boundary conditions which are specific for this application. As a typical example the costs for the production of an acid, i.e. HCl and a base, i.e. NaOH from the corresponding salt, i.e. NaCl shall be calculated. For this calculation a number of approximations and assumptions concerning the investment costs have been made. Therefore, the calculation does not represent a true practically verified application. The calculation is described in more detail elsewhere [H. Strathmann et al., 2000].

3.4.1 Example of Designing a Bipolar Membrane Electrodialysis Plant

As an example the production of 1 m^3 h^{-1} of a 2 n acid and base from a 2.5 n NaCl solution in the stack composed of three cell units will be discussed. The scheme of the process is shown in Figure 5.17. The cell dimensions and the flow velocities of the different solutions in the stack are assumed to be identical. The flow of all three solutions is cocurrent. It is furthermore assumed that the process is continuous without feed and bleed of the salt solution or the acid and base. Based on the mass balance illustrated in Figure 5.19 and the stack dimensions and other input data for the membrane properties and the applied current density and current utilization the required membrane area, the process path length, and the energy consumption for the process can be calculated.

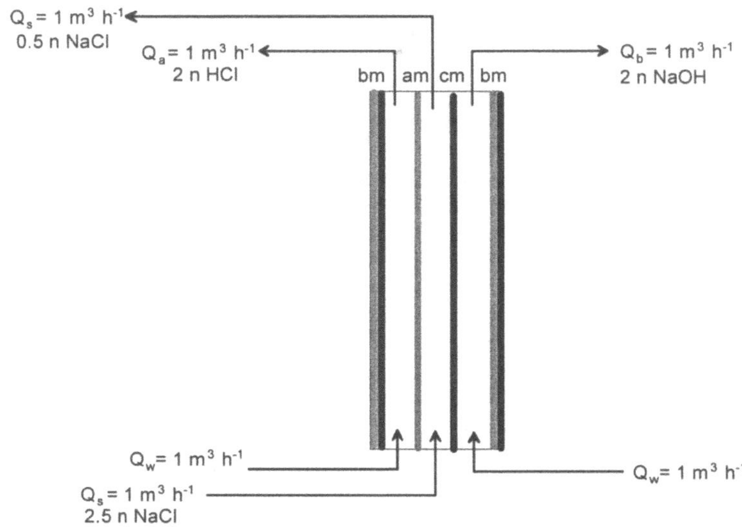

Fig. 5.19 Process flow scheme of a bipolar membrane electrodialysis unit for the production of 1 m^3 2n HCl and 1 m^3 2n NaOH per hour from a 2.5 n NaCl solution.

The calculation of the design parameters and production costs of acid and base of an electrodialytic water dissociation plant with bipolar membranes is based on a mass balance between the different flow streams in the electrodialysis stack. The input data for the calculation consist of values related to the plant capacity, membrane and solution properties, electrodialysis stack design, stack operating strategies, etc. Output data are process path length, required membrane area, cell units per stack number of stacks, voltage drop per stack, limiting current density, total current, and energy consumption. Based on cost relevant data such as membrane costs, energy costs, charges on investment costs, etc. the final product costs can be estimated. The calculation scheme is very similar to that applied for designing an electrodialysis desalination plant which has been discussed earlier. However, the limiting current density has not the same relevance in bipolar membrane electrodialysis as it has in conventional electrodialysis, since the limiting current density caused by limited water transport into the bipolar membrane is so high that in practical applications it will generally not be exceeded. The same is true for concentration polarization effects. A major problem in electrodialysis however, is the high energy input into the stack which must be dissipated as heat. The design of an bipolar membrane electrodialysis stack must take this into account.

Process input data	
plant capacity (acid and base production)	1 m^3h^{-1}
feed solution	2.5 n NaCl
acid and base concentration	2 n HCl and 2 n NaOH
cation-exchange membrane resistance	2.0 Ω cm^2
anion-exchange membrane resistance	2.0 Ω cm^2
bipolar membrane resistance	3.5 Ω cm^2
transition region resistance	3.6 Ω cm^2
applied current density	1000 A m^{-2}
cell thickness	1×10^{-3} m
cell width	0.5 m
cell length	0.5 m
flow velocity in the stack	2×10^{-2} ms^{-1}
ion-exchange membrane costs	US \$ 100.- m^{-2}
bipolar membrane costs	US \$ 500.- m^{-2}
current utilization	0.7
maintenance costs	10 % of investment costs
stack costs	1.5 × membrane costs
peripheral equipment costs	0.5 × stack costs
membrane and peripheral equipment life	2 years
energy costs	0.1 US \$ kWh^{-1}
plant operation	24 h d^{-1}, 320 d y^{-1}
Basic output data	
required cell unit area	77 m^2
process path length	5.5 m
number of stacks in series	11
cell units per stack	28
energy consumption for acid and base production	161 kWh m^{-3}
additional energy for pumping and controls	2 kWh m^{-3}
voltage drop per stack & electrodes	65 V
average current through a stack	250 A
Cost data	
membrane costs	53 900.- US \$
stack costs	80 850.- US \$
peripheral equipment costs	40 425.- US \$
total investment	121 275.- US \$
charges on investment costs	60 6375.- US \$ y^{-1}
maintenance costs	12 127.-US \$ y^{-1}
energy costs	125 184 US \$ y^{-1}
total costs for the production of 2n HCl and 2n NaOH	24.20 US \$ m^{-3}

The example calculation shows that the total production costs are dominated by the energy costs and by the costs of the bipolar membranes. In a commercial plant the applied current density must be determined by a cost from a plot of the total costs versus the applied current density.

3.5 Operational Problems in Bipolar Membrane Electrodialysis

A major problem in the practical application of bipolar membrane electrodialysis for the production of acids and bases from the corresponding salts is caused by the limited permselectivity of the bipolar membrane which can result in a significant salt leakage through the bipolar membrane into the product as indicated in Figure 5.20, which shows the salt concentration in the produced acid and base as a function of product concentration. The experimental data are determined in a laboratory test cell with a 1 molar Na_2SO_4 solution as feed and an experimental bipolar membrane [H.J. Rapp, 1995]. At high product concentrations the permselectivity of the bipolar membrane is drastically decreased due to the decreased Donnan exclusion, and the contamination of the acid as well as the base with salt is significant and is a problem when high purity products are required.

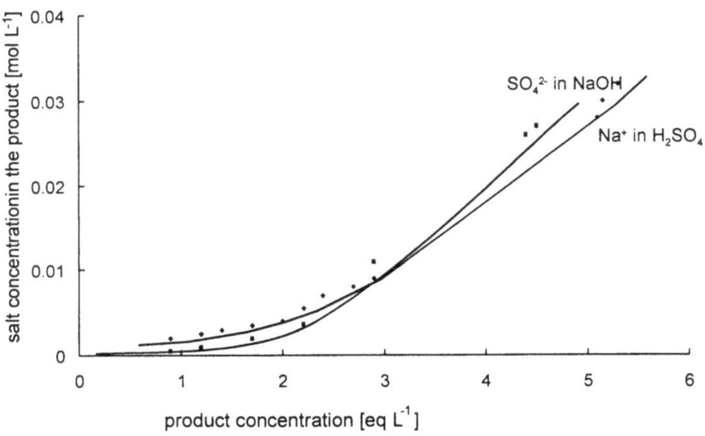

Fig. 5.20 Experimentally determined salt contamination as a function of the acid and base concentration produced by electrodialytic water dissociation with a bipolar membrane and a 1 m Na_2SO_4 feed solution [H.J. Rapp, 1995].

Another problem in the practical application of bipolar membranes for the production of acids and bases is scaling and fouling. Any di- or multivalent cations will precipitate in the alkaline solution and must therefore be removed from the feed solution. The same is true for organic components which can precipitate at extreme pH-values.

Another problem is the poor current utilization which is mainly caused by the proton and hydroxide ion leakage through the monopolar membranes. At high acid and base concentrations the retention of the anion-exchange membrane for protons and that of the cation-exchange membrane for the hydroxide ions is very poor and the current utilization reaches an economically unacceptable low value. However, the development of the so-called acid and base blocker membranes have resulted in a substantial improvement of the current utilization in bipolar membrane electrodialysis [G. Pourcelly et al., 1994].

4 DIFFUSION DIALYSIS PROCESS AND EQUIPMENT DESIGN

Today the main application of diffusion dialysis is the recovery of acids or bases from spent pickling solutions which are generated in the metal refining and electroplating industry. The principles of the process have been discussed earlier. The concept is based on diffusion of components through a semipermeable membrane under the driving force or a chemical potential gradient. Due to the permeation selectivity of the membrane certain feed solution components will be removed from a feed and transferred into a receiving or stripping solution while other components are retained by the membrane.

If ion-exchange membranes are used to separate ions from a feed solution the process is referred to as diffusion dialysis. The transport mechanism in diffusion dialysis is generally more complex than in conventional dialysis due to the electrostatic interaction between positive and negative charges and the electroneutrality requirement. The process and equipment design, however, is very similar to that used in conventional dialysis. Since the driving forces in the main application of diffusion dialysis, i.e. the recovery of acid from a mixture with salts, for the diffusion of the components to be removed are generally rather low, the process is slow and requires a relatively large membrane area for a certain plant capacity. But operating costs are very low. Thus, the economy of diffusion dialysis is dominated by the investment costs and the properties of the membranes, and the stack design play a dominant role in the practical application of diffusion dialysis.

4.1 System Components and Dialyser Design Aspects

A typical diffusion dialysis plant is shown in the process flow diagram of Figure 5.21.

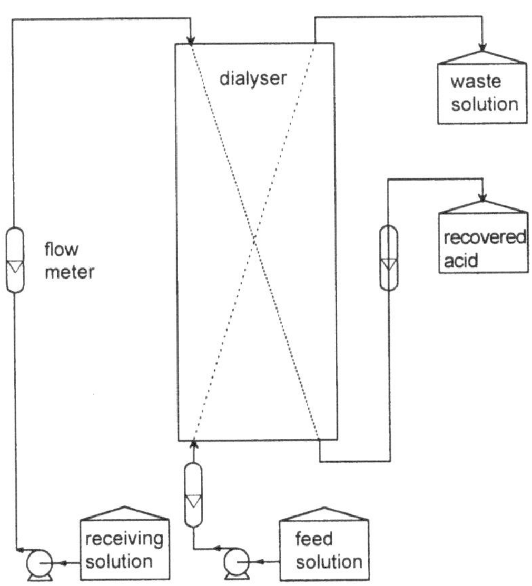

Fig. 5.21 Flow diagram of the diffusion dialysis process.

Depending on the specific application the diffusion dialysis plant consists of a pretreatment unit and the actual diffusion dialysis stack with pumps, filter, flow control devices, and concentration monitoring instrumentation.

The main component in the process is the dialysis stack. The design of a dialyser depends upon the intended use. In conventional dialysers such as the artificial kidney mainly hollow fiber type membranes are used today because of packing density and production costs. In industrial type diffusion dialysis usually flat sheet membranes are installed in a stack as indicated in Figure 5.22.

The construction of a diffusion dialyser stack is very similar to a conventional electrodialysis stack. However, the membranes used in the stack and the mode of operation are different. As indicated in Figure 5.22, which shows a diffusion dialysis stack to be used for the recovery of an acid from a salt mixture, only anion-exchange membranes are installed in the stack and the flow of the receiving and the feed solutions is counter-current. The linear flow velocities in a diffusion dialysis stack are usually quite low, i.e. generally less than a few cm s^{-1}. To design a diffusion dialysis stack in terms of the process-path length, flow velocities, feed solution concentration, product recovery rate, etc. a complete mass balance taking into account all mass transport between the feed and stripping solution is necessary as indicated by Equations 4.128 and 4.129 or Equations 4.135 to 4.140.

4.1.1 The Diffusion Dialysis Stack

A diffusion dialyser consists of a number of stacks depending on the required process path length to achieve a certain acid recovery rate. Each stack has between 200 and 400 anion-exchange membranes arranged in parallel which form individual cells. A feed containing, e.g. a mixture of an acid and a salt, and a receiving solution, generally pure water, are pumped through a manifold into alternating cells as indicated in Figure 5.22.

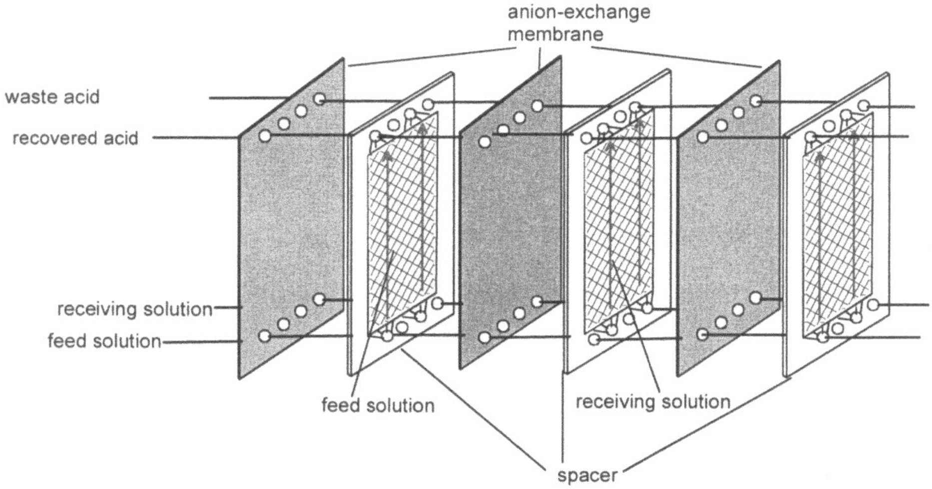

Fig. 5.22 Schematic drawing illustrating the design of a diffusion dialysis stack.

Due to chemical potential gradients acid is transported from the feed solution into the receiving solution. At the exit of the final stack a certain amount of the acid is recovered from the feed solution for re-use while the salts are retained and discharged as waste.

For a system operated in counter-current flow of the feed and stripping solutions in a stack, where all cells have the same geometry and identical flow velocities, a more simple relation can be used to describe the mass transport in diffusion dialysis instead of a complete mass and momentum balance [S.J. Oh et al., 2000]. When a pure solvent is used as a stripping solution, all osmotic effects are neglected and the concentrations are assumed to be identical with the concentration of the individual components. The material balance for a given component is given by:

$$C_i^f = C_i^d + C_i^p \tag{5.50}$$

Here C_i^f, and C_i^d are the concentrations of the component i in the feed at the cell entrance and the cell exit, C_i^p is the concentration of the component i in the product solution, i.e. the concentration in the stripping solution at the cell exit.

The amount of the component i removed from the feed solution and transferred across the membrane is equal to the amount of the component i received in the stripping solution. Thus is:

$$N_i = k^* A^m \Delta C_i = -Q^m C_i^p \tag{5.51}$$

Here k* is a mass transfer coefficient describing the mass transport through the membrane and the adjacent laminar boundary layers in the bulk solutions at the membrane surface, A^m is the membrane area in the diffusion cell, and ΔC_i is the concentration difference between the two solutions, C_i^p is the product concentration, and Q^m is the volume flow rate through the membrane.

Since the concentrations in the bulk solutions change during the passage through the cell ΔC_i also changes with the length of the cell. To calculate the mass transport between the two solutions an average $\Delta\overline{C}_i$ must be introduced which under the above stated conditions, i.e. counter-current flow in identical cells at equal velocities, is given by the arithmetic mean concentration difference:

$$\Delta\overline{C}_i = \frac{C_i^f + C_i^d - C_i^p}{2} \tag{5.52}$$

Applying the material balance leads to:

$$\Delta C_i = C_i^f - C_i^p \tag{5.53}$$

Introducing Equation 5.53 into Equation 5.51 and rearranging gives a relation for the fractional recovery of a component i in the stripping solution:

$$k^* A^m \left(C_i^f - C_i^p\right) = -Q^m C_i^p \tag{5.54}$$

and

$$\Delta = \frac{C_i^p}{C_i^f} = \frac{k^*}{k^* + \dfrac{Q^m}{A^m}} \tag{5.55}$$

Here C_i^f and C_i^p are the concentrations of the component i in the feed and the product, k* is a mass transfer coefficient, A^m is the membrane area, Q^m is the volume flow rate through the membrane, and Δ is the product recovery rate.

Equation 5.55 gives a simple relation between the concentration of a component i in the feed and stripping solution, the recovery rate, the total amount of the component i transferred through the membrane, and the required membrane area for a given dialysis constant.

4.1.2 The Dialyser Membrane

A key component in diffusion dialysis for recovering acids from a feed solution is the anion-exchange membrane. Many commercially available anion-exchange membranes have the required acid/salt separation properties but lack the chemical stability. A membrane especially developed for this application is the Néosepta AFN. This membrane provides the required chemical stability in strong acid solutions [Y. Kobuchi et al., 1984]. The thickness of the membrane is ca. 160 mm, its ion-exchange capacity is 2.0 m eq g^{-1}. This membrane not only shows a relatively high dialysis coefficient but also good separation factors between acids and salts as indicated in Table 5.1.

Table 5.1 Dialysis coefficients of an AFN-membrane for various salts and acids and the separation factor between various acids and salts [Y. Kobuchi et al., 1984].

salt-acid mixture	acid concentration [mol m^{-3}]	salt concentration [mol m^{-3}]	$k^*_{(acid)}$ [m s^{-1}]	$k^*_{(salt)}$ [m s^{-1}]	$k^*_{(salt)}/k^*_{(acid)}$
HCl-NaCl	2.0	2.0	2.4 10^{-6}	1.3 10^{-7}	0.054
HCl-$FeCl_2$	2.0	0.5	2.4 10^{-6}	4.7 10^{-8}	0.02
H_2SO_4-Na_2SO_4	1.0	0.5	9.7 10^{-7}	3.9 10^{-8}	0.04
H_2SO_4-$FeSO_4$	1.0	0.5	1.0 10^{-6}	1.0 10^{-8}	0.01
HNO_3-$Cu(NO_3)_2$	1.5	0.8	2.6 10^{-6}	4.7 10^{-8}	0.018
H_3PO_4-$MgHPO_4$	0.5	0.1	2.4 10^{-7}	5.0 10^{-9}	0.021

The dialysis coefficient of the acid as well as the salt is temperature and concentration dependent and increase with increasing temperature and overall electrolyte concentration. The selectivity of a diffusion dialysis membrane is also concentration dependent and decreases with increasing electrolyte concentration.

4.2 Problems Related to the Design of a Diffusion Dialysis Plant

Modeling the diffusion dialysis process and designing a plant for a typical practical industrial effluent is complicated due to the complex composition of the feed solution which generally contains a multitude of different ions in a relatively high concentration. The activity of the various ions in the solution is generally rather different from their concentration. These activities are usually not readily available in the literature and difficult to measure or calculate. However, the activities provide not only the driving force for the transport of the different components but they also determine the osmotic water flux which can have a drastic effect on the overall process performance of a plant. Furthermore, the anion-exchange membrane has different permeabilities for different anions. Since the transport rates of all ions are coupled by the electroneutrality requirement it is difficult to determine the recovery rates and concentrations in the receiving and the depleted solution on a theoretical base for a feed solution containing more than one acid. Since salts and acids generally have common anions the concentration of the acid in the receiving solution can exceed that of the feed solution significantly. For a single acid in the feed solution the recovery rate of the acid cannot exceed 1. However, for a feed solution containing more than one acid the recovery rate of the different acids is often different when the anions of the acids have different permeabilities in the anion-exchange membrane. In this case, the recovery rate of one acid might exceed that of the others significantly and may be higher than 1. [Y. Kobuchi et al., 1984]. Equations 5.50 to 5.55 can only be used as a first approximation for designing a diffusion dialysis process.

4.3 Practical Example of a Diffusion Dialysis Plant

Diffusion dialysis is not nearly as widely used as conventional electrodialysis. There are only very few industrial scale plants described in the literature. One of the first large scale application of diffusion dialysis has been in the treatment of spent pickling solutions used in the steel industry. Many plants are used to recover HNO_3 and HF from a waste solution containing these acids in a mixture with Fe^{2+}-ions. The flow scheme and mass balance of a plant with a capacity of $1 m^3 h^{-1}$ spent pickling solution is shown in the simplified diagram of Figure 5.23 as a typical example of a commercial size plant [Y. Kobuchi et al., 1986].

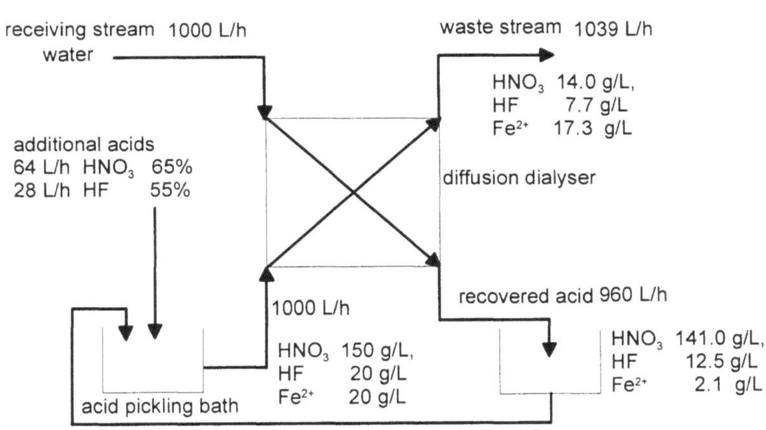

Fig.5.23 Mass balance flow diagram of a diffusion dialysis process to recover HNO_3 and HF from a spent steel pickling solution.

The main data concerning the 1 $m^3 h^{-1}$ capacity plant are summarized as follows:

feed solution concentrations	
HNO_3	150 $g L^{-1}$
HF	20 $g L^{-1}$
Fe^{2+}	20 $g L^{-1}$
receiving solution (water)	1 $m^3 h^{-1}$
recovered acids	0.960 m^3 h
recovery rates	
HNO_3	90 %
HF	60 %
waste solution discharge rate and composition	1.040 $m^3 h^{-1}$
Fe^{2+}	17.3 $g L^{-1}$
HNO_3	14 $g L^{-1}$
HF	7.7 $g L^{-1}$
required membrane area	1000 m^2

4.3.1 Diffusion Dialysis Process Costs

As in any other process the total costs in diffusion dialysis are the sum of charges related to capital investment and operating and energy costs. However, the energy costs are very low since electrical energy is required only for pumping the solutions through the stack. Since the flow velocities are extremely low, i.e. in the range of several cm per minute the pressure losses and thus the required pumping energy is very low. The operation of the process is also quite simple and operating costs are minimal. The main cost factor are charges related to the capital investment which are quite high due to the fact that the diffusion of the acids are slow and a large membrane area is required for a relatively low capacity plant. The useful membrane life under operating condition in an environment of strong acids is limited to ca. 1-2 years. The process costs are, therefore, determined by membrane costs and membrane life.

In some cases the feed solution coming from a pickling bath may be contaminated with suspended solids and large particles which must be removed to values of less than 2-3 mg L^{-1} to avoid a blocking of the membranes and passage of the solution through the cells of the stack. For the pretreatment cartridge filters and in some cases settling tanks are used to precipitate solid materials. Also inhibitors such as fats and oils which are often contained in the pickling waste solution can cause severe membrane fouling. In such cases pre-treatment with active carbon may be required. For the proper function of a diffusion dialysis stack an extensive pretreatment may be necessary, and costs for the pretreatment can become significant for the overall acid recovery costs.

4.3.2 Operational Problems and Limitations in Diffusion Dialyis

A major concern in operating an industrial diffusion dialyser is the flow distribution of the solutions in the dialysis stack. Because of the low flow velocities it is difficult to achieve uniform flow distribution and to avoid concentration polarization effect. A potential problem also exists due to gas bubbles trapped in the downward flowing solution which are not removed by the slowly flowing solution and which are eventually trapped in the upper part of a cell. To avoid the problems caused by gas bubbles the solutions can be degassed either by heating, by hydrophobic mirofiltration membranes, or by reversing the flow directions in the stack for certain time periods [N. Shigekuni et al., 1979]. Membrane fouling due to adsorption of organic material can be a problem which makes periodical rinsing and cleaning cycles of the entire stack necessary.

Another problem sometimes encountered in diffusion dialysis of acids is a rise in temperature due to the heat of dilution of the acids. This heat can lead to a damage of the membranes [Y. Aoki et al., 1991].

The formation of complexes is a problem in the presence of certain metal ions. For example, Zn^{2+}-ions form $ZnCl_3^-$- and $ZnCl_4^-$-complexes with HCl that behave as anions and permeate an anion-exchange membrane much faster than the Zn^{2+}-ions. Negatively charged metal complexes can lead to a significant metal ion transfer into the recovered acid and can impair the separation efficiency of the process.

There are also some intrinsic limitations of the diffusion dialysis process. A single acid in a mixture with its salt cannot be completely recovered because a certain concentration difference between the feed solution and the receiving solution is needed as driving force for the acid transport through the membrane. Also, the salt will diffuse into the acid because of the limited permselectivity of the anion-exchange membranes at high ion concentrations. This means that the recovered acid will always be contaminated with salt. This, however, is no problem when the acid can directly be recycled in the pickling process.

5 CONTINUOUS ELECTRODEIONIZATION PROCESS DESIGN

Continuous electrodeionization is widely used today for the preparation of high quality deionized water for the preparation of ultra pure water in the electronic industry or in analytical laboratories. The process is described in some detail in the patent literature and company brochures. There are also some variations of the basic design as far as the distribution of the ion-exchange resin is concerned. In some cases the diluate cell is filled with a mixed bed ion-exchange resin, in other cases the cation- and anion-exchange resins are placed in series in the cell. More recently also bipolar membranes are used in the process. A performance comparison of the different stack constructions is described in the literature [S. Thate, 2002].

5.1 System Components and Design Aspects

The process design and the different hardware components needed in electrodeionization are very similar to that used in conventional electrodialysis. The main difference is the stack construction. The diluate cell and some times also the concentrate cell is filled with an ion-exchange resin which effects the flow distribution and the pressure drop of the solution in the cell drastically. Therefore, the dimensions of the diluate and concentrate cells are generally quite different. The different concepts used for the distribution of the cat- and anions in the cell is illustrated in Figure 5.22. In the conventional electrodeionization process the diluate cell is filled with a mixed bed ion-exchange resin with a ratio of cation- to anion-exchange resin being close to 1 as shown in Figure 5.22 a). The cation- and anion-exchange resins can also be placed in separate beds in series in a stack as is illustrated in Figure 5.22 b). Finally, the bipolar membranes can be placed between the cation- and anion-exchange resin containing cells to provide protons and hydroxide ions for the regeneration of the ion-exchange resins. This stack arrangement is illustrated in Figure 5.22 c)

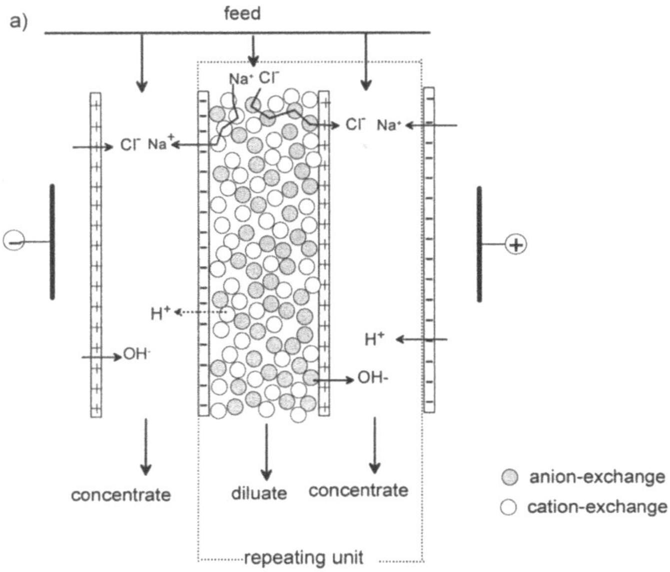

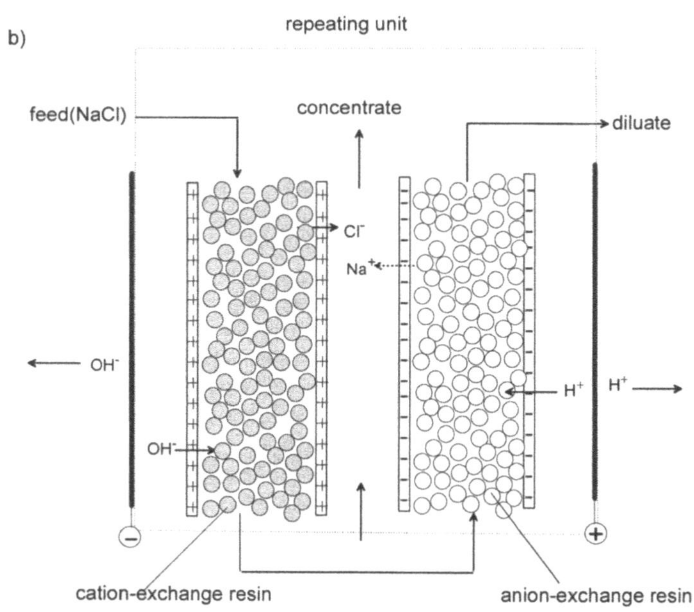

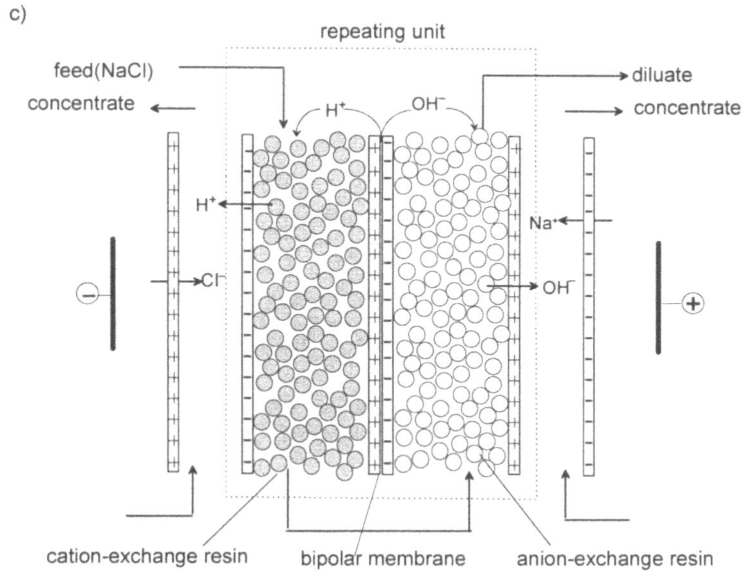

Fig 5.24 Schematic drawing illustrating different stack concepts used in continuous electrodeionization; a conventional stack with diluate cells filled with a mixed bed ion-exchange resin, b) stack with cation-exchange and anion-exchange resins in different diluate cells and generation of the ion-exchange resins by H^+- and OH^- -ions generated at the anode and cathode, c) stack with cation-exchange and anion-exchange resins in different diluate cells and generation of the ion-exchange resins by H^+- and OH^- -ions generated in a bipolar membrane.

The main difference between the electrodeionization system with the mixed bed ion-exchange resin and the system with separate beds is that in mixed bed electrodeionization systems anions and cations are simultaneously removed from the feed while the solution leaving the diluate cell is neutral. In the electrodeionization system with separated cation- and anion-exchange beds as indicated in Figure 5.24 b) and c) the cations will first be exchanged by the protons generated at the anode or the bipolar membrane, and the solution leaving the cation-exchange bed is acidic. This solution is then passed through the cell with the anion-exchange resin where the anions are exchanged by the OH^--ions generated at the cathode or in the bipolar membrane, and at the exit of the anion-exchange filled cell the solution is also neutral. The only difference between the systems shown in Figure 5.24 b) and c) is that the H^+- and OH^--ions needed for the regeneration of the ion-exchange resins in the system illustrated in b) are produced electrolytically. while in the system shown in c) the H^+- and OH^--ions are the result of water dissociation within a bipolar membrane.

All three continuous electrodeionization systems have been studied in detail on a laboratory scale and their advantages and disadvantages clearly pointed out [S. Thate, 2002]. The mixed bed electrodeionization system is characterized by the simultaneous removal of anions and cations from the feed in one cell. The advantage of this system is that both cations and anions, which migrate from the concentrate cell back into the diluate cell due to incomplete membrane permselectivity, will be adsorbed by the corresponding ion-exchange resin in the diluate cell and transported to the adjacent concentrate cell with the results that the achievable conductivity in a mixed bed system is lower than in a system with separated ion-exchange beds. In an electrodeionization system with separated ion-exchange beds coions migrating back from the concentrate cell will not be adsorbed by the ion-exchange resin of the diluate cell and will, therefore, not completely be removed from the diluate stream.

In a system with separate ion-exchange resins, on the other hand, is the diluate in the cell which is first contacted with the feed solution either acidic or basic depending whether the first cell is filled with an anion- or a cation-exchange resin. The pH-value shift in the first ion-exchange bed has some advantage for the removal of weak acids or bases [E.J. Parsi, 1966]. Since the degree of dissociation of weak acids or bases depends on the pH-value they are better adsorbed and thus removed from the feed solution at high or low ph-value, respectively, than in a neutral environment. Therefore, the separate bed ion-exchange electro deionization system provides a better removal of weak acids or bases while in the system with the mixed bed ion-exchange resin the achievable conductivity is lower.

In the systems with separate ion-exchange resins, the H^+- and OH^--ions used for the continuous regeneration are produced externally either by electrolysis at the anode and cathode, or by water dissociation with bipolar membranes. The main difference is that the electrolytic generation of H^+- and OH^--ions is connected with the production of H_2 and O_2 gases which can lead to additional operational problems. The production of H^+- and OH^--ions by water dissociation in bipolar membranes is without the formation of gases as by-products which simplifies the stack design considerably.

However, today mostly mixed bed ion-exchange electrodeionization systems are used in industrial scale water purification plants.

5.2 Operational Problems in Practical Applications of Electrodeionization

One of the problems encountered in the electrodeionization system the mixed bed ion-exchange resin is the incomplete removal weak acids or bases. The salt migration is a mjor problem in the system with separate ion-exchange resin beds. But there are several other problems such as the uneven flow distribution in the ion-exchange resin bed which leads to poor utilization of the ion-exchange resins. The fouling of the ion-exchange resins by organic components such as humic acids and bacterial growth on the surface of the resin is a problem which requires a very thorough pretreatment of the feed solution to guarantee a long-term stability of the system. The effect of the cell geometry, i.e. the ratio of its length to width and thickness has been studied extensively and is described in various patents [A.J. Giuffrida, et al., 1990], [K.K. Siu et al., 1988], [E.J. Parsi, 1966].

6 DONNAN DIALYSIS PROCESS AND EQUIPMENT DESIGN

The principle of Donnan dialysis has been discussed before in chapter 4 in some detail. The process is in direct competition with conventional ion-exchange technology. In spite of several advantages it has not yet reached large industrial relevance. Applications that have been studied on the laboratory and pilot plant scale are the recovering and concentration of Cu^{2+}-ions from industrial effluents or from dump leaching of low grade copper oars and the removal of divalent ions for water softening.

6.1 System Components and Design Aspects

The process design and stack construction is very similar to that used in diffusion dialysis. In the case of Cu^{2+} -ion concentration generally concentrated H_2SO_4 is used as a stripping solution. The transport of the Cu^{2+}-ions is achieved by a flux of H^+-ions through a cation-exchange membrane. The flow streams of the feed and the stripping solution are counter-current in a stack similar to that used in diffusion dialysis. In water softening NaCl is often used as electrolyte in the stripping solution exchanging divalent cations such as Ca^{2+}-ions with Na^+-ions or divalent anions such as SO_4^{2-}-ions with Cl^--ions.

6.2 Problems in the Practical Application of Donnan Dialysis

The problems in the practical application of Donnan dialysis are very similar to those encountered in diffusion dialysis. These are slow exchange rates of the counterions, high osmotic water transport into the stripping solution, heat generation due to dissolution of the acids especially when sulfuric acid is used in the stripping solution, and general membrane fouling by organic matter or polyelectrolytes. But presently the main problem seems to be the relatively high investment cost for the large membrane area required in the process.

6.3 Process Costs and Economics

As in diffusion dialysis the process costs are dominated by charges related to the total investment of a given capacity plant which are determined by the costs of the membranes and their life under operating conditions. Unfortunately, there are hardly any large scale commercial units in operation which provide enough data for a reliable cost calculation.

7 OTHER ION-EXCHANGE MEMBRANE PROCESSES

In addition to the processes discussed in this chapter there are other processes utilizing ion-exchange membranes. In these processes transport through ion-exchange membranes is combined with electrochemical reactions [T. Sata, 1991]. Typical applications of electrochemical reactions which include ion-exchange membranes as an ion transporting element are in inorganic and organic synthesis or in the elctrochemical destruction of toxic materials such as polychlorinated bisphenyls. The combination of ion-exchange technology with controlled transport through ion-exchange membranes is used to recover toxic or valuable metal ions from industrial waste streams [J. Johann et.al., 1993].

8 LIST OF SYMBOLS

A	area	$[m^2]$
a	constant in limiting current density	$[A\ s\ mol^{-1}]$
b	constant	[-]
C	concentration	$[mol\ m^{-3}]$
$\overline{C}$	average concentration	$[mol\ m^{-3}]$

$^{\Delta}C_s$	salt concentration due to transport through membrane	[mol m^{-3}]
E	electrical energy	[A V s]
F	Faraday constant	[A s eq^{-1}]
I	current	[A]
i	current density	[A m^{-2}]
k_s	mass transfer coefficient	[m s^{-1}]
$k^{\times}$	mass transfer coefficient	[m s^{-1}]
L_{pp}	process path length	[m]
N	flux	[mol s^{-1}]
N	number	[-]
P	power	[V A]
Q	volume flux	[m^3 s^{-1}]
r	area resistance	[Ω cm^2]
R	gas constant	[J $^oK^{-1}$ mol^{-1}]
s	safety factor	[-]
T	transport number	[-]
T	temperature	[^{o}K]
t	time	[s]
U	voltage	[V]
u	velocity	[m s^{-1}]
x	directional coordinate	[m]
y	directional coordinate	[m]
z	directional coordinate	[m]
X	cell length	[m]
Y	cell width	[m]

Greek symbols

α	coefficient	[-]
β	coefficient	[-]
Δ	cell thickness	[m]
Δ	recovery rate	[-]
Δ	difference	[-]
η	current efficiency	[-]
ξ	current utilization	[-]
Λ	equivalent conductivity	[$\Omega^{-1}eq^{-1}$]
ξ	current utilization	[-]

Superscripts

am	anion-exchange membrane
bml	ion-exchange layers of bipolar membrane
cm	cation-exchange membrane
f	feed
c	concentrate
d	diluate
fc	feed concentrate
fd	feed diluate

i	inlet
m	membrane matrix
p	product
s	sotution
tr	transition region

Subscripts

a	acid
b	base
cell	cell
des	desalination
F	Faraday efficiency
i	component
lim	limiting current density
m	membrane
p	product
prac	practical
prop	prodution
rev	reversal
s	salt
tot	total
spec	specific
st	stack
s	salt
tot	total
theo	theoretical
w	water

LITERATURE

Aoki, Y., Hamano, T., 1991. Method and apparatus for recovering an acid from an acid containing waste liquor, US Patent 5 062 960.

Donnan, F.G., and E.A. Guggenheim, 1932. Exact thermodynamics of membrane equilibrium, Z. Phys. Chem., A 162:346-360.

Giuffrida, A.J. Jha, A.D., Ganzi, G.C., 1990. Electrodeionization apparatus and method, US Patent 4 925 541.

Huffmann, E.L., Lacey, R.E., 1972. Engineering and economic considerations in electromembrane processing, in: Industrial processing with membranes, edts.: Lacey, R.E., Loeb, S., pp. 39-55, John Wiley & Sons, New York.

Johann, J., Eigenberger, G.,1993. Elektrodialytische Regeneration von Ionenaustauscher-harzen, Chem.-Ing. Tech., 65: 75-78.

Katz, W.E., 1979. The electrodialysis reversal (DER) process, Desalination,, 28: 31-40.

Krol, J.J., Jansink, M., Wessling, M., Strathmann, H., 1998. Behavior of bipolar membranes at high current density, Sep. & Pur. Techn., 14: 41-52.

Kobuchi, Y., Motomura,H., Noma, Y., Hanada, F., 1984. Application of ion-exchange membranes to the recovery of acids by diffusion dialysis, presented at Europe-Japan Congress on Membranes and Membrane processes, June 18-22, Stresa, Italy.

Kobuchi, Y., Motomura, H., Noma, Y., Hanada, F., 1986. Application of ion-exchange membranes to the recovery of acids by diffusion dialysis, J. Membrane Sci., 27: 173-179.

Lee, H.J., Safert, F. Strathmann, H., Moon, S.H., 2002. Designing of an Electrodialysis desalination plant, Desalination, 142: 267-286.

Liu, K.J., Chlanda, F.P., Nagasubramanian, K.J., 1977. Use of bipolar membranes for generation of acid and base, J. Membrane. Sci., 2: 109-124.

Mani, K.N., 1991. Electrodialysis water-splitting technology. J. Membrane Sci., 58:117-138.

Meller, F.H., 1984. Electrodialysis(ED) and electrodialysis reversal (EDR), Ionics brochure.

Oh, S.J., Moon, S.H., Davis, T., 2000. Effects of metal ions on diffusion dialysis of inorganic acids, J. Membrane Sci., 169: 95-105.

Parsi, E.J., 1966. Removal of weakly basic substances from solution by electrodeionization, US Patent 3 291 713.

Pourcelly, G., Tugas, I., Gavach, C., 1994. Electrotransport of sulphuric acid in special anion-exchange membranes for the recovery of acids, J. Membrane Sci.,97: 99-107.

Rapp, H.J., 1995. Die Elektrodialyse mit bipolaren Membranen, Theorie und Anwendungen, Ph.D.-Thesis, University of Stuttgart, Stuttgart Germany.

Sata, T., 1991. Ion-exchange membranes and separation processes with chemical reactions, J. Appl. Electrochemistry, 21. 283-294.

Shigekuni, N., Motomura, K., 1979. Diffusion dialysis methode, Japanese Patent 54 136 580:

Schaffer, L.H., Mintz, M.S. , 1966. Electrodialysis, in: Principles of desalination, edt.: Spiegler, K.S., pp. 3-20, Academic Press, New York.

Siu, K.K., Dracut, A.D., Ganzi, G.C., 1988. Depletion compartment and spacer construction for electrodeionization apparatus, US Patent 4 747,929.

Simons, R., 1990. Development of an acid impermeable anion-exchange membrane, Desalination, 78: 297-302.

Siwak, L.R., 1992. Here is how electrodialysis reverses and why DER works, Desalination &Water Reuse, Vol. 2/4, pp. 26-30.

Spiegler K.S., 1956. Electrochemical operations, in: Ion Exchange Technology, edts.: Nachod, F.C., Schubert, J., pp 118-181, Academic Press, New York.

Strathmann, H., Krol, J.J., Rapp, H.J., Eigenberger, G., 1997. Limiting current density and water dissociation in bipolar membranes, J. Membrane Sci., 125: 123-142.

Strathmann, H., Koops, G.H., 2000 Process economics of the electrolytic water dissociation for the production of acid and base in: Bipolar membrane technology, edt.: Kemperman, A.J.B., Twente University Press, Enschede, The Netherlands.

Thate, S., 2002. Untersuchung der electrochemischen Deionisation zur Reinstwasser-herstellung, Ph.D.-Thesis, University of Stuttgart, Germany, ISBN 3-89722-911-0.

Chapter 6

Applications of Ion-Exchange Membrane Separation Processes

1 INTRODUCTION

Ion-exchange membranes separation processes are used today in a large variety of applications from water desalination by conventional electrodialysis to the production of acids and bases by electrodialysis with bipolar membranes, or the elimination of toxic components from industrial effluents. The technical and commercial relevance of the various processes is very different. Some processes such as electrodialysis are used on a large industrial scale for the desalination of water or the removal of certain electrolytes from food and chemical products. Other processes such as Donnan dialysis, bipolar membrane water dissociation or the continuous electrodeionization are used on a much smaller scale for the production of ultra pure water or for the purification and separation of certain food and pharmaceutical components, or for the treatment of industrial waste streams to recover toxic and valuable materials such as heavy metals or radioactive isotopes. Some of the applications can be considered as state-of-the-art technology such as the production of potable water from brackish water sources, or the pre-concentration of NaCl from sea water for the production of table salt, while other applications are still in the pilot plant or even laboratory stage such as the production of acids and bases by electrodialysis with bipolar membranes. Often ion-exchange membrane separation processes are in competition with other mass separation techniques and their application is determined by economic considerations. In some applications ion-exchange membrane processes provide higher quality products or are more environmentally friendly and will therefore be used in spite of a cost disadvantage. Also increasing costs of raw materials and environmental awareness have increased the application of ion-exchange membrane separation processes especially in highly industrialized and densely populated countries.

2 PRACTICAL APPLICATION OF ELECTRODIALYSIS

Conventional electrodialysis is today commercially by far the most relevant ion-exchange membrane separation process which was developed first for the desalination of brackish water to produce potable water. Currently, water desalination and the pre-concentration of NaCl from sea water are still the most important large scale applications of conventional electrodialysis. But other applications in the food and chemical industry such as the demineralization of whey or de-acidification of fruit juice are becoming increasingly more important [T. Kawahara, 1992]. The production of high quality process water for power plants and for the electronic industry by electrodialysis in combination with other processes such as ion-exchange or reverse osmosis has become today a state-of-the-art technology. Electrodialysis is also widely used to recover toxic or valuable components from effluents of galvanic and metal surface finishing processes and from dump leaching waste water. Another large scale application of electrodialysis is the preconcentration of sea water for the production of table salt. This application, however, is limited to countries which do not have mineral salt deposits such as Japan or Korea. In many applications electrodialysis is in competition with reverse osmosis, nanofiltration or conventional ion-exchange. The competitiveness of electrodialysis in a given task depends on the composition of the feed solution composition, the required quality of the product, and on local considerations such as energy or waste disposal costs.

Some of the more important large scale industrial applications of conventional electrodialysis and the stack and process design used in this application as well as the major limitations are listed in Table 6.1.

Table 6.1 Industrial application of conventional electrodialysis.

industrial applications	stack and process design	status of application	limitations	key problems
brackish water desalination	sheet flow, tortuous path stack, reverse polarity	commercial	concentration of feed and costs	scaling, costs
boiler feed water production	sheet flow, tortuous path stack, reverse polarity	commercial	product water quality and costs	costs
waste and process water treatment	sheet flow stack, unidirectional	commercial	membrane properties and costs	membrane fouling
ultra pure water production	sheet flow, tortuous path stack, reverse polarity	commercial	product water quality and costs	membrane biofouling
demineralization of food products	sheet flow, or tortuous path stack, unidirectional	commercial or pilot phase	membrane selectivity and costs	membrane fouling, product loss
table salt production	sheet flow stack, unidirectional	commercial	costs	membrane fouling
concentration of reverse osmosis brine	sheet flow stack, unidirectional	pilot phase	costs	waste disposal

2.1 Brackish Water Desalination by Electrodialysis

The production of potable water from brackish water is the largest single application of electrodialysis. The world-wide installed capacity of electrodialysis plants for brackish water desalination is well in excess of 10^6 m^3 d^{-1} [K., Wangnick, 1992]. The majority of these plants is operated in reversed polarity mode which is referred to as electrodialysis reversal. In brackish water desalination, electrodialysis is competing directly with reverse osmosis. However, in a certain range of feed water salt composition electrodialysis has a clear economic advantage over other desalination processes. Electrodialysis is mainly used in small to medium size plants with capacities of less than a few 100 m^3 d^{-1} to more than 20 000 m^3 d^{-1}, and brackish water salinity of 1000 to 5000 mg L^{-1} total dissolved solids. Since in electrodialysis both the energy consumption and the required membrane area are strongly increasing with increasing feed water concentration reverse osmosis is considered to have an economic advantage for the desalination of brackish water with total dissolved salts in excess of 10 000 mg L^{-1}. Nevertheless, electrodialysis has been used successfully to desalinate well water with a total salt concentration of 36 000 mg L^{-1} [H.C. Valcour, 1985], [M. Turek, 2002]. The advantages of electrodialysis compared to reverse osmosis are:

- high water recovery rates even for raw water with high sulfate content,
- long useful life of membranes due tohigher chemical and mechanical stability,
- operation at elevated temperatures up to 50°C possible,
- less membrane fouling or scaling due toprocess reversal,
- less raw water pretreatment,
- ion-exchange membranes tolerate higher level of chlorine and extreme pH-values,
- the process can easily be adjusted to varying feed water quality,
- easy start-up and shut-down of the process for intermittent operation.

The desalination of brackish water by electrodialysis has another disadvantage compared to reverse osmosis desalination because neutral toxic components such as viruses or bacteria are not removed, and the product water may require a post-treatment procedure when used as potable water supply. Furthermore, the generation of chlorine gas at the anode can lead to corrosion problems in the surrounding of the plant if the venting is insufficient. In spite of the fact that electrodialysis reversal is significantly less sensitive to membrane fouling than reverse osmosis some pretreatment of the feed water is required to achieve trouble free operation. Especially, the iron and manganese concentrations must be kept below 0.3 and 0.05 mgL^{-1}. The removal of iron and manganese is achieved by oxidation of the dissolved iron and manganese with potassium permanganate prior to precipitation and filtration. Brackish water desalination plants used for the production of potable water can be rather different in size. Container mounted units with a capacity of 50 to 300 m^3d^{-1} are often used to supply isolated hotels, small islands or the labor force in camps in isolated dessert areas with potable water. Larger installations with capacities between 10,000 to 30,000 m^3d^{-1} are used for municipal water supply or in industrial applications. A typical large scale electrodialysis water desalination plant is shown in the photograph of Figure 6.1. The plant is built by Ionics Incorporated, Watertown, MA, USA using the Aquamite XX EDR system.

Fig. 6.1 Industrial size electrodialysis reversal (EDR) drinking water plant (photograph courtesy of Ionics Incorporated).

2.2 Production of Industrial Water by Electrodialysis

Depending on the application industrial process water must meet certain quality standards in terms of total dissolved solids and colloidal material. Generally, surface or well water must be purified and demineralized before it can be used in certain production processes or as boiler feed water. Since the amount of water used in industrial processes is often quite large cost effective treatment procedures are mandatory. Traditionally, precipitation, filtration, and ion-exchange are used in the production of industrial water. Today, these processes are replaced or complimented more and more by microfiltration, reverse osmosis, and electrodialysis. Major applications of electrodialysis in industrial water processing include:

- predemineralization of boiler feed water,
- desalination of cooling water for reuse,
- desalination of process water for chemical production processes,
- desalination of contaminated industrial water for reuse.

In most of these applications electrodialysis is in competition with conventional ion-exchange techniques and reverse osmosis. However, for certain raw water compositions electrodialysis has a clear cost advantage over the competing processes.

2.2.1 Electrodialysis Predemineralization of Boiler Feed Water

For the preparation of boiler feed water large quantities of raw water must be demineralized. The level of deionization may require multiple or redundant ion-exchange techniques to insure that the required quality standard is met on a consistent basis. The costs associated with the deionization depend on a number of factors such as the feed water composition and variability, the extent of the required deionization, and the operation of the ion-exchange system. The most significant costs are attributed to the regeneration of the ion-exchange resin. With pre-demineralization of the feed water by electrodialysis the regeneration costs can drastically be reduced. This is illustrated in Table 6.2 which shows the production costs of 1000 m^3 d^{-1} boiler feed water based on the use of ion-exchange only and on a combination of ion-exchange and electrodialysis.

Table 6.2 Cost comparison of a water deionization plant with a capacity of 1000 m^3 d^{-1} operated by conventional ion-exchange and by ion-exchange with electrodialysis pretreatment.

feed salt concentration [mg L^{-1}]	300	600	900
without electrodialysis [US \$ m^{-3}]	0.36	0.72	1.08
electrodialysis operating costs [US \$ m^{-3}]	0.14	0.20	0.25
costs for ion-exchange regeneration with electrodialysis [US \$ m^{-3}]	0.04	0.08	0.12
regeneration costs saving with electodialysis predemineralization [US \$ m^{-3}]	0.18	0.44	0.71
annual saving for a plant capacity of 1000 m^3 d^{-1} and 360 d y^{-1} operation [US \$]	64800.-	158400.-	255600.-

The cost data listed in Table 6.2 are calculated for a given set of conditions and prices for the regeneration chemicals and energy, which may vary from location to location. Therefore, substantial variations in cost calculations can be expected for different plants. However, in spite of the uncertainties of the calculation substantial cost savings can be expected when in the preparation of deionized water electrodialysis is used as pre-demineralization step prior to the ion-exchange procedure.

Another benefit of the predemineralization by electrodialysis is a significant reduction in waste disposal. A simplified flow diagram of a typical boiler feed water treatment plant with an electrodialysis predemineralization unit is shown in Figure 6.2. A cost analysis shows that the boiler feed water treatment with pre-demineralization by electrodialysis leads to significant savings in regeneration chemicals. The savings increase with increasing feed water salinity. In Table 6.2 the annual costs of water demineralization by ion-exchange and by ion-exchange with predemineralization with electrodialysis are listed for a plant capacity of 1000 m^3 per day and 3 different feed water salinities of 300, 600, and 900 mgL^{-1}. It is assumed that an Ionics AQUAMITE XX EDR unit is used for the predemineralization [L.R. Schmauss et al, 1984].

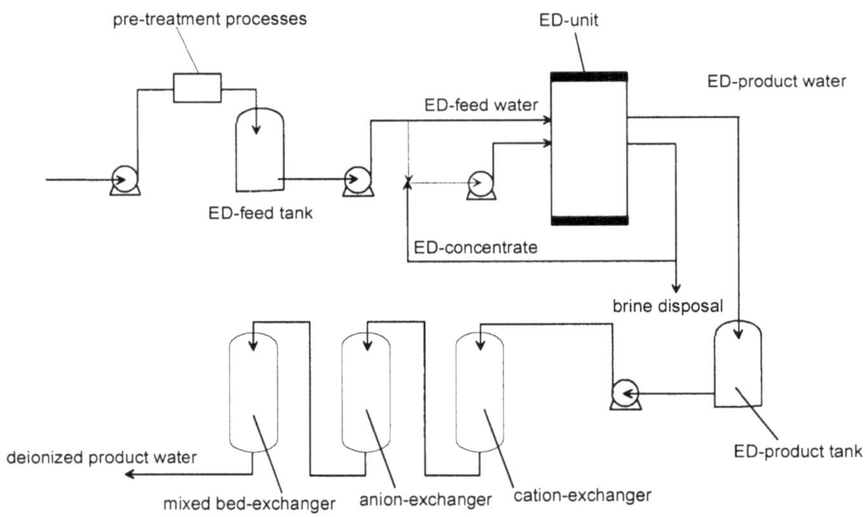

Fig. 6.2 Simplified process flow diagram of an electrodialysis unit integrated predemineralizer in a boiler feed water preparation plant.

During recent years there has been a substantial decrease in the quality of raw water for industrial use with a corresponding increase in water costs. Also, the discharge of certain industrial waste waters is often problematic and costly because of high salt concentrations or toxic water constituents. This has resulted in a trend to conserve water and safe costs by reuse of industrial process water. For recycling waste water back into an industrial process it must undergo certain purification steps which usually include a desalination process. Often electrodialysis can be used very effectively and economically in recycling industrial process water.

A typical application of electrodialysis is the recycling of cooling tower blowdown water. Environmental concerns and raw water costs have led to an increase in the reuse of cooling tower blowdown water. A reduction of raw water intake and waste water discharge have been achieved by increasing the cycles on concentration in cooling towers using acids, anti-precipitants, and side stream clarification and softening. A further increase in water reuse and reduction of waste water can be achieved by desalination of the recycled water. Electrodialysis seems to be particularly suited for this purpose for a couple of reasons:

- In electrodialysis high recovery rates up to 95 % and high brine concentrations up to 100,000 mg L^{-1} can be achieved which result in a reduction in waste water discharge.
- Ion-exchange membranes are stable over a wide pH-range from 1 to 11 and can be operated at temperatures up to 50°C which is in the range of most cooling systems.
- Electrodialysis reversal desalination systems need a minimum of pretreatment.

Another example of a successful application of electrodialysis is the recycling of produced water from crude oil production. In many oil wells steam is being injected into the ground to heat and liquefy the crude oil so that it can be pumped. The in the process the steam is condensed and pumped out of the ground mixed with the oil. The water is then separated from the oil by an oil/water separator. The recovered water usually also contains large quantities of salt and hardness extracted from the ground. To be reused as boiler feed for the generation of steam the produced water must be softened. This is generally done in a conventional one or two stage water softener. Since the salt concentration in the produced water is relatively high, i.e. in the order of 5,000 to 10,000 mg L^{-1} large quantities of salt are used for the regeneration of the ion-exchange resin which makes the process labor intensive and costly. The regeneration effluents contain double the amount of salt than the original water. Their deposition is also quite costly. Substantial savings in regeneration costs can be achieved by using electrodialysis as a pre-softening desalination step. Here again electrodialysis is superior to other desalination processes such as reverse osmosis because of the thermal and chemical stability of the ion-exchange membranes which can operate reliably at temperatures in excess of 50 °C with a minimum of feed water pretreatment. A simplified flow diagram of a produced water treatment system with an integrated electrodialysis unit is shown in Figure 6.3. In the first step of produced water recovery process excess oil is removed by an oil skimmer. The oil content of the water is further removed by a filter system to a level below 2 mg L^{-1}. Before going into the electrodialysis unit the produced water is cooled down from ca. 80°C to about 45°C. In the electrodialysis reversal system 80 to 90% of the total dissolved solids are removed depending on the feed water salt concentration. The partially demineralized water is then further treated in a softener and finally recycled to the steam generator. The brine concentration and with that the recovery rate of the electrodialysis reversal unit was limited by the solubility of the $CaSO_4$.

Operating data from the above described produced water recovering plant are given in the literature [E.P. Geishecker, 1984]. They are summarized in Table 6.3.

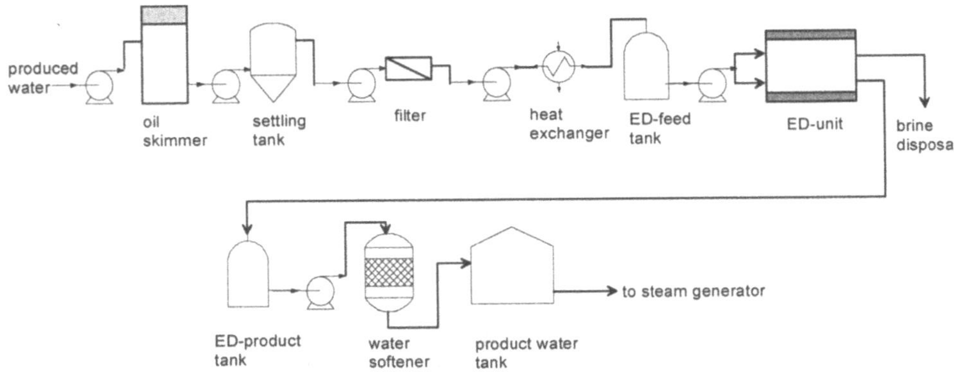

Fig. 6.3 Simplified flow diagram illustrating a produced water recovering plant with an integrated electrodialysis unit.

Table 6.3 Operating data of a produced water recovering plant with an integrated electrodialysis reversal unit.

electrodialysis plant	Aquamite XV-3
electrodialysis feed water salt concentration	2,500-12,000 [mg L^{-1}]
electrodialysis product water salt concentration	350-3,500 [mg L^{-1}]
electrodialysis brine salt concentration	8,000-35,000 [mg L^{-1}]
salt removal rate	70-85 %
feed water volume	680-560 [m^3 d^{-1}]
product water volume	480 [m^3 d^{-1}]
waste water volume	80-200 [m^3 d^{-1}]
water recovery	70-85 %

There are many more examples where electrodialysis has successfully been utilized as an integrated step in a recycling process of salt contaminated industrial process water with the result of substantial cost savings and reduced waste disposal problem. With increasing water and salt disposal costs the electrodialysis will most likely play an increasingly important role in industrial water recycling in the future.

2.2.2 Waste Water Treatment by Electrodialysis

Very closely related to the recycling of industrial water is the treatment of industrial waste waters. In this application the water recovery is not necessarily the main objective. Very often toxic or valuable components must be removed from an industrial effluent to avoid pollution of the environment. Especially in the highly populated industrial areas the pollution of the environment is a rapidly increasing problem. A major source of pollution are effluents of the metal processing industry. Heavy metal ions such as nickel-, zinc-, cadmium-, chromium-, silver-, mercury-, and copper-ions are toxic in relatively low concentrations and often difficult to separate from a mixture with other salts for safe disposal. Today, electrodialysis is playing an increasingly important role in separating, concentrating, and recycling these heavy metal ions. In combination with other processes such as ion-exchange, ultra- and microfiltration the application of electrodialysis not only provides a solution to severe pollution problems but saves substantial production costs by recovering and recycling valuable components and water.

A typical example is the recovery of nickel, cadmium, and copper from rinse solutions of galvanic processes. The units are generally relatively small, processing a few m^3 per day [T.V. Tran et al., 1986], [Tokuyama, 2001], [S. Ito et al., 1980].

Another interesting application is the recovery and recycling of water and H_2SO_4 from the rinse solution of a lead battery production line. To refine the spent rinse water of the electrodes formation of a battery production line a treatment scheme was developed by Osmota GmbH, Germany, which consisted of several of processes including electrodialysis. The flow scheme of the treatment process is indicated in Figure 6.4.

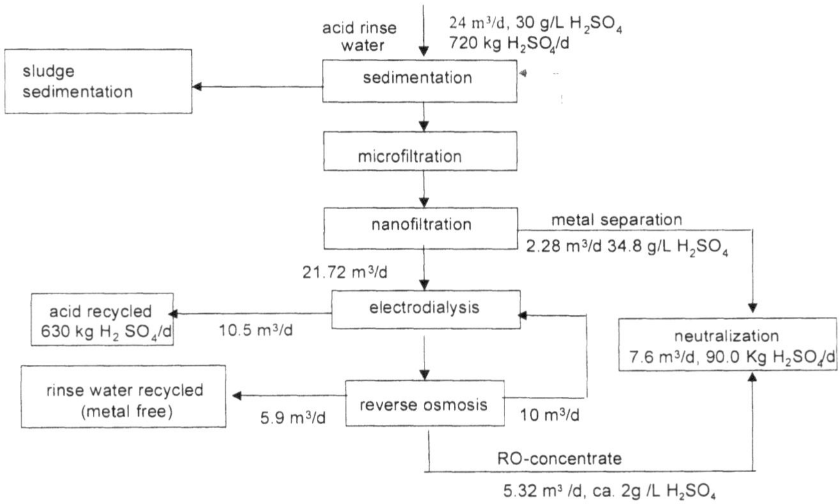

Fig. 6.4 Process scheme illustrating the treatment and recycling of spent rinse water from a lead/acid battery production line [H.J. Rapp, 2002].

Ca. 24 m^3d^{-1} of a rinse solution containing lead sludge, sulfuric acid, and a low level of heavy metals such as chromium, copper, iron, nickel, etc. must be treated. The goal was to recycle as much as possible of the water and the sulfuric acid. However, a certain concentration level of the metals in the recovered water should not be exceeded. The maximum tolerable concentrations are: Cr > 0.1 mg L^{-1}, Cu > 0.1 mg L^{-1}, Fe >1.5 mg L^{-1}, and Ni >0.1 mg L^{-1}.

A photograph of the electrodialysis part of the plant is shown in Figure 6.5.

Fig. 6.5 Photograph showing an electrodialysis unit containing 4 stacks with 180 m^2 membrane area each integrated in a process line to remove heavy metals from spent rinse water of a lead/acid battery production and to recycle the water and recover and concentrate the sulfuric acid for further reuse (photograph courtesy of Osmota GmbH).

The first step in the process scheme of treating the spent rinse water of the electrode production line is the precipitation of lead sludge in a settling tank. The supernatant solution which contains ca. 30 g L^{-1} sulfuric acid and various metal ions and particles is fed to a microfiltration system. The retentate of the microfiltration procedure is recycled to the sedimentation tank. The filtrate is passed through a nanofiltration unit which passes the sulfuric acid but retains the metal ions. The nanofiltration system is operated at a recovery rate of ca. 90%. The 2.28 m^3 d^{-1} retentate which contains the metal ions and 34.8 g L^{-1} sulfuric acid is fed to a neutralization tank. 21.72 m^3 d^{-1} filtrate is further processed in an electrodialysis unit containing 180 m^2 of cell pair area in 4 stacks. The concentrate stream of the electrodialysis unit is 10.5 m^3 d^{-1} and contains ca. 60 g L^{-1} sulfuric acid. The diluate stream which still contains a small amount of sulfuric acid is fed to a reverse osmosis unit which retains the sulfuric acid almost completely. In the reverse osmosis unit 5.9 m^3 d^{-1} are obtained as filtrate which is free of sulfuric acid and metal ions and is directly recycled to the rinsing tank. The concentrate of the reverse osmosis unit is partially recycled into the electrodialysis unit and 5.32 m^3 d^{-1} containing 2 g L^{-1} H_2SO_4 is fed to the neutralization tank.

The overall mass balance shows that 88 wt % of the sulfuric acid and 25 wt % of the rinse water are recovered. This results in reduction of costs for water and sulfuric acid and substantial saving in neutralization chemicals and sludge disposal costs.

A large number of interesting applications for electrodialysis are in the galvanic industry. The recovery of valuable or toxic heavy metals such as nickel, copper, zinc or cadmium from effluents of electroplating processes is of prime importance for the environmental pollution control and for economic reasons [E. Korngold et al., 1978]. An interesting application of electrodialysis in the galvanic industry is the recycling of metal ions from electroplating rinse solutions. A typical example is the recovery and concentration of nickel the still rinse.

Metal parts that have been plated in a galvanic bath are rinsed in a series of rinse tanks to remove metal ions carried out of the plating bath. The procedure is illustrated in Figure 6.6.

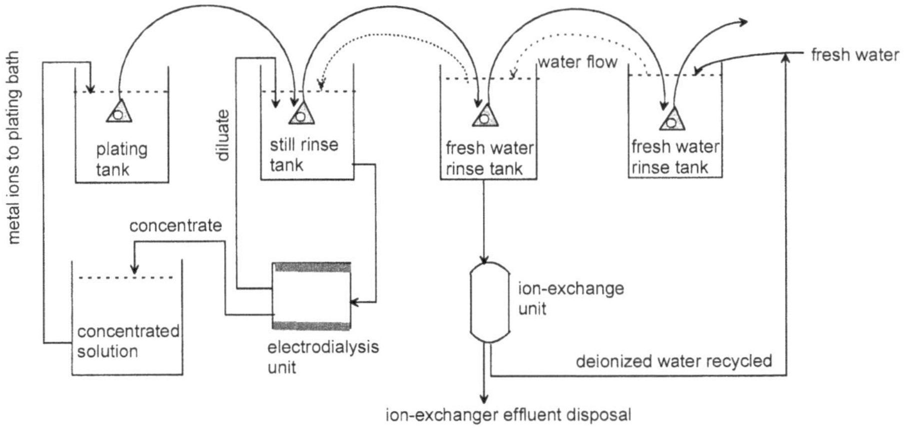

Fig. 6.6 Simplified flow diagram illustrating the integration of an electrodialysis unit in the rinsing process of plated metal parts to recycle rinse water and recover metal ions.

The figure shows the pathway of plated metal parts removed from the plating tank to a series of rinse tanks. The first tank is referred to as still rinse. In this tank most of the components carried out of the plating bath are rinsed off leading to a rapid increase in the concentration of metal ions dragged out of the plating bath. The ion concentration in the still rinse tank determines the amount of material carried into the next two tanks which are continuously supplied with fresh deionized water which flows from the last tank to the second last tank. To control the water level in this tank a small amount flows further into the still rinse tank while the rest is fed to an ion-exchange bed where the metal ions are removed. The deionized water is recycled to the last rinse tank while the effluent of the ion-exchange regeneration goes to the waste disposal. The amount of ions which are carried from the still rinse to the other tanks and thus the amount of ions that must be removed by the ion-exchange process, depends on the concentration in the still rinse tank. When a certain concentration in this tank is reached it must be discharged to limit the amount of dragged out metal ions. The concentration of the still rinse tank, however, can be kept on a low level for a long time period by installing an electrodialysis unit between the still rinse tank and the plating bath as illustrated in Figure 6.6. The solution of the still rinse tank is fed to the electrodialysis unit. The obtained diluate is directly recycled to the still rinse tank while the concentrate goes to a concentrate tank and is either directly or after a proper treatment recycled to the plating bath. In this process arrangement the electrodialysis has a "kidney function". By continuously removing the metal ions from the still rinse tank the concentration in this tank can be kept low, i.e. to 10 – 20 wt % of the plating bath concentration, and the ion drag-over into the two fresh water rinse tanks is kept correspondingly low. Water that is carried out of the still rinse tank with plated metal parts and removed as concentrate by the electrodialysis process is replenished from the fresh water tanks. The integration of the electrodialysis unit into the plated metal rinse process results in a substantial reduction of ions that must be removed by the ion-exchange resin. Due to the water recycling, the recovery of the metal ions, and the reduction of sludge and waste water treatment substantial cost reduction is achieved. Typical concentrations of the nickel plating bath, the still rinse tank, and the diluate and concentrate of an electrodialysis unit installed between the still rinse tank and the plating bath are given in Table 6.4 [Tokuyama Soda, 2001].

Tab. 6.4 Example showing salt concentrations of the plating bath, the still rinse, and the diluate and concentrate solutions of an electrodialysis unit integrated in a nickel plating process to recycle metal ions from the still rinse.

component	concentrations [g L^{-1}]			
	galvanic bath solution	still rinse solution	diluate solution	concentrate solution
$NiSO_4 \times 6\,H_2O$	300	45	13	288
$NiCl_2 \times 6\,H_2O$	50	12	5	66
H_3BO_3	50	12	11	13

Other ions that can be recovered and recycled from metal plating rinse solutions by integration of an electrodialysis process are zink, cadmium, silver, and copper.

The metal plating and the metal surface treatment industry provides a multitude of economically very interesting applications of electrodialysis. In many of these applications electrodialysis is combined with other processes. A typical example is the combination of retardation with electrodialysis which is applied to recover acids from pickling bathes used in metal surface treatment. The process scheme is illustrated in Figure 6.7.

The prevailing method of metal surface treatment is pickling with acids. The acids used are H_2SO_4, HCl, HNO_3, HF, etc. The pickling solution, however, deteriorates with time and becomes uneffective. Since the spent pickling solution contains various heavy metals and substantial amounts of acids it must be neutralized and the obtained sludge must be deposited safely. For economic reasons it is interesting to recover the acids from the spent pickling solution and reduce the sludge disposal. One of the methods used today to recover and recycle acids is retardation. However, the effluent of the retardation unit contains still a substantial amount of acids. These acids can be recovered by electrodialysis [H.J. Rapp, 2002]. The schematic diagram in Figure 6.7 shows the flow scheme of an electrodialysis process to recover acids from the effluent of a retardation unit which is used to recycle HNO_3 and HF from a stainless steel pickling bath. 30 m^3d^{-1} effluent from the retardation unit which contains 25 to 35 gL^{-1} HNO_3, 8 gL^{-1} HF, and 15 to 20 gL^{-1} of metal salts are treated in an electrodialysis unit having a total 500 m^2 membrane area. The diluate from the electrodialysis unit contains 2 to 5 gL^{-1} HNO_3 to the neutralization process while the concentrate which contains ca. 176 g L^{-1} HNO_3 is directly recycled into the pickling bath.

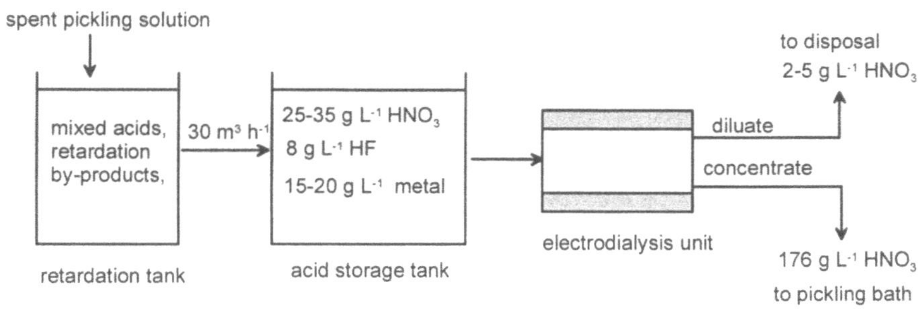

Fig. 6.7 Simplified flow scheme illustrating the recycling of HNO_3 by electrodialysis from the retardation effluent obtained from a spent pickling solution treatment system.

A photograph of the retardation acid recovery plant with the integrated electrodialysis unit is shown in Figure 6.8.

There are many more interesting applications of electrodialysis in waste water treatment. Some of these applications require only relatively small electrodialysis units, as e.g. the treatment of small amounts of effluents from chemical processes containing highly toxic compounds. In other applications rather large quantities of water must be treated. This is the case in the paper and pulp industry. Here electrodialysis has been successfully used to remove NaCl selectively from the chemical recovery cycle of Kraft Pulp production. The accumulation of chloride is a serious problem in the Kraft pulping process. It was shown that chloride can be selectively removed from the dissolved electrostatic precipitator dust obtained in a chemical recovery process by electrodialysis using anion-exchange membranes with a high selectivity for monovalent ions [H.J. Rapp et al.,1998]. The removal of radioactive components from nuclear power stations has been studied extensively in the laboratory. But there are until today no large scale commercial installations. The same is true for the removal of nitrate from potable water sources. Due to nitrate fertilizers in intensively used agricultural land the nitrate level in ground water is high above the drinking water standard level. The removal of nitrate from polluted water is an option that has been extensively tested on a pilot plant scale [K. Kneifel et al. 1988].

Fig. 6.8 Photograph showing an electrodialysis unit containing 500 m^2 membrane area integrated into a retardation process to recover HF and HNO_3 from a spent pickling bath (photograph courtesy of Osmota GmbH).

2.3 Food Processing by Electrodialysis

In the food and beverage industry but also in biotechnology and the pharmaceutical industry electrodialysis has found a large number of applications. Some of these applications can be considered as state-of-the-art processes such as the deionization of whey or the de-acidification of fruit juices. Other applications such as the potassium tartrate or the de-ashing of sugar or the removal of salts from protein solutions have been tested and evaluated on a pilot plant or laboratory scale. The removal of salts and organic acids such as lactic acid and certain amino acids from fermentation broth by electrodialysis has been tested extensively in the laboratory and on a pilot plant scale. Many applications of conventional ion-exchange resins in the food industry can be replaced by electrodialysis providing better economics, simpler operation, and less waste material which must be deposited. Typical applications of electrodialysis in the food industry are summarized in Table 6.5.

Table 6.5 Examples for the application of electrodialysis in the food and beverage industry.

industry	application of electrodialysis
dairy industry	demineralization of cheese whey, demineralization of non-fat milk
sugar industry	demineralization of molasses in sugar production, demineralization of waste molasses, demineralization of polysaccharide
fermentation industry	demineralization of soy sauce, desalination of amino acids, recovering of organic acids
wine industry	removal of tartrate
beverage industry	de-acidification of fruit juice

2.3.1 Demineralization of cheese whey and skimmed milk

In the dairy industry large amounts of whey are obtained as in the production of various types of cheese. In the past and even today most of the whey is discharged and often causes a severe pollution problem. On the other hand whey provides an excellent source of protein, lactose, vitamins, and minerals. But in its normal form it is considered as not suited as food material because of the high salt content of whey. When the salts in whey are substantially reduced its composition is close to that of human milk. With the proteins and vitamins in the natural functional form, whey is a valuable product to be used in baby food as additive to cheese products, ice cream, candies etc. Thus, demineralization of whey under condition that do not cause a decomposition of the proteins and vitamins would be very desirable. Early attempts to demineralize whey with the use of ion-exchange media were not successful because of severe fouling of the ion-exchange resins and a substantial pH-shift which caused denaturation of the proteins. Although, the demineralization of whey by electrodialysis was not without problems in the beginning because of severe fouling of the anion-exchange membranes and insufficient chemical stability of the membranes [R.M. Ahlgreen, 1972]. With the development of ion-exchange membranes which were stable in solutions with very high and low pH-values, and which showed a better resistance to organic fouling, and due to a better electrodialysis stack design which tolerated higher internal pressure caused by the high required feed flow velocity in the stack that prevented the precipitation of proteins and calcium phosphate on the surface of the membranes, electrodialysis was successfully used for the demineralization of whey and skimmed milk [L.J. Andres et al., 1995], [M. Bleha et al., 1992], [M. Greiter et al., 2002]. The operation of an electrodialysis plant used by Snow Brand Milk Products Corporation for the demineralization of whey and skimmed milk are described in the literature [K. Kobuchi et al., 1983], [Tokuyama Soda Co. Technical Information, 1987].

The main characteristics of the 4-stage electrodialysis unit is used in process are:

- continuous operation for an extended time period,
- high degree of demineralization,
- short liquid holding time avoids excessive growth of microorganisms,
- clean in place after extended period of operation,
- compact system design,
- economical operation due to high current efficiency and long membrane life.

The system is illustrated in Figure 6.9 showing the flow diagram of the 4-stage demineralization unit.

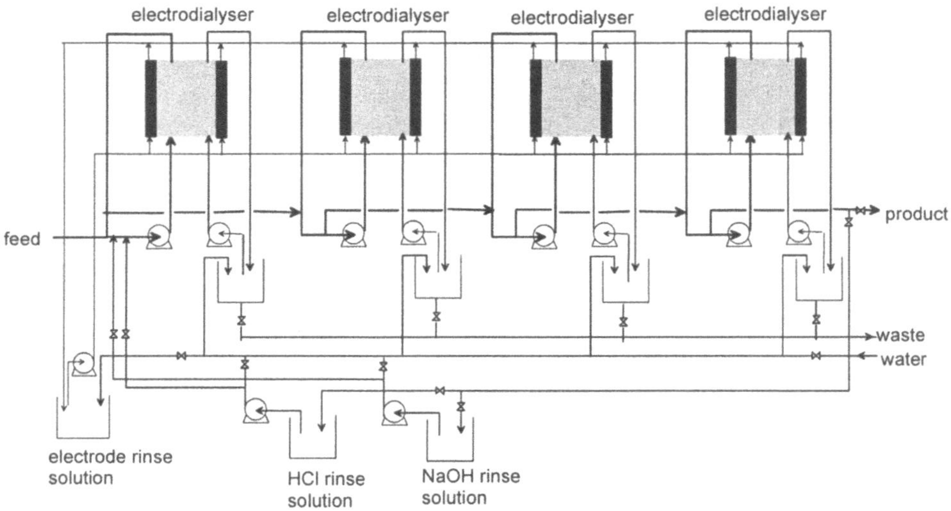

Fig. 6.9 Simplified flow diagram of the 4-stage electrodialysis unit used for the demineralization of cheese whey and skimmed milk.

In the 4-stage demineralization unit the raw solution, i.e. the cheese whey or the skimmed milk is fed through the unit, and in each stage it is demineralized to a certain extent reaching the final degree of demineralization at the exit of the last stage. The feed solution and the concentrate is circulated in each stage by individual pumps while a separate solution is pumped through the electrode cells. In the demineralization of whey or skimmed milk it is very important to keep the product from being contaminated by microorganisms. Therefore, cleaning and sanitation of the electrodialysis stack is very important. To achieve an effective sanitation the entire system is cleaned in place and rinsed by HCl and NaOH solutions.

The degree of demineralization depends on the individual salts as indicated in Table 6.6 which shows the composition of cheese whey and skimmed milk that was demineralized in the plant. The total removal of salts is ca. 90 wt % for cheese whey and 70 wt % for skimmed milk [Tokuyama Soda Technical Information, 1987].

Table 6.6 Typical composition of cheese whey and skimmed milk before and after electrodialysis demineralization based on dry product.

feed constituents	cheese whey			skimmed milk		
	feed [g L^{-1}]	product [g L^{-1}]	demineralization [%]	feed [g L^{-1}]	product [g L^{-1}]	demineralization [%]
protein	140	144		352	369	
fat	41	45		7.5	10	
lactose	734	800		554	593	
sodium	9	0.7	92	8	1.3	83
potassium	27	0.4	98	21	0.4	95
magnesium	1	0.5	50	2	0.6	60
calcium	6	2.2	63	11	7	36
phosphate	7	1.7	76	11	6	65
chloride	18	>0.01	>99	12	0.2	98
others	15	4.5	70	21	10.5	50
total ash	84	10	88	86	26	70

The ions such as K^+-, Na^+-, and Cl^--ions, permeate first, then Ca^{2+}-, Mg^{2+}, and PO_3^{3-}-ions are removed. Therefore, the demineralization rate for monovalent ions is higher than for the multivalent ions. Furthermore, the demineralization of cheese whey is generally slightly higher than that of skimmed milk.

2.3.2 Other Applications of Electrodialysis in the Food Industry

In other applications of electrodialysis in the food industry such as the demineralization of soy sauce, sugar molasses, and amino acids or the de-acidification of fruit juice, the same membranes and system design are used as in the demineralization of cheese whey. However, these applications have not yet reached the same industrial relevance. In most cases only pilot plant studies have been carried out and only a few applications have been commercialized. A process that has reached some industrial relevance is the removal of salts from fermentation broth. For example, soy-sauce produced by fermentation contains 16 to 18 wt% of NaCl. For human consumption the salt content must be reduced to 5 – 10 wt%. This can be done effectively be electrodialysis without loosing fragrance, color or taste [Tokuyama Soda, 2001]. Since the electrodialysis unit is generally operated as batch-type process the adjustment of the desalination time allows to produce soy sauce with arbitrary salt concentration. This is indicated in Table 6.7 which shows examples of desalination of soy sauce.

Table 6.7 Results of a soy sauce desalination by electrodialysis operated in a batch process.

soy sauce feed	raw soy sauce	desalted solution		
operating time [h]	0	4	7	9
product quantity [L]	10	9.2	8.1	7.1
NaCl [wt %]	16.8	12.4	8.9	5.9
total N [wt %]	1.53	1.61	1.72	1.80
pH-value	4.7	4.81	4.85	4.87
desalination rate [%]		32.1	57.1	75.1
N-loss [%]		3.2	8.9	16.5

The data of Table 6.7 show that with increasing operating time of a given size electrodialysis batch process from 4 to 7 to 9 hours the degree of desalination is increasing from 32.1 to 57.1 to 75.1 %. Since a part of the solution is removed as concentrate the loss of protein is also increasing with operating time, i.e. the degree of desalination, as the analysis of total nitrogen indicates.

The desalination of amino acid containing feed solutions is an interesting application of electrodialysis that has been commercialized. Electrodialysis is used to remove NaCl from the mother liquid for the separation of amino acids. This not only increases the yield of amino acid but also reduces the costs of waste water treatment. Since amino acids are amphoteric electrolytes having carboxylic acid and amino groups in the molecule their loss during the electrodialysis procedure can be minimized by adjusting the pH-value of the feed to the isoelectric point of the amino acid.

The use of electrodialysis in the production process of natural essences is an application that has been studied extensively on a laboratory and pilot plant scale.

Natural essences are produced from meat, seafood, and vegetables and fruits. They are used mainly as natural seasonings, and raw material for drugs and health food. In many cases they are extracted from the raw material by a saline solution. The products are usually provided in powder or granular form. However, because of the high salt concentration of the extracted solution the salt content in the dried seasoning powder is generally too high for human consumption and must be reduced. This can easily and efficiently be done by electrodialysis without significant loss of flavor. The process design is in principle the same as applied to the desalination of soy sauce or cheese whey. The ion-exchange membranes used in this application must also meet the requirements for food hygiene and the system design must provide an easy in place cleaning and sanitation by HCl and NaOH solutions.

De-acidification of fruit juice and the removal of tartaric acid from wine is another interesting application that has been explored mainly in Europe [H. Voss, 1986]. Especially in the production of bottled champagne it is necessary to avoid the formation of crystalline tartar formation which occurs when the wine is cooled and the solubility of potassium-sodium salt of the tartaric acid exceeds its solubility limits. Crystalline tartar on the bottom of a champagne bottle enhances the release of dissolved CO_2 which results in an instantaneous and uncontrolled foam formation when the champagne bottle is opened. To avoid the crystallization of tartar salts in the wine, tartaric acid and its salts must be reduced to a level that does not exceed their solubility in the wine at the temperature the wine is processed stored. This can be done efficiently by electrodialysis.

Several other applications of electrodialysis in the pharmaceutical industry such as recovery of lactic acid from fermentation broth have been studied on a laboratory scale [P.B. Reed, 1984], [H.J., Lee et al., 2002], [D.A. Glassner et al., 1990]. Most of these applications are concerned with the desalting of solutions containing valuable components such as certain proteins or dextrans . Here, electrodialysis is often competing with dialysis and ion-exchange. But in many cases electrodialysis is the superior process as far as economics and the quality of the products is concerned. Especially in down stream processing in biotechnology electrodialysis is more and more replacing conventional ion-exchange techniques [E.G. Lee et al., 1998].

2.4 Pre-Concentration of Salts by Electrodialysis

The concentration of dilute salt solutions by electrodialysis is used in the treatment of certain industrial effluents to recover valuable or toxic waste water constituents or to avoid excessive salt discharge into waste water treatment plants. It is also of interest for the production of certain salts such as KBr or KI from certain raw water sources. Of very special interest is the concentration of NaCl from sea water prior to evaporation for the production of table salt. Because of the unique circumstances this application has been developed exclusively in Japan. Japan has no native salt deposits. The large amounts of rainfall make solar evaporation difficult and conventional evaporation is rather costly, and with evaporation all sea water constituents are equally concentrated. For the production of table salt and as raw material for the chlorine alkaline production it is desirable to have NaCl as pure as possible. Using electrodialysis as a pre-concentration step not only substantial energy savings were achieved but with the development of low resistance monovalent ion selective membranes a more effective concentration of NaCl was achieved. Further improvements of the sea water pretreatment process have eliminated the extensive cleaning procedures. Operation for several years and cleaning without disassembling the stack are now possible [M.Yamamoto et al., 2000]. Large plants with a capacity of 20,000 to more than 200,000 tons of table salt per year are now in operation in Japan.

A typical plant for the pre-concentration of sea water is installed in the Sanuki Salt Manufacturing Co. The plant has a production capacity of 230,000 tons per year table salt using 6 TSX and 1 TSW electrodialysers manufactured by Tokuyama Corporation. The stacks used in this plant are of the sheet flow type developed already in 1967 [T. Nishiwaki, 1972], [R. Yamane et al., 1969]. A simplified flow diagram of the plant as it is operated today is shown in Figure 6.10.

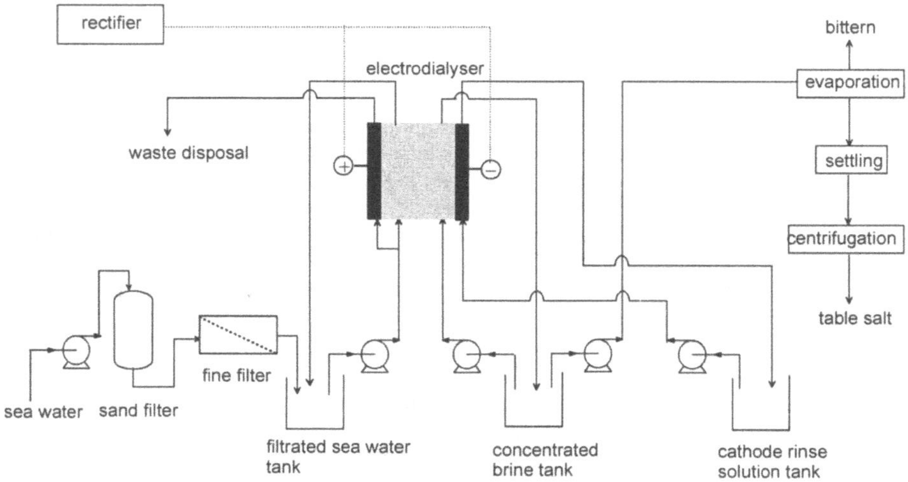

Fig. 6.10 Simplified flow diagram of a table salt production plant using a TSX electrodialyser of Tokuyama Soda.

The specifications of a TSX-200 electrodialyser and the performance data are summarized in Table 6.8. Both the stack design and the membranes have significantly been improved over the years resulting in substantial energy savings.

Table 6.8 Specification and design parameters of the TSX-200 electrodialyser of Tokuyama Soda Corporation.

electrodialyser specifications:	
cation-exchange membrane	CIMS
anion-exchange membrane	ACS-3
effective cell pair area	2 [m^2]
cell thickness	0.4 [mm]
number of cell pairs per stack	3,500
operating conditions:	
current density	300 [A m^{-2}]
feed solution flow velocity	6 [cm s^{-1}]
average sea water temperature	25°C
electrodialyser performance	
production capacity	30,000 [tons NaCl per year]
NaCl concentration in brine	205 [g L^{-1}]
power consumption by electrodialyser	155 [kWh per ton NaCl]

A typical plant is shown in the photograph of Figure 6.11.

Fig. 6.11 230,000 tons per year table salt production plant using 6 electrodialysis TSX units to concentrate sea water from ca. 3.5 wt% salt to ca. 20 wt% NaCl (photograph courtesy of Tokuyama Corporation).

The production of table salt from sea water, however, is limited to only a few countries which do not have large salt deposits. However, concentrating and recovering other electrolytes from brines generated by reverse osmosis or from salt lakes such as the dead sea by electrodialysis seems economically feasible.

2.5 Electrodialysis in Hybrid Processes

The application of electrodialysis in desalination is most efficient in a certain range of concentration as far as the feed solution, the diluate, and the concentrate is concerned. If extremely low diluate concentrations are required from feed solutions having low salt concentration, electrodialysis has to be operated at very low current densities and often becomes uneconomic because very large membrane areas are required for a given capacity plant, and therefore ion-exchange is the more efficient process. If very high concentrations are required in the concentrate electrodialysis is impaired by low current utilization and high energy consumption, and evaporation often is the more economic process. To utilize the most effective range of operation of electrodialysis in desalination as far as the concentrations of feed, diluate and concentrate is concerned a combination with other separation processes has proved to be very effective. Especially in the production of ultra pure water the combination of electrodialysis with ion-exchange is very effective and has finally resulted in the development of the continuous electrodeionization process and to the continuous regeneration of ion-exchange resins [J. Johann et al., 1993].

Another interesting application which is presently studied on a laboratory is the integration of ion-exchange membrane in the so-called membrane reactors. In many chemical and biochemical reactions the reaction products or the reaction by-products inhibit the reaction when a certain concentration is exceeded. This often limits the achievable product concentration and requires additional separation and concentration steps. A continuous removal of the reaction inhibiting components often makes a continuous more economic production possible.

A substantial effort has also been concentrated on reversing electrodialysis to recover energy from mixing a concentrated salt solution such as sea water with river or surface water in an electrodialysis stack. Although the process is technical feasible it seems very unlikely that it ever will be economic.

3 APPLICATIONS OF BIPOLAR MEMBRANE ELECTRODIALYSIS

Since bipolar membranes became available as commercial products in 1977 a very large number of potential applications of electrodialysis with bipolar membranes has been identified and has been studied extensively on a laboratory or pilot plant scale [K.J. Liu, et al., 1977 and 1978], [K.N. Mani, 1991], [G. Pourcelly et al. 2000]. However, in spite of the obvious technical and economical advantages of the technology large scale industrial plants are still quite rare. The main reasons for the reluctant use of bipolar membrane electrodialysis are shortcomings of the available bipolar and monopolar membranes which often do not have the required chemical stability and permselectivity which result in a short useful life of the membranes under operating conditions, poor current utilization, and high product contamination. While the intrinsic limitation of the presently commercially available bipolar and monopolar membranes will limit their large scale application in certain areas such as the production of mineral acids and bases from the corresponding salts, there are a number of smaller scale applications in the chemical process industry, in biotechnology, in food processing and especially in pollution control, in waste water treatment and in recycling of constituents in industrial effluents. Some of the potential applications of electrodialysis are summarized in Table 6.9. The list of possible applications in Table 6.9 is far from being complete. Some of the more relevant applications will be discussed later in more detail. A substantial number of studies are carried out in industrial enterprises and their results are not published in the open literature. Today, unsatisfactory membrane properties and lag of application know-how and long term experience are still a major drawback for the large scale industrial application of bipolar membranes. However, with the development of new and better bipolar and monopolar membranes the number of potential applications of bipolar membrane electrodialysis on a larger industrial scale will most like increase substantially.

Table 6.9 Potential applications of electrodialysis with bipolar membranes, their state of development and possible advantages and experienced problems.

application	state of process development	potential advantages	problems related to application
production of mineral acids and bases from corresponding salts	pilot plant operation	lower energy consumption	contamination of products and low current utilization due to poor membrane permselectivity
recovering of organic acids from fermentation processes	commercial and pilot plant operation	simple integrated process, lower costs	unsatisfactory membrane stability and fouling
pH-control in chemical and biochemical processes or reverse osmosis feed solutions	laboratory tests	fewer by-products, less chemicals, and salt production and disposal	application experience, process costs
removal of SO_2 from flue gas	extensive pilot plant test	decreased salt production, reduced salt disposal costs	high investment costs, long term membrane stability
recovering and recycling of H_2SO_4 and NaOH from waste waters such as the rayon production effluent	laboratory and pilot plant tests, some commercial plants	purity of the recovered products is not critical, savings in chemicals and sludge disposal costs	no long term experience, membrane stability under operating conditions, membrane fouling, high investment costs
recycling of HF and HNO_3 from steel pickling solutions	commercial plants	cost savings due to recovered acids and decreased salt disposal	relatively complex process, high investment costs
ion-exchange resin regeneration	pilot plant tests	decreased salt disposal	high investment costs
high purity water production by electrodeionization	laboratory tests	better removal of weak acids and bases	no long term experience,
production of sodium methoxide from methanol	laboratory tests	more economic than conventional production process	no long term experience
energy storage and conversion	only theoretical considerations	eventually economical advantages	no experimental verification

3.1 Production of Acids and Bases by Bipolar Membrane Electrodialysis

The largest potential application of bipolar membrane electrodialysis is the production of acids and bases from the corresponding salts. Presently, caustic soda is produced as a co-product with chlorine by electrolysis. Because of environmental problems caused by many chlorinated hydrocarbons products the worldwide demand for chlorine is growing less than that for caustic soda, and chlorine may become a less valuable by-product. Therefore, the bipolar membrane electrodialysis would be an interesting alternative process to produce caustic soda and acids such as hydrochloric acid from the corresponding salts. However, presently the process is not yet economic for a large scale industrial production for several reasons as has been pointed out before. Extensive laboratory and pilot plant tests have shown that the maximum acid and base concentration that can be reached is too low to be economic because today's available bipolar as well as monopolar membranes have insufficient permselectivity at high ion concentrations resulting in substantial product salt contamination and low current utilization. The useful life of the ion-exchange membranes under operating conditions, i.e. at high acid and base concentrations and temperature in excess of 40°C is insufficient leading to high membrane replacement costs. Problem free operation of the electrodialysis unit also requires a substantial pretreatment of the salt solution as indicated in the flow diagram of Figure 6.12. Multivalent cations must be kept well below 1 mg L^{-1} to avoid precipitation of metal hydroxide in the membranes. Organic pollutants such as aromatic compounds can cause swelling of the membranes and must be removed.

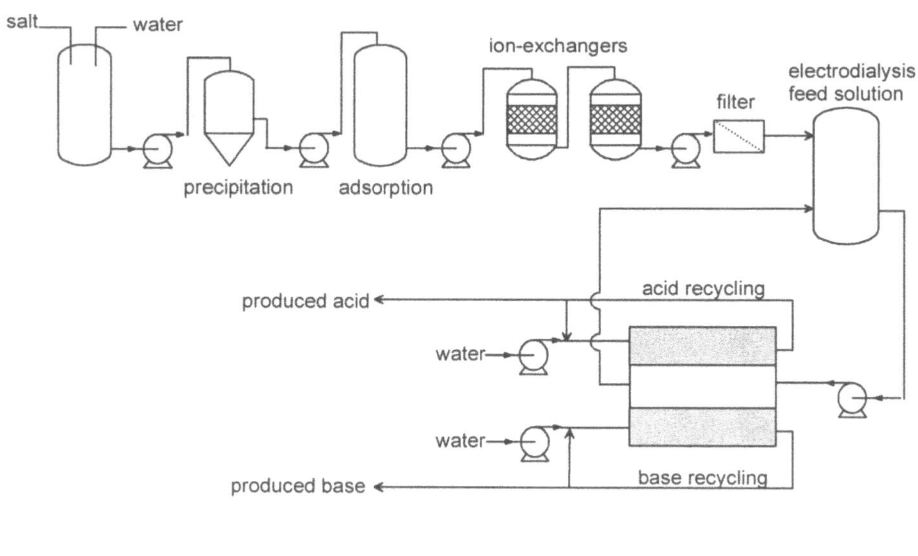

Fig. 6.12 Schematic drawing illustrating the pretreatment and production of a mineral acid and base by electrodialysis with bipolar membranes.

In spite of extensive pilot plant tests bipolar membrane electrodialysis is not used today on a large commercial scale for the production of high purity concentrated mineral acids and bases. The main problems effecting the overall feasibility of the process are [H.J. Rapp, 1995]:

- insufficient membrane permselectivity at high acid and base concentrations,
- insufficient permselectivity of the bipolar membranes causes contamination of the products with salt and lowers the current utilization,
- insufficient permselectivity of the monopolar membranes causes a substantial decrease in the current utilization,
- large water transport from the salt solution into the product due to osmosis and electroosmosis effects the current utilization and the maximum achievable product concentration,
- poor chemical stability of bipolar and monopolar membranes in highly concentrated acids and bases.

The useful life of the ion-exchange membranes under operating conditions in concentrated acids and bases is too low. The bipolar membranes are relatively expensive leading to high investment related costs.

The overall result of the laboratory tests is that the production of mineral acids and bases by bipolar membrane electrodialysis does not meet the product quality requirements under economic conditions.

However, the situation is quite different when acids or bases can be recovered from salts of chemical reactions or neutralization processes. In these cases the requirements for the concentration and the purity of the recovered acids or bases are not as stringent as in the production high quality commercial products and bipolar membrane electrodialysis can be used effectively to recycle acids and bases recovered from the corresponding salts from a waste stream.

3.2 Applications of Bipolar Membranes in Pollution Control Processes

Recovering acids and bases from their salts generated in neutralization reactions to control pollution and minimize waste disposal is one of the most promising applications of bipolar membrane electrodialysis. One of the first commercial plants was installed 1987 at Washington Steel Corporation in Pennsylvania [J.C. McArdle et al., 1991], [G. Pourcelly et al., 2000] to recover hydrofluoric and nitric acid from an effluent stream containing potassium fluoride and nitrite generated by neutralization of a steel pickling bath. Operating and performance data of the plant are published in the literature. The economic analysis shows that that the plant provides substantial savings in the operating cost of the pickling process with a relative short pay-back time of the investment which is less than 4 years. The bipolar membrane electrodialysis plant is designed and build by Aqualitics [K.N. Mani, 1991].

Table 6.10 Operating data of a bipolar membrane electrodialysis plant operated by Washington Steel Corporation to recover HF and HNO_3 from a waste stream generated by neutralization of a steel pickling bath.

plant capacity	27 m^3 per day
pickling solution composition	HF: 1.8 wt % HNO_3: 9 wt % FeF_3: 3.6 wt % Ni- and Cr-complexes
electrodialysis feed solution	ca. 10 wt % total salt
electrodialysis product concentration	HF: ca. 2 to 3 wt% HNO_3: ca. 8 to 10 wt % KOH: ca. 1.5 mol L^{-1}
product recovery	HF: ca. 90 % HNO_3: ca. 95 %
installed membrane area	3 x 150 m^2
expected membrane life	bipolar membranes 2 years monopolar membranes 5 years
pretreatment	filtration to remove metal hydroxides preconcentration by electrodialysis
economics	
total investment	US $ 2,950,000.-
operating costs	
membrane replacement	US $ 300,000.-
power consumption 3,000,000 kWh	US $ 150,000.-
general maintenance	US $ 500,000.-
total operating costs	US $ 750,000.-
operating profits	
saved acids	US $ 770,000.-
saved neutralization costs	US $ 750,000.-
saved sludge disposal costs	US $ 100,000.-
total operating savings	US $ 1,620,000.-
profit from operation	US $ 870,000.-
pay-back time of investment	ca 3.5 years

The process is illustrated in the simplified flow diagram of Figure 6.13. The spent pickling acid is neutralized with potassium hydroxide. The solution is then filtered and the precipitated heavy metal hydroxides are removed. The neutral potassium fluoride and nitrite containing solution is fed to the bipolar membrane electrodialysis unit in which the salts are converted to the corresponding acids and potassium hydroxide. The potassium hydroxide is recycled to the neutralization tank and the acids to the pickling bath. The depleted salt solution from the bipolar membrane electrodialysis unit is concentrated in a conventional electrodialysis system and recycled directly to the bipolar membrane unit. The diluate is used for rinsing and cleaning of the filter.

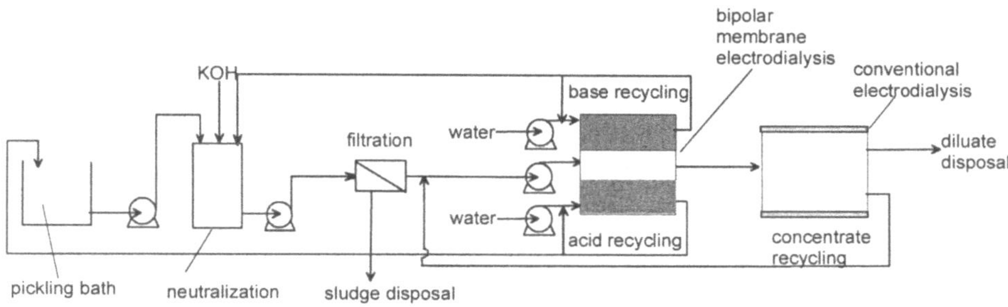

Fig. 6.13 Simplified flow diagram of acid recovery and recycling from steel pickling neutralization bath.

The treatment of alkaline or acid scrubbers that are used to remove components which are harmful to the environment such as NO_x, SO_2 or certain amines from waste air streams is another interesting application for bipolar membrane electrodialysis. In alkaline and acid scrubbers large amounts of acids or bases are consumed and salts are produced. The disposal of the produced salts is often difficult due to contamination with toxic materials such as heavy metal ions or organic compounds. The recovery of the base used in the scrubber to remove SO_2 from flue gas of coal burning power plants by bipolar membrane electrodialysis is a large scale application that has been studied extensively. There are basically two processes for the removal of SO_2 from flue gas. In the first process NaOH is used as an adsorbent which forms with oxygen which is always available in the flue gas, Na_2SO_4. The base is then regenerated from the salt solution in a three cell bipolar membrane electrodialyser and H_2SO_4 is obtained as by-product. The process is illustrated in Figure 6.14 which shows a simplified flow diagram of a SO_2 adsorber combined with a bipolar membrane electrodialysis unit for the regeneration and recycling of NaOH.

In the first step of the process SO_2 is adsorbed by NaOH forming $NaHSO_3$ which is further oxidized to Na_2SO_4 according to the following reaction:

$$SO_2 + NaOH = NaHSO_3 \quad \text{and} \quad 2\,NaHSO_3 + O_2 = Na_2SO_4 + H_2O$$

The Na_2SO_4 is then fed to the middle compartment of a three cell bipolar membrane electrodialyser. Here, the Na^+-ions migrate through the cation-exchange membrane towards the cathode into the adjacent cell and form NaOH with the OH^--ions generated in the bipolar membrane. The SO_4^{2-}-ions migrate through the anion-exchange membrane towards the anode into the adjacent cell and form H_2SO_4 with the protons generated in the bipolar membrane. NaOH is recycled to the adsorber.

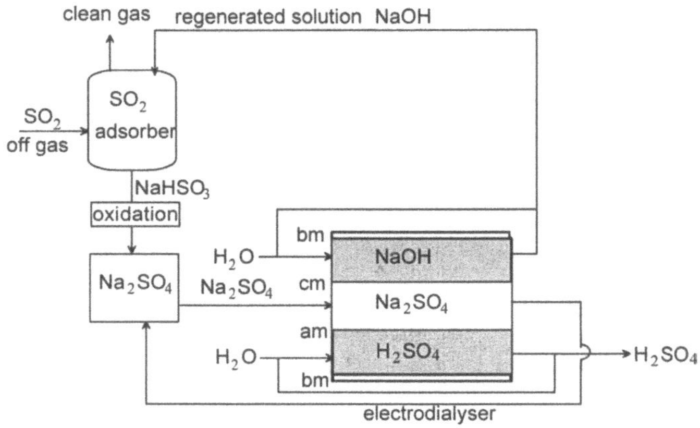

Fig. 6.14 Simplified flow diagram of the flue gas desulfurization process with NaOH as adsorber and utilizing bipolar membrane electrodialysis to regenerate and recycle the NaOH.

In the second process, the so-called Soxal™-process, SO_2 is removed from flue gas by a regenerative scrubbing system using a two cell bipolar membrane electrodialysis unit. In this case a concentrated SO_2 stream is obtained as by-product [K.N. Mani et al., 1986]. The process is illustrated in the simplified flow diagram of Figure 6.15.

The first step in the process is the adsorption of SO_2 from the flue gas by Na_2SO_3 forming $NaHSO_3$ according to the following reaction:

$$Na_2SO_3 + H_2O + SO_2 \rightarrow 2\,NaHSO_3$$

There is also oxygen in the flue gas which reacts with $NaHSO_3$ to Na_2SO_4:

$$2\,NaHSO_3 + O_2 \rightarrow 2\,Na_2SO_4 + H_2O$$

The spent solution leaving the adsorber is composed of a mixture of $NaHSO_3$ and Na_2SO_4. This mixture is fed to the bipolar membrane electrodialyser with a two cell system in a repeating unit. The two cells are separated by a cation-exchange membrane. In the cell facing the anion-exchange layer of the bipolar membrane hydroxide ions are generated and Na_2SO_3 is formed according to the following reaction:

$$Na_2SO_4 + NaHSO_3 + OH^- + Na^+ \rightarrow 2Na_2SO_3 \rightarrow H_2O$$

In the cell facing the cation-exchange layer of the bipolar membrane protons are generated and SO_2 is formed according to the following reaction:

$$Na_2SO_4 + NaHSO_3 + 3H^+ - 3Na^+ \rightarrow 2H_2SO_3 \rightarrow SO_2$$

The sodium sulfite is recycled into the flue gas adsorber. The SO_2 is stripped at high vacuum and recovered as major product in high concentration up to 95 %. It is further processed to sulfuric acid or elementary sulfur.

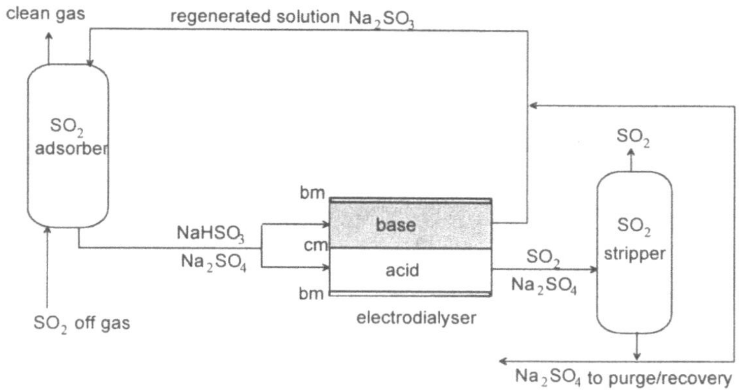

Fig.6.15 Simplified flow diagram of the Soxal™ flue gas desulfurization process utilizing bipolar membrane electrodialysis.

Comparing the Soxal™-process with SO_2 adsorption by NaOH it seems that the Soxal™-process has an economic advantage. In both processes the adsorption media, i.e. Na_2SO_3 and NaOH can be recovered and recycled by bipolar membrane electrodialysis. However, to recover the Na_2SO_3 a two cell system with one monopolar membrane is used in a repeating unit while for the recovery of NaOH a three cell repeating unit is used. Therefore, the energy consumption in the bipolar membrane unit in the Saxol™-prosses is most likely lower that in the NaOH recovery system. Furthermore, the by-product of the NaOH recovery system is a rather dilute H_2SO_4 which is also contaminated with salts and therefore of little commercial value. The by-product of the Saxol™-process is fairly pure SO_2 which can further be converted to elementary sulfur or high quality H_2SO_4 and is probably of higher value than the dilute H_2SO_4 obtained in the NaOH regeneration process.

An example for recovering and recycling an acid from the salt of a spent liquor bath which seems to be economically promising is the treatment the effluent of the rayon spinning bath. The process which is described in detail in the literature [K.N. Mani, 1991] involves bipolar membrane electrodialysis to convert Na_2SO_4 into H_2SO_4 and NaOH which is recycled to the cellulose dissolution process. By recovering the spent acid and base from the spinning bath liquor substantial savings in chemical and water costs are achieved in addition to a reduction in waste salt disposal.

Another interesting application of the bipolar membrane electrodialysis that has been studied on a pilot plant scale [H.J. Rapp, 1995] is the recovery of dimethylisopropylamine removed from a waste air stream by a sulfuric acid scrubber. The waste air stream is generated when aluminium casting moulds are made from a sand/epoxy resin mixture by injecting dimethylisopropylamine in a mixture with air as catalyst to cure the resin instantaneously. The amine is not consumed in the process and emitted in a waste air stream containing ca. 0.5 g amine per m^3 waste air. The amine can be recovered as amine sulfate in an acid scrubber as indicated in Figure 6.16 which shows the conventional process of using an acid scrubber to clean an amine contaminated air stream and to recover the amine for recycling.

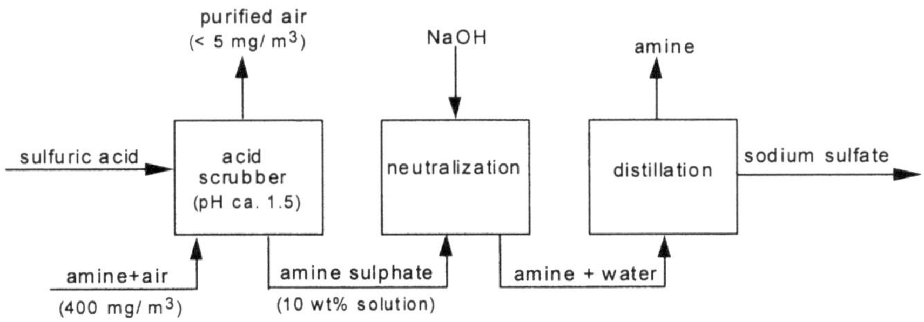

Fig. 6.16 Schematic diagram illustrating a conventional process for recovering an amine from a contaminated air stream using an acid scrubber.

The reaction in the scrubber is given by:

$$2(CH_3)_2C_3H_7N + H_2SO_4 \rightarrow [(CH_3)_2C_3H_7NH]_2SO_4$$

Thus, amine and H_2SO_4 react to amine sulfate, a water soluble salt. At a pH-value of less than 1.5 virtually all amine is in the solution as amine sulfate according to the equilibrium between the free amine and the amine sulfate. By increasing the pH-value the free amine can be released due to the equilibrium between the free or undissociated amine and the dissociated amine according to:

$$NR_3 + H_2O \Leftrightarrow R_3NH^+ + OH^-$$

The ratio of the free amine to the total amine is given by:

$$\frac{[R_3N]}{[R_3N]+[R_3NH^+]} = \frac{1}{1+\frac{[R_3NH^+]}{[R_3N]}}$$

Further is:

$$\frac{[R_3NH^+][OH^-]}{[R_3N]} = 10^{-pK_b},$$

$$[OH^-] = 10^{-(14-pH)}$$

and

$$\frac{[R_3NH^+]}{[R_3N]} = \frac{10^{-pK_b}}{10^{-(14-pH)}}.$$

The ratio of the free amine as function of the pH-value is given by:

$$\frac{[R_3N]}{[R_3N]+[R_3NH^+]} = \frac{1}{1+\frac{10^{-pK_{bb}}}{10^{-(14-pH)}}}$$

In Figure 6.17 free amine is shown as a function of the pH-value of the solution. Increasing the pH-value to ca. 12 by addition of NaOH virtually all amine in the solution is in the form of the free amine and can be recovered by distillation. The amine sulfate reacts with the NaOH to free amine and Na_2SO_4. The net result of the process, however, is the production of large amounts of sodium sulfate.

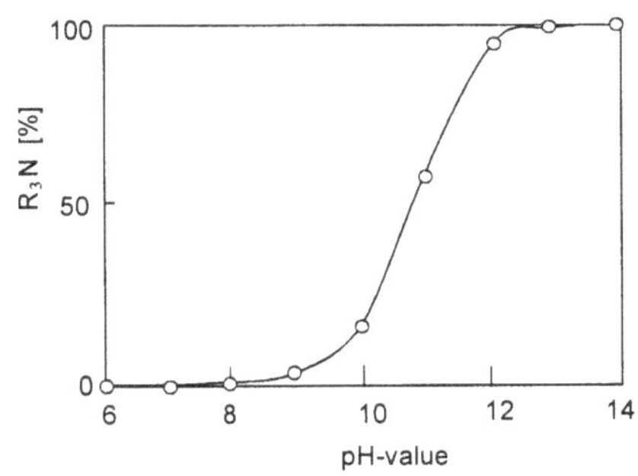

Fig. 6.17 Diagram showing the ratio of not dissociated, i.e. R_3N to the total amine as a function of the pH-value of the solution.

A complete recycling of the amine, the sulfuric acid, and water is achieved without the production of a salt by combining the bipolar membrane electrodialysis with distillation. The process is illustrated in Figure 6.18. The waste air stream containing the amine is fed into an acid scrubber where the free amine is converted into amine sulfate. The effluent of the acid scrubber containing ca. 10% amine sulfate in a mixture with sulfuric acid is then fed into a three cell bipolar membrane electrodialysis apparatus which contains bipolar membranes and anion- and cation-exchange membranes in alternating series in a repeating unit between two electrodes. Here, the amine sulfate is converted back into the free amine and sulfuric acid which is recycled to the acid scrubber. The amine water mixture is distilled to recover the amine and the water is recycled to the electrodialysis unit. Thus, the process allows a complete recovery of the amine from a waste air stream by combination of an acid scrubber and electrodialytic water dissociation. In pilot plant test it was shown that the current utilization strongly depends on amine sulfate concentration and the pH-value in the acid compartment because of the proton leakage through the anion-exchange membrane. At an amine sulfate concentration of 0.5 molL-1 and a pH-value of 1 the current utilization is ca 70% [H.J. Rapp, 1995]. Due to the relatively high price of the amine and savings in sludge disposal costs the regeneration and recovery of the amine from the amine sulfate of the scrubber seems economically very interesting.

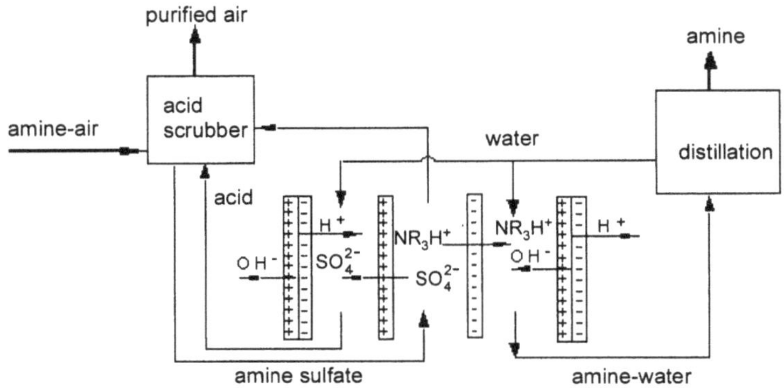

Fig. 6.18 Schematic diagram illustrating the recovery of dimethylisopropylamine from a waste air stream by combination of an acid scrubber, electrodialysis with bipolar membranes, and distillation.

Other applications of bipolar electrodialysis are in the regeneration and recycling of acids or bases from industrial effluents including the sodium hydroxide recycling in pulping and bleaching processes and in aluminum etching processes. The regeneration of ion-exchanger effluents is another potential application of bipolar membrane electrodialysis. Further applications of bipolar membranes in the chemical process industry are tested on the laboratory and pilot plant scale. Some of these applications are treated as company know-how and are not published in the open literature.

3.3 Applications of Bipolar Membrane Electrodialysis in Biotechnology

A very promising application of electrodialysis with bipolar membranes is the recovery of organic acids from fermentation processes. The process is illustrated in Figure 6.19 a) and b) which show the as an example the production of itaconic acid by a conventional batch-type fermentation process and by continuous fermentation with a integrated product recovery process based on bipolar membrane electrodialysis.

During the fermentation of itaconic acid the pH in the fermentation broth shifts to lower values due to the production of the acid. To avoid product inhibition the pH-value of the fermentation broth is maintained at a certain level by addition of sodium or ammonium hydroxide which reacts with the itaconic acid and forms a soluble salt. In the conventional batch type fermentation process illustrated in Figure 6.19 a) the spent medium is separated from the biomass by filtration and the free acid is recovered by lowering the pH-value. The pH-value adjustment in the fermenter as well as in the spent medium is and creates a substantial amount of salts mixed with the product, which complicates the final purification of the itaconic acid and creates an additional waste disposal problem. By integrating bipolar membrane electrodialysis the production of additional salts in the fermentation broth can be eliminated and the fermentation and thus the production of itaconic acid can be carried out more efficiently in a continuous process as illustrated in the simplified flow diagram of Figure 6.19 b).

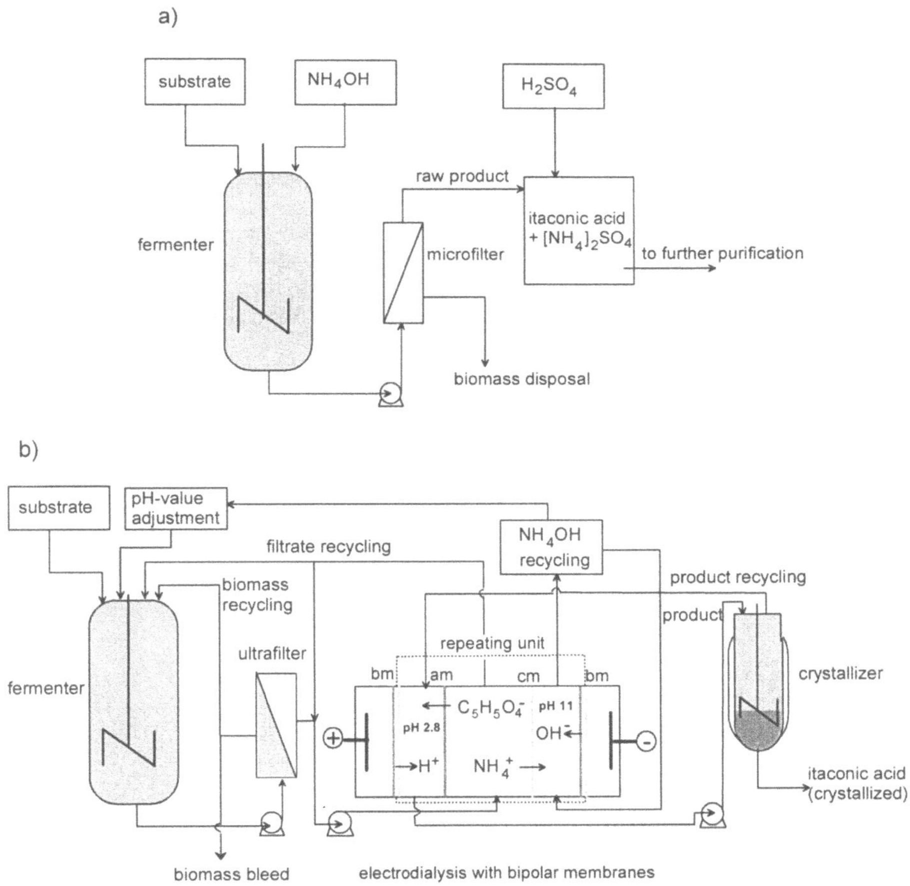

Fig. 6.19 Simplified flow diagram of the production of itaconic acid a) by batch fermentation and b) continuous fermentation with integrated electrodialysis with bipolar membranes.

The fermenter is continuously fed with substrate. Its constituents are passed through an ultrafiltration unit where the biomass is separated from the product containing solution. The most part of the biomass, i.e. ca. 95% is recycled into the fermenter and ca. 5% is bled for disposal to remove dead cell tissue. The product containing filtrate is fed to the middle cell of the three compartment electrodialysis stack repeating unit. In this cell the ions will be removed from the solution by migration due to the applied potential gradient. The cation, i.e. the NH_4^+-ions permeate the cation-exchange membrane and form, with the OH^--ions generated by the bipolar membrane NH_4OH which is concentrated and then fed back to the reactor to adjust the pH-value in the fermentation broth. The anion, i.e. the itaconate migrates through the anion-exchange membrane and forms, with the protons generated by the bipolar membrane, the itaconic acid which is concentrated and precipitated by cooling. Thus, itaconic acid is produced continuously by fermentation with integrated electrodialysis without the addition of bases or acids and the production of additional salt. The continuous production of itaconic acid which was studied extensively on a laboratory scale is just an example for the production of organic acid by fermentation and integrated electrodialysis with bipolar membranes.

There are many more acids that can be produced by this technique [K.N. Mani, 1991]. Of special interest is the production of lactic acid [D.A. Glassner et al., 1990], [Y.H. Kim et al., 2001]. A commercially operated industrial plant has been installed recently by Eurodia Industry [G. Pourcelly et al., 2000]. The simplified flow diagram of the conventional lactic acid production and the production with integrated electrodialysis is given in Figure 6.20 a) and b). In the conventional lactic acid production process which is shown in Figure 20 a) the separation and purification of the lactic acid is achieved mainly by ion-exchange and requires different ion-exchange steps resulting in a large volume of waste water from regeneration salts. The production process with integrated electrodialysis which is shown in Figure 6.20 b) a minimum of ion-exchange resin is needed in a final purification step. The concentration of the lactate salt is achieved by conventional electrodialysis and the conversion of the lactate into lactic acid by bipolar membrane electrodialysis. The simultaneously produced base is recycled to the fermenter to control the pH-value.

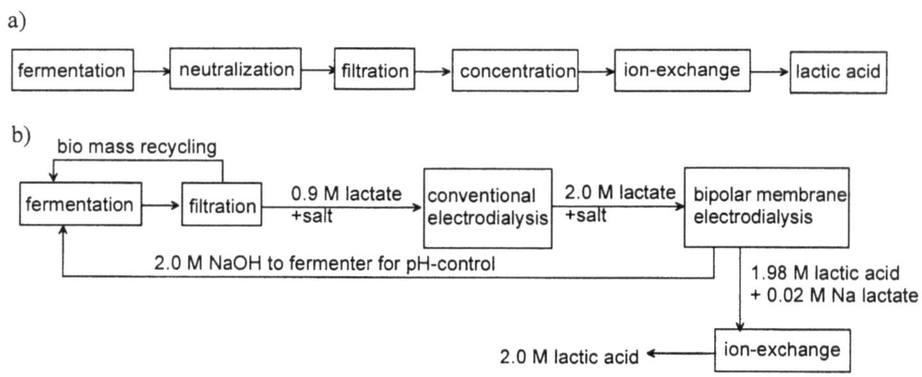

Fig. 6.20 Simplified flow diagram illustrating a) the conventional lactic acid production process and b) the production process with integrated electrodialysis.

Relevant technical data of the plant build by Eurodia Industry, France as given by the authors [G. Pourcelly et al., 2000] are summarized in the following Table:

Table 6.11 Operating data and costs of a bipolar membrane electrodialysis plant used for recovering lactic acid from a fermentation process.

plant capacity	5,000 tons lactic acid per year	
plant design	2 unit system with 900 m^2 membranes in the conventional electrodialysis system and 280 m^2 membranes in electrodialysis system with bipolar membranes	
process characteristics	current utilization 60 %	
	conversion rate 96 %	
pretreatment	cross-flow microfiltration (500 m^2 membrane area)	
	conventional electrodialysis (900 m^2 membrane area)	
post treatment	ion-exchange (9000 L)	
economics	membrane life and costs:	
	mono polar membranes, 7000 h,	0.05 US $ kg^{-1} lactic acid
	bipolar membranes,10,000 h,	0.11 US $ kg^{-1} lactic acid
	energy consumption:	1.43 kWh kg^{-1} lactic acid

Other typical applications of bipolar membrane electrodialysis in biotechnology are the recovery of gluconic acid from sodium gluconate and the production of ascorbic acid from sodium ascorbate. For organic acids with an anion of high molecular weight the migration of the ion through conventional anion-exchange membranes is very slow and the membrane exhibits a high electrical resistance. In this case, the use of a bipolar membrane unit with two repeating units seem to be better suited than the three cell unit in spite of an addition purification step that might be required. The two cell system is shown in the diagram of Figure 6.21.

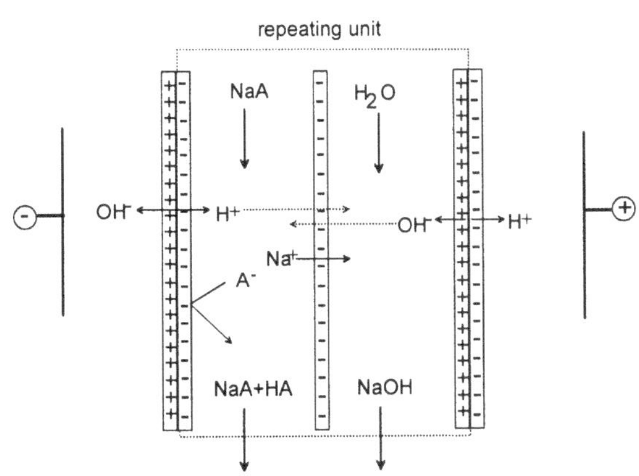

Fig. 6.21 Schematic diagram illustrating the two cell bipolar membrane electrodialysis unit to be used in the recovery of organic acids with large anions.

Whether a three cell or two cell bipolar membrane electrodialysis unit will be used depends on the molecular weight of the organic acid. For low molecular weight components the tree cell system seems to be better suited while for higher molecular weight acid the two cell system seems to have economical advantages. Many applications of bipolar membrane electrodialysis for other separation problems in biotechnology have been tested on a laboratory scale [Th. Lehmann et al., 1995]. Some of these applications are quite promising, others are presently not economic or provide not the same quality products as competitive conventional processes.

An industrial scale bipolar membrane electrodialysis plants are available for various applications today. A photograph of typical industrial size commercial plant is shown in Figure 6.22.

Fig. 6.22 Photograph of an industrial scale bipolar electrodialysis unit (photograph courtesy of Eurodia Industry, S.A., France).

3.4 Applications of Bipolar Membranes in Non-Aqueous Solutions

Bipolar membranes may not only be used for the electrodialytical dissociation of water. They can also be applied for the electrodialytical dissociation of alcohols and thus for the production of alcoholates [S. Shridhar, 1996], [T. Chou et al., 1998]. Methanol, e.g. is both a weak acid and weak base. Like water it can be converted to its conjugated base by a conjugated base of a weaker acid like ammonia. However, the reaction is difficult to perform and requires the complete absence of water.

An alternative route for the production of methanol salt such as sodium methoxide is based on the electrodialytic alcohol dissociation with bipolar membranes. The production of sodium methoxide with bipolar membrane eledctrodialysis is illustrated in the schematic diagram of Figure 6.23 which shows an electrodialysis stack with a bipolar membrane in a repeating unit consisting of two compartments between two electrodes. Water-free methanol and sodium acetate are fed into the cell formed by the bipolar membrane and the cation-exchange membrane which is directed towards the cathode while water-free methanol is fed into the other cell directed towards the anode. Due to an electrical potential gradient between the electrodes methanol is dissociated in the bipolar membrane into protons and CH_3O^--ions. The proton forms acetic acid with the acetate ions wile the CH_3O^--ions react with the Na^+-ions which migrate from the adjacent sodium acetate containing cell and form CH_3ONa. Thus, sodium acetate and methanol are converted to sodium methoxide and acetic acid.

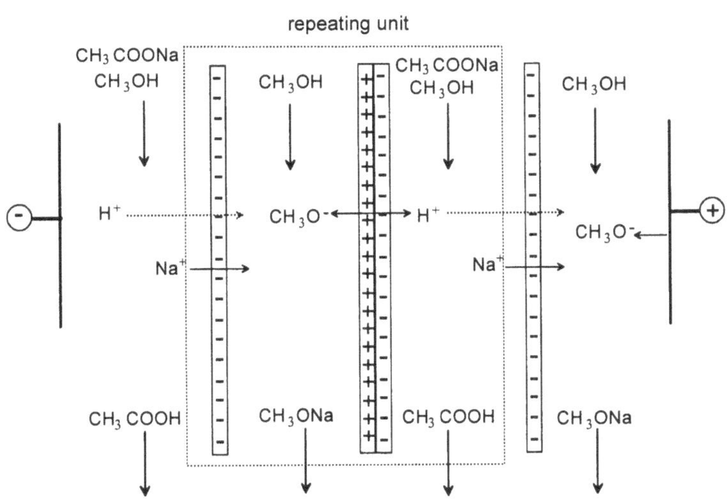

Fig. 6.23 Schematic drawing illustrating the production of sodium methoxide by electrodialysis with bipolar membranes from methanol using sodium acetate as electrolyte.

The current utilization is rather low and decreases with increasing methylate and acetic acid concentration, respectively, because of proton transport from the acetic acid compartment into the sodium methoxide containing compartment. The electrical resistance of the methanol solution is also significantly higher that in an aqueous electrolyte solution which results in higher energy consumption. However, at least technically the process seems to be feasible.

3.5 Other Processes Involving Bipolar Membrane Electrodialysis

Bipolar membrane electrodialysis in combination with ion-exchange resins can also be utilized in the continuous electrodeionization process for the production of high purity water as has been mentioned before. The process has been studied extensively and its advantages as well disadvantages have been pointed out [S. Thate, 2002]. A similar process is described to purify water with ion-exchange resins filled cells in a system having individual anion and cation depletion compartments. The anion depletion cell is composed of a bipolar and an anion-exchange membrane and contains the anion-exchange resin, while the cation depletion cell is composed of a bipolar membrane and a cation-exchange membrane and contains the cation-exchange resin. Between the two depletion cells is the ion concentration cell [A.J. Giuffrida et al.,1993].

4 PRACTICAL APPLICATIONS OF DIFFUSION DIALYSIS

Diffusion dialysis is well established today as a relevant commercial process with a number of applications mainly in the treatment of industrial effluents from the metal processing industry. There are several potential applications in the chemical process industry for the purification of acids or bases or in the food industry for deacidification of fruit juice. But these applications are presently of minor commercial relevance.

4.1 Waste Water Treatment by Diffusion Dialysis

Typical applications are:

- recovery of an acid from the ion-exchange regeneration process,
- recovery of acids such as HF, HNO_3, and H_2SO_4 from spent pickling baths,
- recovery of HCl from spent aluminum foil etching baths of a capacitor production,
- recovery of NaOH from spent aluminum etching baths,
- recovery of H_2SO_4 from spent aluminum anodizing baths.

The major commercial application of diffusion dialysis is the recovery of acids from pickling solutions used to remove metal oxides from steel parts prior to surface finishing. A large number of diffusion dialysis plants is in operation since many years in Japan and many are being presently installed in the USA and Europe. The recovery of HF and HNO_3 from spent pickling solutions seems to be particularly attractive because these acids are relatively expensive and are causing severe pollution problems. But the recovery of HCl from aluminum etching rinse water or H_2SO_4 from steel pickling processes is also technically and economically feasible.

As an example for a successful large scale application of diffusion dialysis is a plant located in a Nippon Stainless Co. Ltd. In the plant 0.9 m^3 h^{-1} of a spent stainless steel pickling bath is treated [Y. Kobuchi et al., 1983, 1986, 1987]. A simplified flow scheme of the process is shown in Figure 6.24.

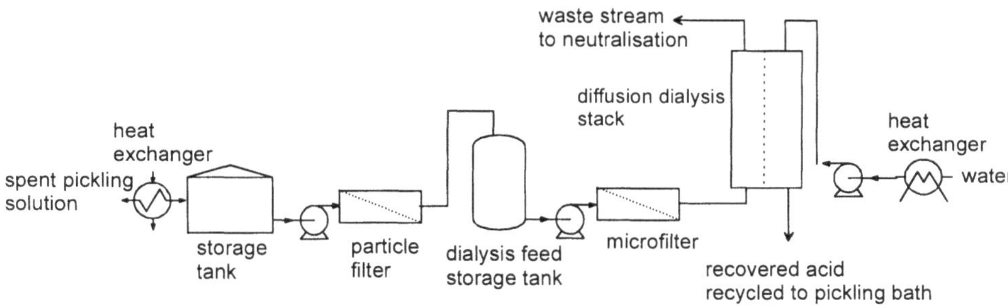

Fig. 6.24 Flow scheme of a diffusion dialysis plant to recover HNO_3 and HF from 0.9 m^3 d^{-1} of a spent stainless steel pickling bath.

After being cooled down to 45 to 50°C the spent pickling bath solution is stored in a holding tank. The spent pickling solution contains dissolved ferric, nickel, and chromium ions in addition to a substantial amount of suspended metal particles derived from surface scale. The suspended solids are removed in a non-woven fiber filter and the solution is stored in the dialysis feed tank. Then the solution is passed through a cartridge filter and fed to the diffusion dialysis stack. Water heated to ca. 40°C is pumped in counter-current flow as acid stripping solution through the stack. The heating is necessary to enhance the recovery of the acid and to reduce the air release from the water during the dialysis process. The actual commercial diffusion dialysis stack used in this application is a Tokuyama Soda TSD-50-2000 model which has an active cell pair area of 0.5 m^2 and 2000 membranes in one unit. The flow velocity is very low and in the range of 0.5 cm s^{-1}. The slow flow velocity provides the necessary process path length for the continuous operation and avoids pressure differences between the feed and the stripping solution which could lead to convective flow between the two solutions through the membranes. Due to the low flow rate air is very poorly removed from the water in the stripping solution and can block part of the membrane surface. Therefore, the water must be degassed by heating. The recovered acid is adjusted in concentration with fresh acid and recycled to the pickling tank. The results and relevant operating data of the diffusion dialysis unit are summarized in Table 6.12.
The performance of the diffusion dialysis plant satisfactory It has been operated reliably without major problems for several years. The recovered acid is recycled to the pickling bath without further treatment. The waste acid and 95% the Fe^{3+}-ions in the stripped feed solution go directly to the waste disposal system. A cost analysis shows that the cost savings from the recovered

acids, and the reduction in neutralization chemicals and sludge disposal make the process extremely attractive with a pay-back time of the investment of less than one year.

Table 6.12 Equipment specification and performance data of a diffusion dialysis unit recovering acids from a spent stainless steel pickling bath solution.

diffusion dialyser	TSD 50-2000
membrane	Neosepta AFN
plant capacity	0.9 $m^3\ h^{-1}$
membrane area	950 m^2
composition of spent pickling solution	
HNO_3	55 [g L^{-1}]
HF	11 [g L^{-1}]
Fe^{3+}-ion	21 [g L^{-1}]
acid recovery rate	
HNO_3	97%
HF	50%
Fe^{3+}-ion elimination rate	95%

5 APPLICATIONS OF CONTINUOUS ELECTRODEIONIZATION

The demand by industry for high purity water is rapidly increasing. Especially the electronic industry but also analytical laboratories and power stations need water which is free of particles and organic matter, and has a conductivity of less than 0.06 μS cm^{-1}. Generally, well or surface water is purified in a series of processes that include water softening, microfiltration, reverse osmosis, ultrafiltration, UV-sterilization, and mixed bed ion-exchange. A typical pure water production line is shown in the simplified diagram of Figure 6.25.

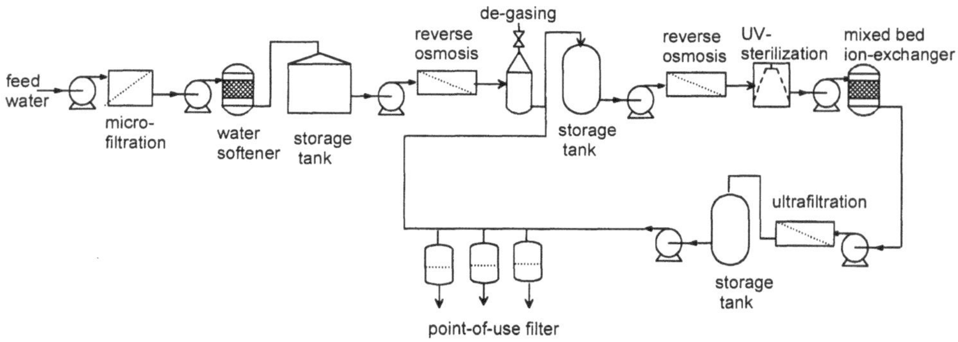

Fig. 6.25 Schematic diagram illustrating a conventional ultra pure water production process line.

The process line indicated in Figure 6.25 may be changed depending on the quality of the raw water and might include also a precipitation or a carbon adsorption step. But the key processes are the reverse osmosis, microfiltration, and mixed bed ion-exchange. While reverse osmosis micro- and ultrafiltration or UV-sterilization could all be operated in a continuous mode, the mixed bed ion-exchanger, however, which is necessary to reach the required low conductivity, must be regenerated in certain time intervals. This regeneration is not only labor intensive and costly but it requires extremely long rinse down times to remove traces of regeneration chemicals. It is furthermore effected by biological fouling due to micro organisms brought into the resin during the regeneration process and an additional ultrafiltration step is required before the ultra pure water can be stored for final use. By replacing the mixed bed ion-exchanger in the ultra pure water production line the process can be substantially simplified producing consistently high quality water in a completely continuous process. The most important advantage of electrodeionization is that no chemical regeneration of the ion-exchange resins is required. Thus, avoiding contamination of the product water and the need of chemicals is drastically reduced. As a result, electrodeionization provides a continuous process with a reduction in system components, regeneration chemicals, and leads to a substantial reduction in the overall water production costs.

The electrodeionization is used today on a very large scale for the production of boiler feed water used in power generation plants. Here, the requirements for the water purity are not as stringent as in the electronic industry. Generally, a conductivity below 0.1 $\mu S\ cm^{-1}$ is acceptable. The demand for high quality water in the steam generation plants is enormous and readily increasing. Large power plants have a demand for pure water of more than 400 $m^3\ h^{-1}$. Pure water production plants based on microfiltration, reverse osmosis, and electrodeionization with a capacity of a few 100 $m^3\ h^{-1}$ which guaranty a water conductivity of less than 0.06 $\mu S\ cm^{-1}$ are now commercially available [J. D'Ailly, 2000]. For the production of high quality boiler feed water a relatively simple water treatment train of processes, which includes an electrodeionization unit, is used today as shown in the schematic drawing of Figure 6.26.

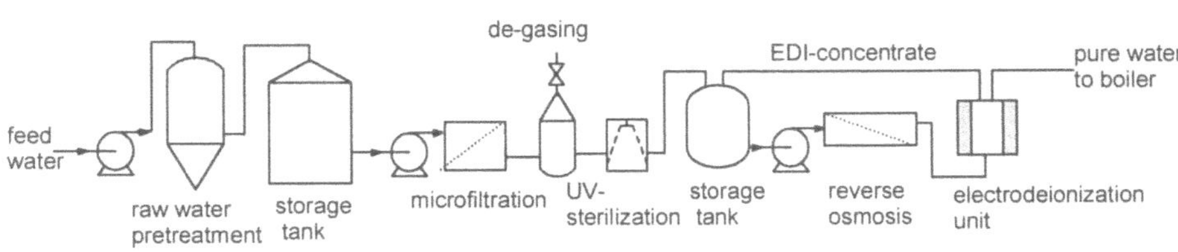

Fig. 6.26 Ultra pure water production line used for the boiler feed water preparation.

The advantages of the ultra pure water production with an integrated electrodeionization unit compared to the use of a mixed bed ion-exchanger are a simpler process, no need of regeneration chemicals, less raw water consumption, and a substantial reduction in costs.

6 PRACTICAL APPLICATIONS OF DONNAN DIALYSIS

So far, Donnan dialysis is not used on a large industrial scale. Most data concerning the performance and process cost data have been generated with laboratory or pilot plant scale equipment in relatively short term tests. The major applications of Donnan dialysis would be in water softening and the removal of certain heavy metal ions from waste water. But apparently here the process has no significant technical or economical advantage to the conventional ion-exchange process.

LITERATURE

Ahlgreen, R.M., 1972. Electromembrane processing of cheese whey, in: Industrial processing with membranes, edts.: Lacey, E.R., Loeb, S., pp 57-69, Wiley Interscience, New York.

Andres, Z., Riera, F.A., Alvarez, R., 1995. Skimmed milk demineralization by electrodialysis: Conventional versus selective membranes, J. Food Eng., 26: 57-66.

Bleha, M., Tishenko, G., Sumberova, V., Kudela, V., 1992. Characterization of the critical state of membranes in electrodialysis desalination of milk whey, Desalination, 86: 173-186.

Chou, T., Tanioka, A., 1998. Current voltage curves of composite bipolar membrane in alcohol-water solutions, J. Physic. Chem. B, 102: 7866-7870.

D'Ailly, J., 2000. E-cell promises chemical-free future, Membrane Technology, 125: pp. 12-14, ISSN 0958-2118, Elsevier, Amsterdam.

Geishecker, E.P., 1986. Reuse of contaminated water, First Biennial Conference National Water Supply Improvement Association, Washington, D.C. June 8-12.

Glassner, D.A., Datta, R., 1990. Process for production and purification of lactic acid, European Patent 393818 A1.

Greiter, M., Novalin, S., Wendland, M., Kulbe, K.D., Fischer, J., 2002. Desalination of whey by electrodialysis and ion-exchange resins: Analysis of both processes with regard to sustainability by calculating their cumulative energy demand, J. of Membrane Sci., 210: 91-102.

Giuffrida, A.J., Ganzi, G.C., 1993. Process for purifying resins utilizing bipolar interface, US Patent 4,871,431.

Ito, S., Nakamura, I., Kawahara, T., 1980. Electrodialytic recovery process of metal finishing waste water, Desalination, 32: 383-389.

Johann, J., Eigenberger, G., 1993. Electrodialytische Regenerierung von Ionenaustauscherharzen, Chem.-Ing.-Tech., 65: 75-78.

Kawahara, T., 1992. Industrial applications of ion-exchange membranes, part 1: from Application of ED technology from desalination to concentration, Desalination &Water Reuse, Vol. 2/4, pp. 26-30.

Kim, H.Y., Moon, S.H., 2001. Lactic acid recovery from fermentation broth using one-stage electrodialysis, J. Chem. Technol. Biotechnol., 76: 169-178.

Kneifel, K., Luhrs, G., Wagner, G., 1988. Nitrate removal by electrodialysis from brewing water, Desalination, 68: 203-209.

Kobuchi, Y., Motomura, H., 1983. Demineralization of whey by multistage continuous electrodialysis system, Presented at the International Membrane Technology Conference, November 8-10, 1983, Sydney Australia.

Kobuchi, Y., Matsunaga, Y., Noma, Y., Hanada, F., 1987. Application of ion-exchange membranes to the recovery of acids by diffusion dialysis and electrodialysis, in: Synthetic Polymeric Membranes, edts.: Sedlacek, B., Kahovec, J.,pp. 412-428, W. de Gruyter, Berlin, Germany.

Kobuchi, Y., Motomura, H., Noma, Y., Hanada, F., 1986. Application of ion-exchange membranes to the recovery of acids by diffusion dialysis, J. Membrane Sci., 27: 173-179.

Korngold, E. , Kock, K., Strathmann, H., 1978. Electrodialysis in advanced waste water treatment, Desalination, 24: 129-139.

Lee, E.G., Moon, S.H., Chang, Y.K., Yoo, I, Chang, H-N., 1998. Lactic acid recovery using two-stage electrodialysis and its modeling, J. Membrane Sci., 145: 53-66.

Lee, H.J., Oh, S.J., Moon, S.H., 2002. Removal of hardness in fermentaion broth by electrodialysis, J. Chem. Technol. Biotechnol., 77: 1005-1012.

Lehmann, Th., Engel, D., 1995. Evaluation of industrial electrodialysis for the processing of amino acids, Conference on: Electrochemical processing, Toulouse, France, April 24-27.

Liu, K.J., Chlanda, F.P., Nagasubramanian, K.J., 1977. Use of bipolae membranes for generation of acid and base: an engineering and economic analysis, J. Membrane Sci., 2: 109-124.

Liu, K.J., Chlanda, F.P., Mani, K.N., 1978. Application of bipolare membrane technology: a novel process for control sulfur dioxide from flue gases, J. Membrane Sci., 3: 57-70.

Mani, K.N., 1991. Electrodialysis water-splitting technology, J. Membrane Sci., 58: 117-138

Mani, K.N., Chlanda, F.P., 1986. Electrodialytic water splitter and process for removal of SO_2 from gases containing same for the recovery of SO_2 and NaOH, US Patent 4,629,545.

McArdle, J.C., Piccari, J.A., Thornburg, G.G., 1991. Aquatech System's pickling liquor recovery process Washington steel reduces waste disposal costs and liability, Iron and Steel Engineer, 68: 39.

Pourcelly, G., Gavach, G.,2000. Electrodialysis water splitting- Application of electrodialysis with bipolar membranes, in: handbook bipolar membrane technology, edt.: A.J.B. Kemperman, Twente University Press, Enschede, The Netherlands.

Rapp, H.J., 1995. Die Elektrodialyse mit bipolaren Membranen, Theorie und Anwendungen, Ph.D.-Thesis, University of Stuttgart, Germany.

Rapp, H.J., Pfromm, P.H., 1998. Electrodialysis for chloride removal from the chemical recovery cycle of a Kraft pulp mill, J. Membrane Sci., 49: 1-13.

Rapp, H.J., Pfromm, P.H., 1998. Electrodialysis field test for selective chloride removal from the chemical recovery cycle of a Kraft pulp mill, Ind. Eng. Chem. Res., 37: 4761-4767.

Rapp, H.J., 2001. Säureaufbereitung in kombinierten Prozessen mit Elektrodialyse, Proceedings 8. Aachener Membrane Kolloquium, March, 27-29.

Reed, P.B., 1984. Electrodialysis for the purification of protein solutions, Chem. Eng. Progress, Dec.1984, pp.47-50.

Schmauss, L.R., Aiken, J.W., 1984. Economics of the electrodialysis reversal process in conjunction with ion-exchange, Ultrapure Water, pp. 32-35, November/December.

Sridhar, S., 1996. Electrodialysis in non-aqueous medium: production of sodium methoxide, Journal of Membrane Science, 113: 73-79.

Thate, S., 2002. Untersuchung der elektrochemischen Deionisation zur Reinstwasserherstellung, PhD.-Thesis, ISBN 3-89722-911-0, University of Stuttgart, Germany.

Tran, T.V., Clemens, P.B., 1986. Recovery of nickel salts by electrodialysis reversal process, Presented at 73rd annual AESF-Conference, Philadephia, PA June 23-26.

Tokuyama Soda Technical Information, 1987.

Tokuyama Soda Brochure 2001.

Turek, M. 2002., Cost effective electrodialytic seawater desalination, Desalination, 153: 371- 376.

Valcour, H.C., 1985. Recent application of EDR, 2nd World Congress on Desalination and Water Reuse, Hamilton , Bermuda, November 17-22.

Voss, H., 1987. Deacidification of citric acid solutions by electrodialysis, J. Membrane Sci., 27: 165-171.

Wangnik, K., 2000. Desalination plants inventory report, published by: The International Desalting Association.

Yamane, R., Ichikawa, M., Mitzutani, Y., Onoue, Y., 1969. Concentrated Brine production from sea water by electrodialysis using ion-exchange membranes, I&EC Process design and Development, 8: 159-565.

Yamamoto, M., Hanada, F., Futaki, S., Takashima, K., 2000. A new electrodialyser technique for the salt production by ion-exchange membrane, in: Eighth Symposium on Salt, edt.: Geertman, R.M., Vol. I , pp. 647-652, Elsevier, Amsterdam.

Appendix I

Table Properties of selected commercial membranes

trade name	type	resistance[1)] $[\Omega cm^2]$	transport number[2)]	water content[3)] [wt %]	thickness[4)] [mm]	remarks
Aciplex[R] (Asahi Chemical Industry)						
K 192	strongly acidic	1.5 – 2.0	0.99	35 - 40	0.13 -0.17	monovalent cation selective, sea water concentration
K 501	strongly acidic	2 - 3	0.99	30 - 40	0.18 - 20	desalting of food and bio-products
K 521	strongly acidic	2 - 3	0.99	25 - 30	0.14 - 0.18	low diffusion rates, desalting of bio-products
A 192	strongly basic	1.8 - 2.0	0.99	25 - 35	0.12 - 0.15	monovalent anion selective, sea water concentration
A 201	strongly basic	3.6 - 4.0	0.99	25 - 28	0.20 - 0.26	desalting of food and bio-products
A 221	strongly basic	1.4 - 1.7	0.99	24 - 30	0.17 - 0.2	high acid diffusion, diffusion dialysis
A 700	strongly basic	3.5 - 5.5	0.99	13 - 18	0.10 - 0.13	high acid retention, concentration of acids
Acidiplex-F[R] (Asahi Chemical Industry)						
F 2204	laminated membrane	2.5 - 2.7	0.97	10 -20		chlor-alkali electrolysis, production of < 24 % NaOH

Table continued

F 3201	laminated membrane	2.9 - 3.0	0.97	10 -20		chlor-alkali electrolysis, production of < 26 % NaOH
F 4202	laminated membrane	2.0 - 2.2	0.97	10 -20		chlor-alkali electrolysis, production of < 34 % NaOH
Neosepta[R] (Tokuyama Soda)						
CM-1	strongly acidic	1.0 - 2.0	0.98	35 - 40	0.13 - 0.15	low electrical resistance, general desalination
CM-2	strongly acidic	2.0 - 3.0	0.98	25 - 35	0.13 - 0.15	low diffusion rates, desalting of bio-products
CMS	strongly acidic	1.5 - 1.8	0.98	35 - 40	0.15 - 017	monovalent cation selective, sea water concentration
CMX	strongly acidic	2.5 - 3.5	0.98	25 - 35	0.17 - 0.2	high mechanical strength, general desalination
AM-1	strongly basic	1.3 - 1.8	0.98	25 - 35	0.12 - 0.16	low electrical resistance, general desalination
AM-3	strongly basic	2.0 - 3.0	0.98	15 - 25	0.12 - 0.16	low diffusion rates, general desalting
ACS-3	strongly basic	1.5 - 2.0	0.98	20 - 30	0.10 - 0.12	monovalent anion selective, sea water concentration
AMX	strongly basic	2.0 - 3.5	0.98	25 - 35	0.16 - 0.18	high mechanical strength, general desalination

Table continued

Nepton^R (Ionics Incorporated)						
CR61CZL-386	strongly acidic	6 - 9	0.95	40 - 45	0.6 - 0.7	high mechanical strength, general desalination
CR67HMR-412	strongly acidic	4 - 6	0.95	40 - 45	0.5 - 0.6	high mechanical strength, general desalination
AR103PZL-386	strongly basic	10 - 12	0.95	45 - 48	0.6 - 0.7	high mechanical strength, general desalination
AR103QZL-389	strongly basic	8 - 9	0.97	35 - 37	0.6 - 0.7	high mechanical strength, general desalination
AR204SZRA-412	strongly basic	5 - 6	0.97	45 - 48	0.5 - 06	low organic fouling, general desalination
Pall RAI, Inc.						
R-5010-L	strongly acidic	2 - 4	0.90	40 - 45	0.2 - 0.3	good thermal stability, general desalination
R-5010-H	strongly acidic	8 - 10	0.95	20 - 25	0.2 - 0.3	good thermal stability, general desalination
R-5030-L	strongly basic	4 - 7	0.85	30 -35	0.2 - 0.3	good thermal stability, general desalination
R-5030-H	strongly basic	11 -15	0.90	20 - 25	0.2 - 0.3	good thermal stability, general desalination

Table continued

Ionac (Sybron Corp)						
MC 3470	strongly acidic	6 - 9	0.75	30 - 35	0.5 - 0.7	mechanically strong, general desalination
MA 3475	strongly basic	5 - 10	0.80	30 - 32	0.5 - 0.7	mechanically strong, general desalination
Selemion[R] (Asahi Glass)						
CMV	strongly acidic	2 - 3	0.92	20 - 25	0.1 - 0.2	low resistance, general desalination
AMV	strongly basic	2 - 3	0.92	20 - 25	0.1 - 0.2	low resistance, general desalination
DSV	strongly basic	1 - 2	0.92	18 - 20	0.1 - 0.2	high proton permeability, diffusion dialysis
ASV	strongly basic	3 - 4	0.95	20 - 25	0.1 - 0.2	monovalent anion selective, sea water concentration
AAV	strongly basic	4 - 6	0.95	20 - 25	0.1 - 0.2	high proton retention, acids concentration
Flemion[R] (Asahi Glass)						
F 795	laminated membrane		0.95	10 - 20	0.3 - 0.4	chlor-alkali electrolysis, production of < 30 % NaOH

Table I (continued)

F 893	laminated membrane		0.95	10 - 20	0.3 - 0.4	chlor-alkali electrolysis, production of < 30 % NaOH

Table continued

Nafion[R] (Dupont)						
N-117	strongly acidic	2 - 3	0.96	15 - 20	0.2 - 0.3	highly chemically stable, fuel cell separator
N981	bi-layer membrane	4 - 5	0.96	10 - 15	0.4 - 0.5	chlor-alkali electrolysis, production of < 35 % NaOH

1) Determined with direct current in 0.5 n NaCl-solution, 2) determined by potential measurement between 0.5/1 n NaCl-solutions, 3) determined 0.5 nNaCl-solution, 4) determined 0.5 n NaCl-solution, or taken from manufacturers data lists

Appendix II

Table 1 Important Constants (Data from: F.W. Küster, A. Thiel, K. Fischbeck, Logarithmische Rechentafeln, De Gruyter, Berlin 1962)

quantity	symbol	value
Avogadro number	N_A	6.02214×10^{23} [mol^{-1}]
elementary charge	e	1.60218×10^{-19} [C]
Faraday constant	$F = e N_A$	96485 [$C\ mol^{-1}$], [$C\ val^{-1}$]
speed of light (in vacuum)	c	2.9979×10^{8} [$m\ s^{-1}$]
Boltzmann constant	k	1.38066×10^{-23} [$J\ K^{-1}$]
gas constant	$R = k N_A$	8.3145 [$J\ K^{-1}\ mol^{-1}$]
vacuum permittivity	ε_o	8.8542×10^{-12} [$C^2\ J^{-1}\ m^{-1}$]
proton mass	m_p	1.6726×10^{-27} [kg]
electron mass	m_e	9.1094×10^{-31} [kg]
standard acceleration	g	9.8066 [$m\ s^{-2}$]

Table 2 Mechanical Units (Data from: F.W. Küster, A. Thiel, K. Fischbeck, Logarithmische Rechentafeln, De Gruyter, Berlin 1962)

quantity	symbol	definition	unit (m kg s)	unit (cm g s)
length	l	basic quantity	meter [m]	centimeter [cm]
time	t	basic unit	second [s]	second [s]
mass	m	basic unit	kilogram [kg]	gram [g]
area	A		m^2	cm^2
volume	V		m^3	cm^3
velocity	u	dl/dt	$m\ s^{-1}$	$cm\ s^{-1}$
acceleration	g	du/dt	$m\ s^{-2}$	$cm\ s^{-2}$
force	F	mass × acceleration	Newton [$N = m\ kg\ s^{-2}$]	dyn [$cm\ g\ s^{-2}$]
work	W	force × distance	Joule, Watt x second	erg [$cm^2\ g\ s^{-2}$]
power	P	work/time	Watt [$W = m^2\ kg\ s^{-3}$]	[$cm^2\ g\ s^{-3}$]

Table 3 Electrical Units (Data from: F.W. Küster, A. Thiel, K. Fischbeck, Logarithmische Rechentafeln, De Gruyter, Berlin 1962)

quantity	symbol	definition	unit (m kg s)	other units
electric energy	W s	Watt second	m^2 kg s^{-2}	J = V A s = N m
electric power	W	Watt	m^2 kg s^{-3}	VA
electrical potential	U	Volt [V]	m^2 kg s^{-3} A	Ω A
electric current	I	Ampere [A]	A	C s^{-1}
electric charge	q	Coulomb [C]	A s	J V^{-1}
electric field	E		m kg s^{-3} A^{-1}	V m^{-1}
electric resistance	Ω	Ohm [Ω]	m^2 kg s^{-3} A^{-2}	V A^{-1}
electric conductivity	S	Siemens [S]	m^{-2} kg^{-1} s^3 A^2	Ω^{-1}
molar conductivity	Λ_m		S m^2 mol^{-1}	m^2 mol^{-1} Ω^{-1}
ion mobility	u		m^2 s^{-1} V^{-1}	C kg^{-1} s
capacitance	C	Farad [F]	m^{-2} kg^{-1} s^4 A^2	A s V^{-1}

Table 4 Conversion of units used in the USA and theUK into metric units (Data from: F.W. Küster, A. Thiel, K. Fischbeck, Logarithmische Rechentafeln, De Gruyter, Berlin (1962)

USA units	multiplied by	metric units
mil	25.40	μm
inch [in]	2.540	cm
foot [ft]	0.3047	m
yard [yd]	0.9144	m
mile [mi]	1.609	km
square inch	6.452	cm^2
square foot	0.09290	m^2
acre	4046	m^2
square mile	2.950	km^2
cubic inch	16.387	cm^3
cubic foot	28.317	dm^3, L

Table 4 (continued)

USA units	multiplied by	metric units
US gallon [gal]	3.785	dm^3, L
UK gallon [gal]	4.546	dm^3, L
pound [lb]	0.45359	kg
ounce [oz]	28.35	g
pounds per square inch [psi]	0.0689	bar, 10^5 Pa
gallons per square foot day [gsfd]	0.0407	$m^3\ m^{-2}\ day^{-1}$
British thermal units [BTU]	1055	J, $m^2\ kg\ s^{-2}$
Fahrenheit [°F]	0.555 x °F –17.8	Celsius [°C]

Table 5 Conversion of Energy Units (Data from: F.W. Küster, A. Thiel, K. Fischbeck, Logarithmische Rechentafeln, De Gruyter, Berlin 1962)

	Joule	liter atmosphere	kilowatt hour	calorie	British thermal unit
[J]	1	0.9868×10^{-2}	2.7778×10^{-7}	0.2388	9.478×10^{-4}
L atm	1.013×10^{2}	1	2.8146×10^{-5}	24.20	9.353×10^{-6}
kWh	3.600×10^{6}	3.5528×10^{4}	1	8.598×10^{5}	3.448×10^{3}
cal	4.1840	4.1319×10^{-2}	1.1622×10^{-6}	1	3.965×10^{-3}
BTU	1.055×10^{3}	10.421	2.930×10^{-4}	2.522×10^{2}	1

Table 6 Conversion of Pressure Units (Data from: F.W. Küster, A. Thiel, K. Fischbeck, Logarithmische Rechentafeln, De Gruyter, Berlin 1962)

	pounds per square inch [psi]	atmosphere [atm] (760 Torr)	atmosphere [at] (kp cm^{-2})	bar [bar]	Pascal [Pa]	mm Hg (20°C)
psi	1	0.06805	0.0703	0.06895	6.895×10^{3}	51.71
atm	14.7	1	1.033	1.013	1.013×10^{5}	762,8
at	14.2	0.968	1	0.981	0.981×10^{5}	738,2
bar	14.504	0.9869	1.0097	1	10^{5}	752.8
Pa	0.0001450	0.9869×10^{-5}	1.009×10^{-5}	10^{-5}	1	0.00753
torr	0.1934	1.3158×10^{-3}	0.001359	0.001333	133.3	1.0036

Table 7 The Water Ion Product Constant K_w and the pH-value of Water at Different Temperatures(Data from: F.W. Küster, A. Thiel, K. Fischbeck, Logarithmische Rechentafeln, De Gruyter, Berlin 1962)

Temperature [° C]	K_w [mol^2 L^{-2}]	pH
0	0.114×10^{-14}	7.472
10	0.292×10^{-14}	7.267
15	0.451×10^{-14}	7.173
20	0.681×10^{-14}	7.083
25	1.008×10^{-14}	6.998
30	1.469×10^{-14}	6.916
35	2.089×10^{-14}	6.840
40	2.919×10^{-14}	6,767
45	4.018×10^{-14}	6.698
50	5.474×10^{-14}	6.631
60	9.610×10^{-14}	6.508

Table 8 Ionic Radii of Various Ions (Data from: Electrolyte solutions, R.A. Robinson and R.H. Stokes, Butterworth Publishers London, 1970)

ion	radius [nm]	ion	radius [nm]
Li^+	0.06	F^-	0.133
Na^+	0.095	Cl^-	0.181
K^+	0.133	Br^-	0.196
NH_4^+	0.145	J^-	0.219
Ag^+	0.126	OH^-	0.145
Ca^{++}	0.099		
Mg^{++}	0.065		
Fe^{++}	0.075		

Table 9 Limiting Equivalent Conductivities of Ions at 25°C in Water (Data from: Electrolyte solutions, R.A. Robinson and R.H. Stokes, Butterworth Publishers London, 1970)

cations	λ^o_{equiv} [$cm^2\Omega^{-1} equiv^{-1}$]	anions	λ^o_{equiv} [$cm^2\Omega^{-1} equiv^{-1}$]
H^+	349.8	OH^-	199.1
Li^+	38.6	F^-	55.4
Na^+	50.1	Cl^-	76.4
K^+	73.5	Br^-	78.1
Ag^+	61.9	NO_3^-	71.5
NH_4^+	73.5	HCO_3^-	44.5
Mg^{2+}	53.0	CO_3^{2-}	69.3
Ca^{2+}	59.5	ClO_3^-	64.6
Ba^{2+}	63.6	ClO_4-	67.3
Cu^{2+}	53.6	SO_4^{2-}	80.0
Zn^{2+}	52.8	$Fe(CN)_6^{3-}$	100.9
Co^{2+}	55.0	formeate	54.6
Pb^{2+}	69.5	acetate	40.9
La^{3+}	69.7	$P_3O_9^{3-}$	83.6
Ce^{3+}	69.8	$Fe(CN)_6^{4-}$	110.1

Table 10 Limiting Equivalent Conductivities of Ions in Water at Different Temperatures (Data from: Electrolyte solutions, R.A. Robinson and R.H. Stokes, Butterworth Publishers London, 1970)

ion	0°C	5°C	15°C	18°C	25°C	35°C	45°C	55°C	100°C
H^+	225.0	250.1	300.6	315.0	349.8	397.0	441.4	483.1	630.0
OH^-	105.0		165.9	175.8	199.1	233.0	267.2	301.4	450.0
Li^+	19.4	22.7	30.2	32.8	38.6	48.0	58.0	68.7	115.0
Na^+	26.5	30.3	39.7	42.8	50.1	61.5	73.7	86.8	145.0
K^+	40.7	46.7	59.6	63.9	73.5	88.2	103.4	119.2	195.0

Table 10 (continued)

ion	0°C	5°C	15°C	18°C	25°C	35°C	45°C	55°C	100°C
Mg^{2+}	28.9			44.9	53.0				165.0
Ca^{2+}	31.2		46.9	50.7	59.5	73.2	88.2		180.0
F^-				47.3	55.4				
Cl^-	41.0	47.5	61.4	66.0	76.35	92.2	108.9	126.4	212.0
Br^-	42.6	49.2	63.1	68.0	78.1	94.0	110.6	127.8	
NO_3^-	40.0			62.3	71.5	85.4			195.0
SO_4^{2-}	41.0			68.4	80.0				260

Table 11 Equivalent Conductivities of NaCl, KCl and BaCl as a Function of Concentration at 25°C (Data from: Electrolyte solutions, R.A. Robinson and R.H. Stokes, Butterworth Publishers London, 1970)

Concentration [mol L^{-1}]	NaCl	KCl	$BaCl_2$
0.0	126.5	149.9	140.0
0.0005	124.5	147.8	134.3
0.001	123.7	146.9	132.3
0.005	120.7	143.6	132.27
0.01	118.5	141.3	119.1
0.05	111.1	133.4	105.2
0.1	106.7	128.9	98.7
0.2	101.7	124.1	
0.5	93.6	117.3	80.6
1.0	85.8	111.9	68.0
1.5	79.9	108.3	
2.0	74.7	105.2	
3.0	65.6	99.5	
4.0	57.2	93.5	

Table 12 Activity coefficients of Electrolytes at 25°C at Various Concentrations (Data from: Electrolyte solutions, R.A. Robinson and R.H. Stokes, Butterworth Publishers London, 1970)

Concentration [mol kg^{-1}]	HCl	HBr	HNO_3	$HClO_4$	H_2SO_4	NaOH	KOH
0.1	0.796	0.805	0.791	0.803	0.265	0.764	0.776
0.2	0.767	0.782	0.754	0.778	0.209	0.725	0.739
0.3	0.756	0.777	0.735	0.768	0.183	0.706	0.721
0.4	0.755	0.781	0.725	0.766	0.167	0.695	0.713
0.5	0.757	0.789	0.720	0.769	0.155	0.688	0.712
0.6	0.763	0.801	0.717	0.776	0.148	0.683	0.712
0.7	0.772	0.815	0.717	0.785	0.142	0.680	0.715
0.8	0.783	0.832	0.718	0.795	0.137	0.677	0.721
1.0	0.809	0.871	0.724	0.823	0.136	0.677	0.735
1.2	0.840	0.917	0.734	0.858		0.679	0.754
1.4	0.876	0.969	0.745	0.900		0.684	0.778
1.6	0.916	1.029	0.758	0.947		0.690	0.804
1.8	0.960	1.094	0.775	0.998		0.698	0.832
2.0	1.009	1.168	0.793	1.055	0.128	0.707	0.863
2.5	1.147	1.389	0.846	1.227	0.133	0.741	0.947
3.0	1.316	1.674	0.909	1.448	0.142	0.782	1.051
3.5	1.518			1.726	0.155	0.833	1.181
4.0	1.762	2.415	0.950	2.08	0.170	0.901	1.314
4.5	2.04			2.53	0.188	0.982	1.49
5.0	2.38	3.503	1.054	3.11	0.208	1.074	1.67
5.5	2.77			3.83	0.231	1.178	1.90
6.0	3.22	5.10	1.287	4.76	0.257	1.296	2.14
8.0					0.386		
10					0.559		
20					1.940		

Table 13 Activity coefficients of Electrolytes at 25°C at Various Concentrations (Data from: Electrolyte solutions, R.A. Robinson and R.H. Stokes, Butterworth Publishers London, 1970)

Concentration [mol kg^{-1}]	NaCl	KCl	$NaNO_3$	Na_2SO_4	$CaCl_2$	$MgCl_2$	$Ca(NO_3)_2$
0.1	0.778	0.770	0.762	0.452	0.518	0.528	0.486
0.2	0.735	0.718	0.703	0.371	0.472	0.488	0.429
0.3	0.710	0.688	0.666	0.325	0.455	0.476	0.397
0.4	0.693	0.666	0.638	0.294	0.448	0.474	0.378
0.5	0.681	0.649	0.617	0.270	0.448	0.480	0.365
0.6	0.673	0.637	0.599	0.252	0.453	0.490	0.356
0.7	0.667	0.626	0.583	0.237	0.460	0.505	0.349
0.8	0.662	0.618	0.570	0.225	0.470	0.521	0.344
1.0	0.657	0.604	0.548	0.2.04	0.500	0.569	0.338
1.2	0.654	0.593	0.530	0.189	0.539	0.630	0.337
1.4	0.655	0.586	0.514	0.177	0.587	0.708	0.337
1.6	0.657	0.580	0. 501	0.168	0.644	0.802	0.339
1.8	0.662	0.576	0.489	0.161	0.712	0.914	0.342
2.0	0.668	0.573	0.478	0.154	0.792	1.051	0.347
2.5	0.688	0.569	0.455	0.144	1.063	1.538	0.362
3.0	0.714	0.569	0.437	0.139	1.483	2.32	0.382
3.5	0.746	0.572	0.422	0.137	2.08	3.55	0.407
4.0	0.783	0.577	0.408	0.138	2.93	5.53	0.438
4.5	0.826	0.583	0.396	2.53	4.17	8.72	0.472
5.0	0.874		0.386		5.89	13.92	0.510
5.5	0.928		0.378		8.18		0.551
6.0	0.986		0.371		11.11		0.596

Subject Index: